Y0-AQW-722

Ecotoxicology

Ecotoxicology

Pesticides and beneficial organisms

Edited by

Peter T. Haskell

and

Peter McEwen

Insect Investigations Ltd
School of Pure and Applied Biology
University of Wales
Cardiff, UK

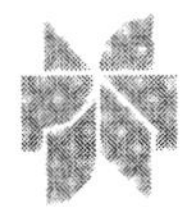

KLUWER ACADEMIC PUBLISHERS

DORDRECHT / BOSTON / LONDON

Library of Congress Cataloging in Publication Card Number: 98-70270

ISBN 0 412 81290 8

Published by Kluwer Academic Publishers,
P.O. Box 17, 3300 AA Dordrecht, The Netherlands.

Sold and distributed in North, Central and South America
by Kluwer Academic Publishers,
101 Philip Drive, Norwell, MA 02061, U.S.A.

In all other countries, sold and distributed
by Kluwer Academic Publishers Group,
P.O. Box 322, 3300 AH Dordrecht, The Netherlands.

Printed in Great Britain

Contents

Contributors

Zulkifli Ayob
Crop Protection Division
Department of Agriculture
Jalan Gallagher
50632 Kuala Lumpur
Malaysia

B. Baier
Biologische Bundesanstalt für Land- und Forstwirtschaft
Institut für Ökotoxikologie im Pflanzenschutz
Stahnsdorfer Damm 81
D-14532 Kleinmachnow
Germany

Frank M. Bakker
MITOX Stichting Duurzame Plaagbestrijding
PO Box 92260
NL-1090 AG
Amsterdam
The Netherlands

Julie A.O. Banken
Washington State University
Puyallup Research and Extension Center
Puyallup
Washington 98371
USA

Katie Barrett
Huntingdon Life Sciences
PO Box 2
Huntingdon
Cambs PE18 6ES
UK

Peter Beaumont
The Pesticides Trust
Eurolink Centre
49 Effra Road
London SW2 1BZ
UK

U. Bienert
IBACON GmbH
Industriestrasse 1
D-64380 Rossdorf
Germany

Franz Bigler
HoechstSchering AgrEvo GmbH
Hoechtstwerke H872
D-65926 Frankfurt
Germany

Linzi Birnie
IACR Rothamsted
Harpenden
Herts AL5 2JQ
UK

Sylvia Blümel
Federal Office and Research Center of Agriculture
Institut für Phytomedizin
Spargelfeldstr. 191
A-1226 Wien
Austria

Kevin Brown
Ecotox Ltd
PO Box 1
Tavistock
Devon PL19 0YU
UK

Richard A. Brown
Zeneca Agrochemicals Ltd
Jealott's Hill Research Station
Bracknell
Berks RG12 6EY
UK

Peter J. Campbell
Zeneca Agrochemicals Ltd
Jealott's Hill Research Station
Bracknell
Berks RG12 6EY
UK

Marco Candolfi
Springborn Laboratories AG
Seestrasse 21
Horn CH-9326
Switzerland

Michael Coulson
Zeneca Agrochemicals Ltd
Jealott's Hill Research Station
Bracknell
Berks RG12 6EY
UK

Brian A. Croft
Department of Entomology
Oregon State University
Cordley Hall
Corvallis
Oregon 97331
USA

Paulo Degrande
Federal Biological Research Centre for Agriculture and Forestry (BBA)
Institute for Biological Control
Heinrichstr. 243
D-64287 Darmstadt
Germany

Ian Denholm
IACR Rothamsted
Harpenden
Herts AL5 2JQ
UK

Gerhard P. Dohmen
BASF-AG Landwirtschaftliche Versuchsstation
Postfach 120
D-67114 Limburgerhof
Germany

Steve A. Ellis
ADAS High Mowthorpe
Duggleby
Malton
North Yorkshire YO17 8BP
UK

Rolf Forster
Biologische Bundesanstalt für Land- und Forstwirtschaft (BBA)
Biol Mittelprufung
Messeweg 11/12
D-38104 Braunschweig
Germany

Geoff Frampton
Biodiversity and Ecology Division
School of Biological Sciences
University of Southampton
Bassett Crescent East
Southampton SO16 7PX
UK

Steen Gyldenkaerne
Danish Institute of Plant and Soil Sciences
Department of Plant Pathology and Pest Management
Lottenborgvej 2
DK-2800 Lyngby
Denmark

Barbara Hackett
IACR Rothamsted
Harpenden
Herts AL5 2JQ
UK

Peter T. Haskell
School of Pure and Applied Biology
University of Wales
PO Box 915
Main Building
Park Place
Cardiff CF1 3TL
UK

Sherif A. Hassan
Institute for Biological Pest Control
Federal Biological Research Centre

Heinrichstr. 243
D-64287 Darmstadt
Germany

Udo Heimbach
Federal Biological Research Centre for Agriculture and Forestry
Institute for Plant Protection in Arable Crops and Grassland
Messeweg 11/12
D-38104 Braunschweig
Germany

Philip A. Heneghan
Department of Entomology
Oregon State University
Cordley Hall
Corvallis
Oregon 97331
USA

S. Hennig-Gizewski
Bayer AG
Crop Protection Research
Institute for Environmental Biology
D-51368 Leverkusen
Germany

Kong Luen Heong
Entomology and Plant Pathology Division
International Rice Research Institute
PO Box 933
1099 Manila
Philippines

John M. Holland
The Game Conservancy Trust
Fordingbridge
Hants SP6 1EF
UK

Paul A. Horne
IPM Technologies Pty Ltd
PO Box 560
Hurstbridge
Victoria 3099
Australia

B. Jäckel
Pflanzenschutzamt Berlin
Mohringer Allee 137
D-12347 Berlin-Neukölln
Germany

Paul C. Jepson
Department of Entomology
Oregon State University
Cordley Hall
Corvallis
Oregon 97331
USA

Jürger Just
Federal Biological Research Centre for Agriculture and Forestry (BBA)
Institute for Plant Protection in Fruit Crops
Schwabenheimer Str. 101
D-69221 Dossenheim
Germany

P. A. Kirsch
IPM Technologies Inc.
4134 N. Vancouver
Portland
Oregon 97068
USA

R. Kleiner
BioChem
D-04451 Cunnersdorf
Germany

Stefanie Klepka
IBACON Institut für Biologische Analytik und Consulting GmbH
Industriestrasse 1
D-64380 Rossford
Germany

Christiane Kühner
GAB Biotechnologie GmbH
Eutingerstrasse 24
D-75223 Niefern-Oeschelbronn
Germany

Alan Lane
ADAS Wolverhampton

Woodthorne
Wolverhampton WV6 8TQ
UK

Gavin B. Lewis
Covance
Otley Road
Harrogate
North Yorkshire HG13 1PY
UK

Martin Longley
Agrochemical Evaluation Unit
School of Biological Sciences
The University of Southampton
Basset Crescent East
Southampton SO16 7PX
UK

F. Louis
SLFA für Landwirtschaft
Weinbau und Gartenbau
Breitenweg 71
D-67435 Neustadt
Germany

Susanne Maise
Novartis Crop Protection AG
4002 Basle
Switzerland

Eddie McIndoe
Zeneca Agrochemicals Ltd
Jealott's Hill Research Station
Bracknell
Berks RG12 6EY
UK

Michael Mead-Briggs
Agrochemical Evaluation Unit
School of Biological Sciences
The University of Southampton
Basset Crescent East
Southampton SO16 7PX
UK

Mark J. Miles
DowElanco Europe
Letcombe Laboratories
Letcombe Regis
Wantage
Oxon OX12 9JT
UK

Albert K. Minks
Research Institute for Plant Protection (IPO-DLO)
PO Box 9060
6700 GW Wageningen
The Netherlands

Derek Morgan
CSL
MAFF
Sand Hutton
York YO4 1LZ
UK

Normah Mustaffa
Crop Protection Division
Department of Agriculture
Jalan Gallagher
50632 Kuala Lumpur
Malaysia

S. Nengel
GAB/IFU
D-75223 Biefern-Öschelbronn
Germany

Christoph Neumann
Novartis Crop Protection AG
4002 Basle
Switzerland

Andrea Nickless
Ecotox Ltd
PO Box 1
Tavistock
Devon PL19 0YU
UK

Pieter A. Oomen
Plant Protection Service

Department of Phytopharmacy
PO Box 9102
6700 HC Wageningen
The Netherlands

B. Reber
Ciba-Geigy AG
Crop Protection/Insect Control/IPM Services
CH-4002 Basel
Switzerland

Jörg Römbke
ECT Ökotoxikologie GmbH
Böttgerstr. 2-14
D-65439 Flörsheim am Main
Germany

David Rosen (the late)
The Hebrew University
Faculty of Agriculture
Rehovot
Israel

S. Schmitzer
IBACON GmbH
Industriestrasse 1
D-64380 Rossdorf
Germany

Richard Schmuck
Bayer AG
AGricultural Centre
D-151368 Leverkusen
Germany

Kenneth G. Schoenly
Entomology Division
International Rice Research Institute
PO Box 993
1099 Manila
Philippines

Burkhard Sechser
Ciba-Geigy AG
Crop Protection/Insect Contro/IPM Services
CH-4002 Basel
Switzerland

Stephen W. Shires
FMC Europe NV
Avenue Louise 480
B-81050 Brussels
Belgium

John D. Stark
Washington State University
Puyallup Research and Extension Center
Puyallup
Washington 98371
USA

J.H. Stevenson
Chairman
IC-PBR Bee Protection Group
16 Old Rectory Close
Harpenden
Herts AL5 2UD
UK

Catherine A. Symington
Department of Zoology
School of Biological Sciences
La Trobe University
Bundoora 3083
Australia

Andreas Ufer
BASF Landwirtschaftliche Versuchsstation
Postfach 120
D-67114 Limburgerhof
Germany

Harold van der Valk
Burg. Jansenstraat 50
5037 NC Tilburg
The Netherlands

Heidrun Vogt
Federal Biological Research Centre for Agriculture and
Forestry (BBA)
Institute for Plant Protection in Fruit Crops
Schwabenheimer Str. 101
D-69221 Dossenheim
Germany

Janny G.M. Vos
CABI-Bioscience
SE Asia Regional Centre
PO Box 210
43409 UPM Serdang
Selangor
Malaysia

Mario Waldburger
Swiss Federal Research Station for Agroecology and Agriculture
Reckenholzstr. 191
CH-8046 Zürich
Switzerland

Keith F.A. Walters
CSL
MAFF
Sand Hutton
York YO4 1LZ
UK

Anna Waltersdorfer
HoechstSchering AgrEvo GmbH
Hoechtstwerke H872
D-65926 Frankfurt
Germany

William K. Walthall
Washington State University
Puyallup Research and Extension Center
Puyallup
Washington 98371
USA

A. Wehling
Springborn Laboratories AG
Seestrasse 21
Horn CH-9326
Switzerland

John A. Wiles
European B&D Center
DuPont Agric Products
Nambsheim
France

H. Wilhelmy
Dr U Noack Laboratorium
Käthe-Paulus-Str. 1
D31157 Sarstedt
Germany

Stephanie F.J. Williamson
CABI-Bioscience Centre, U.K.
Silwood Park
Buckhurst Road
Ascot
Berks SL5 7TA
UK

DAVID ROSEN

Vigevani Professor of Agriculture and Professor of Entomology, The Hebrew University of Jerusalem

David Rosen died a few months after he had delivered the paper which forms Chapter 4 in this book. He was a personal friend of many of us at the Ecotox Conference, and was known to everyone as a world leader in Chalcidoid taxonomy and biological control. He was an active member of IOBC and a member of the IOBC/WPRC Council and played a leading role in the development of integrated pest management both in Israel, and, through his membership of the FAO/UNEP Panel of Experts on Integrated Pest Control, internationally.

The Ecotox Conference was the last international meeting he was to attend. He was his usual witty, helpful and erudite self there, which is how we will remember him.

Index of abbreviations

Agricultural Development and Advisory Service (ADAS)
Agrochemical Evaluation Unit (EAU)
AgroEcosystem Analysis (AESA)
Conventional Farming Practice (CFP)
Days After Treatment (DAT)
Department of Agriculture (DOA)
Enzyme linked immunosorbent assay (ELISA)
Farmer Field Schools (FFS)
Good Agricultural Practice (GAP)
Good Laboratory Practice (GLP)
Insect Growth Regulator (IGR)
Integrated Crop Management (ICM)
Integrated Farming System (IFS)
Integrated Pest Management (IPM)
Joint Pesticide Testing Programme, JTPT
Leaf Area Index (LAI)
Lowest expected environmental concentration (LEEC)
No Observable Effect Concentration (NOEC)
Non government organisations (NGOs)
Pesticide Application Technology (PAT)
Plant Protection Products (PPP)
Predicted Environmental Concentration (PEC)
Predicted initial environmental concentration (PIEC)
Reduced Input Approach (RIA)
Risk Assessment (RA)
Risk Assessment for Beneficials (RAB)
Risk Quotients (RQ)
Seeking Confirmation About Results At Boxworth (SCARAB)
Training Of Trainers (TOT)
World Health Organisation (WHO)

Index of organizations

Advisory Committee on Pesticides (ACP)
Agricultural Development Advisory Service (ADAS)
Asian Networking on Natural Enemies and Pesticides (ANNEP)
Beneficial Arthropod Regulatory Testing Group (BART)
Biologische Bundesanstalt für Land- und Forstwirtschaft/German Federal Biological Research Centre (BBA)
Biopesticide and Pollution Prevention Division (BPPD)
British Agrochemicals Association (BAA)
British Crop Protection Council (BCPC)
Commercial Ecotoxicology Testing Group (COMET)
Common Agricultural Policy (CAP)
Council of Europe (CoE)
Department of the Environment (DoE)
Environmental Protection Agency (EPA)
European and Mediterranean Plant Protection Organisation (EPPO)
European Commission (EC)
European Crop Protection Association (ECPA)
European Standard Characteristics of Beneficial Regulatory Testing (ESCORT)
European Union (EU)
Food and Agriculture Organization of the United Nations (FAO)
Institut tot Aanmoediging van het Wetenschappelijk Onderzoek in Nijverheid en Landbouw (IWONL)
International Commission for Plant-Bee Relationships (ICPBR)
International Institute for Biological Control (IIBC)
International Organisation for Biological and Integrated Control of Noxious Animals and Plants (IOBC)
International Organisation for Biological Control of Noxious Animals and Plants/West Palearctic Regional Section (IOBC/WPRS)
Malaysian Agriculture Research & Development Institute (MARDI)
Ministry of Agriculture, Fisheries and Food – UK (MAFF)
Organisation for Economic Co-operation and Development (OECD)
Pesticides Safety Directorate (PSD)

Research Institute for Plant Protection (IPO-DLO)
Society for Environmental Toxicology and Chemistry (SETAC)
Swiss Development Corporation (SDC)
World Trade Organisation (WTO)

Preface

The international conference on *Ecotoxicology: Pesticides and beneficial organisms* held in Cardiff in October 1996 came about as a result of collaboration between the Welsh Pest Management Forum and a number of other organizations, including the International Organisation for Biological Control (both its global and WPRS sections), the Beneficial Arthropods Regulatory Testing Group (BART), the Commercial Ecotoxicology Testing Group (COMET), an expert group of the European and Mediterranean Plant Protection Organisation (EPPO), the Pesticide Safety Directorate of the UK Ministry of Agriculture, The International Institute for Biological Control and the British Crop Protection Council (BCPC).

A number of research groups in some of these organizations and others in agro-chemical companies have, in the past 15 years, carried out a detailed programme of research on the effects of pesticides on beneficial organisms, mainly beneficial arthropods, which has resulted in the development of highly sensitive and specialized techniques for testing pesticides in the laboratory and in semi-field and field conditions. This work has of course a direct bearing on the testing and labelling of pesticides and on their usage in integrated pest management systems and hence is of considerable scientific and commercial significance.

The conference programme consisted of two parts. The first was a programme of some 36 platform papers designed to provide authoritative reviews by international experts on the current position of the effect of pesticides on beneficial arthropods and in particular the detailed state of development of the testing procedures and possible changes to them in the future. It is these papers, now edited, that comprise the present volume.

The second part of the conference consisted of some 25 poster presentations dealing with a wide range of issues including new types of test, further developments of current testing methodology and the bioavailablity of pesticides. These posters provoked such lively interest and debate at the conference that the Forum decided to publish as a paperback short papers based on these posters. This book entitled *New Studies in Ecotoxicology* was published in 1997, price £20, and is available from the Forum.

Ecotoxicology: Pesticides and beneficial organisms and *New Studies in Ecotoxicology* together constitute a unique record of research on the effects of pesticides on beneficial arthropods. I say unique because, despite the long time span of the research effort and intermediate reviews of certain aspects of it, this conference was the first occasion when all the research groups and workers had assembled to review and discuss the programme in its entirety.

The scientific programme for the conference was developed by an Advisory Committee with the following membership.

Chair:

P.T. Haskell	University of Wales, Cardiff

Members:

D.V. Alford	British Crop Protection Council
A. Armitage	Welsh Development Agency
K.L. Barrett	Huntingdon Life Sciences
M.F. Claridge	University of Wales, Cardiff
J. Dyer	Welsh Development Agency
J. Hemingway	University of Wales, Cardiff
M.S. Hoy	Pesticides Safety Directorate, MAFF
P.C. Jepson	Oregon State University, USA
O.T. Jones	AgriSense BCS Ltd
P.Langley	University of Wales, Cardiff
G.Lewis	Corning Hazelton Ltd (now Covance)
M. Mead Briggs	Agrochemical Evaluation Unit, Southampton
D. Royle	International Organisation for Biological Control (WPRS section)
J. Waage	International Institute of Biological Control

Additional advice and help was received from Dr S. Hassan, Federal Biological Research Centre, Darmstadt, Germany; Dr S. Blümel, Institut für Phytomedizin, Vienna, Austria; Dr P. Oomen, Plant Protection Service, Wageningen, the Netherlands, and the British Crop Protection Council and the Forum acknowledges the help and advice of all these organisations and individuals.

The Welsh Pest Management Forum believes this book will be a powerful aid to the development of integrated pest management systems world wide – a major objective of the Forum – and is glad to acknowledge the continued support of the Technology Transfer Group of the Welsh Development Agency towards this end.

P.T. Haskell CMG, PhD, CBiol, FIBiol
Technical Secretariat, Welsh Pest Management Forum

Part One

The Issues

1

Introduction

Paul Jepson and Brian Croft

The conference from which this volume was generated reported the unique achievements of an international community of researchers, regulators and industrialists with a common interest in reducing the impact of pesticides upon beneficial invertebrates in agriculture. The purpose of the meeting was to evaluate the state of the art with respect to our understanding of side-effects and focus upon future needs in an international context. This volume is itself unique, in that it contains contributions from all the different interest groups as well as presenting a global perspective; it is more than a snapshot of activity at a given moment, but it certainly tapped into an area of activity that is evolving rapidly.

The major achievement of the core group of European scientists, industrialists and regulators at the meeting has been the establishment of testing procedures that evaluate pesticide toxicity against beneficial invertebrates during the pesticide registration process. Although the regulations themselves are still under development and discussion, a body of data has already built up, over two years, concerning the risks posed by new candidates for registration and older products that are being subjected to re-registration. For those who know the somewhat adversarial relationships within the integrated pest management (IPM) debate in some parts of the world and also the inertia built into the process of modifying regulatory procedures, this will doubtless be viewed as a stunning success – so how did it arise? Several factors seem to have conspired together to bring about success, over at least 20 years.

It started with a group, operating under the auspices of the International Organization for Biological Control (IOBC) for over 20 years,

Ecotoxicology: Pesticides and beneficial organisms.
Edited by P.T. Haskell and P. McEwen. Published in 1998 by Chapman & Hall, London. ISBN 0 412 81290 8.

that established a tradition of testing pesticides for their side-effects against a diverse selection of beneficial species from a variety of cropping systems. The *modus operandi* of this group was to be inclusive, accepting as participants anyone from commerce, academe or government research laboratories who had test methodologies to offer or testing services to contribute. The building of trust amongst this community may have been as important to the most recent development of regulatory test procedures as the database that they established, although both are significant. The expanding political union established by the European Community promoted cooperation between member states in areas of pesticide regulatory procedures and established the forum within which the ideas and achievements of the IOBC group could be debated and evaluated. In this arena, cooperation between the European and Mediterranean Plant Protection Organization (EPPO) and the Council of Europe (CoE) was instrumental in moving the debate towards regulatory harmonization. The preparedness of a group of pesticide manufacturers to cooperate and contribute ideas and technology to this evolving debate, rather than simply delay or inhibit discussion in a field that was likely, under all scenarios, to increase the costs of pesticide registration, greatly assisted the process. The Beneficial Arthropods Regulatory Testing (BART) group is the third essential ingredient to the mix that permitted the regulatory procedures to arise. In addition, Europe seems to be the first zone where the principles of ecotoxicology have penetrated IPM and produced a unique blend of ideas and techniques that have moved the discipline through a common bottleneck – simply knowing what to measure and how to measure it. In particular, research in the Netherlands and UK has adapted the ideas of ecological risk assessment to suit the needs of agriculture, dealing with the rather specialized problems associated with the widely distributed but temporary pollution caused by pesticides in the heavily managed ecosystems of agriculture.

Finally, there has been little controversy concerning the nature of the hazards posed by pesticides; and an increasing case history, some of which has been generated by industry itself, has developed concerning the impact of agrochemicals on beneficial arthropods. All participants in these discussions use terms such as 'local extinction' and 'pesticide resurgence' to define the ultimate hazards posed by excessive use of pesticides or the application of excessively broad-spectrum materials.

The diversity of opinion and interest within these groups has of course established an active debate and there is still a need for many issues to be resolved. The greatest challenge to the representative groups within this community continues to be the need to develop techniques that actually result in reduced pesticide impacts against beneficial species in the field. Of greatest priority in this challenge may be the need to involve end-users within this process and build awareness amongst growers about chemical side-effects and how to mitigate them. In this respect, the participants in the meeting were fortunate in being able to hear about experiences in Southeast Asian rice systems, where considerable successes in pesticide impact reduction have emerged through a very different pathway, building from the bottom up. In rice systems, a more proscriptive approach to chemical regulation and use has assisted growers in exploiting their new-found knowledge of pesticide ecotoxicology to reduce chemical impacts against the highly effective beneficial predators and parasites that colonize tropical rice when it is not sprayed by broad-spectrum insecticides. Like the European example referred to above, developments in Asian rice were driven by a long history of research; indeed the sophistication in understanding of the interactions between herbivores and their natural enemies in rice probably exceeds that of any other system. There is enormous scope for this approach to be internationalized, just as there is for the European model to be more widely known and exploited. The important contrast between the two approaches lies, however, in the involvement of growers in the Asian example.

What, then, are the challenges and opportunities that face us in the development of more benign approaches to pesticide use? The list below is in no particular order or priority, but seems to reflect the current trajectories of thinking amongst the participants in this meeting. We need to know more about mechanisms of toxicity, exposure, uptake and effects; undergraduate classes the world over never cease to be amazed at how little we know about the basic toxicology of insecticides to many groups of organisms, including beneficial invertebrates. We also know very little about the relative importance of different routes of chemical uptake or the rates at which pesticides are accumulated by different species from different substrates.

We need to optimize the process of regulatory decision making and achieve a balance with IPM development following registration. Do regulators, for example, have a use for dose-response data for natural enemies, or is it preferable to rely upon trigger values for toxic effects of compounds applied at recommended rates, as a basis for decision making? In a similar vein, how do we achieve an appropriate balance between laboratory testing, which is simple and cost effective, and field testing, which is realistic but expensive and produces results of only local significance? Finally, where does the balance lie between regulatory testing and research following registration that establishes the role of a given compound in a local IPM programme? Technologies and procedures for mitigating chemical effects need to be understood; possible approaches span the development of selective formulation technology, the optimization of dose rate, improvements to sprayer design and even alteration of the pattern of application within or between fields.

There needs to be greater agreement concerning what we are attempting to protect; basic regulatory concern is for environmental protection and the preservation of ecosystem function and maintenance of biodiversity. This can deflect concern away from the cropping zone itself towards field boundaries or so-called off-target areas. This philosophy can divert attention away from the need to protect the beneficial function of natural enemies themselves within the sprayed area where many species complete their life cycles. It is through attention to impacts upon beneficial species within the crop environment that the level of pesticide use will decline in the longer term, protecting off-crop habitats.

There is an urgent need to develop recognition of chemical impacts on natural enemies amongst grower, extension, research, industrial and regulatory communities in many parts of the world; amongst the regions that lag in this thinking are the United States, Japan and parts of the Asia-Pacific Region, the Mediterranean countries and Africa. There are encouraging signs in all of these, but also a need for more rapid progress, building upon the fine examples presented at this meeting. There appears to be excellent scope for learning from achievements in Asian rice where awareness of chemical impacts amongst growers has had a direct influence on chemical use. New educational techniques and approaches are needed to assist this process in other systems.

We are left with an impression of rapid recent progress, though this has been built upon decades of research, but also of an emerging opportunity to develop modes of pesticide invention, development and use which, for the first time, properly exploit what has proved to be one of the most significant (in both positive and negative terms) technologies of the twentieth century. We attach particular significance to natural enemies and the side-effects of pesticides upon them, because their activities substitute for pesticides in so many cropping systems. By cancelling out predation and parasitism, we challenge ecosystems and witness the unsurprising outcome of pest resurgence and increased rates of insecticide resistance development amongst pests, released from limitation by natural enemies. Enlightened self-interest within industry should guide the establishment of types of chemicals and modes of use that minimize side-effects and therefore maximize the market life of these hugely expensive chemical products. Industry, however, needs the assistance of scientists as well as the environmental standards set by society through regulatory agencies to set the context in which it operates. It is in all our interests to maximize knowledge and understanding of side-effects and we hope that this volume will provide an effective introduction or update to many readers.

2

Issues associated with pesticide toxicology and arthropod natural enemies in the pre- and post-registration stages of chemical development

Brian A. Croft, Paul C. Jepson and Philip A. Heneghan

INTRODUCTION

The ecotoxicological impact of pesticides against arthropod natural enemies has a relatively recent history of research and an even shorter history of development and implementation of practical measures that aim to minimize adverse impacts. Some efforts to mitigate the side-effects of toxins (mostly pesticides) to this group of organisms were instituted in the era before synthetic organic pesticides were invented, but these declined with expanding use of mainly neurotoxic, synthetic pesticides (i.e. organochlorines, organophosphates, carbamates, synthetic pyrethroids). In the past 20 years, however, interest in the issue of chemical side-effects against non-target invertebrates has grown. Reasons for this include:

- the increasing diversity of toxins, which now include insect growth regulators (IGRs), attractants (baits, pheromones), neurohormones and natural products;

Ecotoxicology: Pesticides and beneficial organisms.
Edited by P.T. Haskell and P. McEwen. Published in 1998 by Chapman & Hall, London. ISBN 0 412 81290 8.

- increasingly credible and precise documentation of the role of natural enemies in the limitation of pest populations;
- improvements in the means by which natural enemies are mass-produced for augmentative and inundative release;
- increases in the numbers of key target pests exhibiting resistance to pesticides;
- a decline in the tolerance of chemical side-effects.

This chapter discusses what is currently known, and what we still need to know, about side-effects against this important group of organisms, mainly from a biological and ecological perspective. The first two sections emphasize action taken before and after a chemical pest control agent is registered for use. The third section deals with efforts to deliver side-effects information to end users (farmers, pest managers). Finally there is a retrospective analysis to reflect the tone and content of debate during the International Review Conference on Ecotoxicology at which this chapter was first presented as a paper.

RESEARCH AND DECISION MAKING PRIOR TO PESTICIDE REGISTRATION

Physiological selectivity

Research into pesticide toxicology against arthropod natural enemies has seldom been conducted as a part of the pesticide development process, before the registration of new compounds. This is partly because we still only have limited knowledge about selectivity and its possible mechanisms amongst natural enemy groups (Croft, 1990) and also because simple screening tools may not provide an effective methodology for early detection of potential hazards (Stark *et al.*, 1995). It is an emerging perspective however, that testing against natural enemies could be conducted at a much earlier stage in the pesticide development process.

Historically, only a small number of scientists have studied pesticide selectivity and maintained regular testing of new compounds. Normally, this type of programme has supported pest management efforts within high value crops, grown on a large scale and for which well researched integrated pest management (IPM) programmes were already in place (e.g. cotton, citrus and tree fruits). Research has generally aimed to determine chemical effects on local fauna under specific environmental/production conditions and explored the scope for protecting known natural enemies, of established economic importance, by exploiting new chemistry or formulation technology. Seldom has selectivity data been fed back to the development chemist who explores synthetic pathways and toxicological activity within related families of compounds, despite the abundant evidence of wide variability in the natural enemy toxicity of even closely related compounds with equivalent efficacy against the target pest.

An example of the effective operation of this feedback loop between field and laboratory ecotoxicology and developmental chemistry concerns the development of selective acaricides that are used in tree fruit crops (Croft, 1990). Structure/activity relationships have also been explored for honey bees, given their acknowledged economic importance in crops world-wide. There may be new opportunities to expand research into invertebrate side-effects in the earlier stages of pesticide development, triggered by the escalating investment costs for all pesticides (and consequent need to explore potential hazards that would otherwise emerge after large investments of research funds) and increasingly sophisticated developments in plant protection, including transgenic crops containing insecticidal toxins (Jepson *et al.*, 1994), where basic research into ecotoxicology is given a high priority. Symptomatic of this new climate is the establishment, by several pesticide companies, of early stage screening of candidate compounds against selected natural enemy species.

Basic studies on natural enemy physiology and toxicology

Basic studies of physiology and toxicology of beneficials (detoxification, excretion, food metabolism, storage, etc.) have been rare and much data has been extrapolated from phytophagous species (Croft and Mullin, 1990). While the protein-based diet of entomophages confers differences in aspects of metabolism, it is probable that many elements of basic physiology that affect the metabolism of toxins are similar to those of herbivores. The basis for extrapolation from herbivore to carnivore is still not resolved, and the extent to which detoxification mechanisms are general amongst beneficial invertebrates, with their wide taxonomic diversity, is unclear. Research is lacking and it seems that funding to answer these questions comes only from basic science sources.

Optimal pesticide selectivity criteria

As understanding of the biological interactions between natural enemies and pests has improved, there have been parallel improvements in the use of pesticides in physiologically and ecologically selective ways. No longer do we solely consider a single level or type of selectivity, and we increasingly explore pesticides as tools that adjust pest population increase rates or pest/natural enemy ratios in favour of natural enemies. In some circumstances, only a minor adjustment in the predator/prey or parasite/host ratio may be necessary and in others we may need to have more drastic impacts. Sometimes a toxin with more limited pest-killing power may prove extremely effective in this respect, especially where toxicity to natural enemies is limited – for example, propargite, a long-established and apparently uncompetitive organotin compound with considerable promise for IPM use against mites in apple (Croft, 1990). In other cases, reduced doses of broad-spectrum compounds may confer a degree of selectivity (Wiles and Jepson, 1995) or even offer the prospect of reducing parasitism rates upon natural enemies themselves – for example, differential effects on hymenopteran parasitoids and hyperparasitoids (Longley and Jepson, 1996). These examples are extremely rare at present and we have yet to develop the conceptual or methodological framework to manipulate pesticide/invertebrate interactions with multiple pests and beneficials in complex plant/pest systems (Schoenly *et al.*, 1995). In the future, we may establish the prospect for much more broadly based selectivity. Some new pesticides seem to offer specific selectivity at almost order or trophic level. In order to be able to exploit this, advances are needed in the conceptual basis for manipulation of pest population densities in such a way that natural enemy effectiveness is enhanced.

RESEARCH AND DECISION MAKING FOLLOWING PESTICIDE REGISTRATION

The diversity of pesticide modes of action is at an all time high and we are falling further behind in terms of research data concerning toxicity against beneficial species. This is having an adverse impact upon the sustainable use and market life of pesticides, given the importance of natural enemies and evidence of adverse effects against them. Renewed priority must therefore be given to the investigation of chemical impacts on beneficial species. Natural products are sometimes put under fast-track registration procedures and may reach the market-place with less understanding about modes of action or activity against beneficial species than conventional synthetic materials. Attention should therefore be focused on the performance of materials in the early stages of use following registration.

With the emergence of new pesticide chemistry, with more varied target sites and modes of action, comes greater unpredictability in the nature and symptoms of adverse side-effects. To reach the target site, pesticides of the past in fact shared many features of pharmacology, despite an enormous range of chemical structures. Products may have varied in penetration, routes of uptake and the degree of partitioning between body tissues, but the common target site of the nervous system conferred constraints, within which pesticides could be predictably found. These chemicals exhibited a high degree of variability in their toxicity to beneficial invertebrates, but the process of devising toxicological tests and the symptomology of adverse effects was comparatively straightforward when most materials were neurotoxic . With wholly new symptomologies and the likelihood that effects may be stage specific or even delayed between generations, some of these testing systems and ideas may be redundant. Insect growth regulators may, for example, elicit intergenerational effects that would not be detected by conventional assay techniques.

What rules of testing may be affected by these changes? We cannot assume that the ability to extrapolate effects between species, families or even orders with conventional pesticides remains. Differences between species may now arise because of differences in diet/enzyme systems, growth rates and even life histories, with far less predictability than in the past. It is also apparent that too little attention has been paid to sublethal effects, despite the fact that rates of parasitism and predation can be severely affected by the altered behaviours or reduced reproductive rates that pesticides may elicit.

All of the above implies that research following pesticide registration will need to be refined and updated to pick up effects that cannot be detected using conventional methodologies. In systems where the relative importance of specific natural enemies has been quantified, there may have to be a greater emphasis on field validation of laboratory tests because standard test species may exhibit responses that differ from those of key biological control agents. Other improvements to research and monitoring following pesticide registration can be summarized as follows.

Basic toxicology of new (and old) compounds

Immediate benefits would be brought to both pesticide development by industry and the regulatory process if we knew what metabolic, structural, detoxification and even immunological systems were held in common by natural enemy taxa and which systems were different enough to confer significant differences in toxicological risks. By establishing the scientific basis for extrapolation of impacts between organisms, we would reduce the testing burden and increase the likelihood of genuinely selective compounds being developed. This will only arise through an expansion of research in this area, as we have indicated above. Very few research programmes have incorporated detailed mechanistic toxicological studies with field-based experimentation of ecotoxicological impact: perhaps the investigations of Everts (1990) and Jagers op Akkerhuis (1993) are unique.

Improving our ability to predict short-term effects in the field from laboratory testing

Laboratory data may be modified by field exposure estimates to make effective predictions of effects, at least in the short term. The hazard ratio for honey bees (Stevenson: g a.i./ha, divided by LD_{50} in μg/insect) has been quite successful, and there is scope to explore and develop similar indices for natural enemies. Thus, pesticide use in crops that exist as seedlings, where levels of exposure may be greater, might be considered more hazardous than in crops with a well established canopy, where refugia exist that protect certain species (Everts *et al.*, 1991a; Çilgi and Jepson, 1992). More sophisticated models that incorporate substrate-specific influences on bioavailability and the effects of physical factors are also now emerging (Everts *et al.*, 1991b; Jagers op Akkerhuis and Hamers, 1992; Jagers op Akkerhuis, 1993). Similarly in off-crop environments, aspects of exposure and the nature of surfaces receiving pesticide contamination by drift need to be quantified and described to enable potential impacts to be forecast from toxicological data for non-target species (Longley *et al.*, in press a). The scope for development of simple indices that enhance the potential to extrapolate effects between laboratory and field has not really been properly explored (Stark *et al.*, 1995).

Evaluating impacts on behaviour within side-effects testing protocols

There is a long-standing gap in knowledge concerning how invertebrate behaviour affects pesticide uptake (Croft, 1990). Studies with epigeal organisms have shown how movement can influence toxic effects and advances in test methodology may arise through improved understanding of these complex processes. Behavioural analysis is also needed to understand the effects of altered host distributions after spraying (e.g. Duffield *et al.*, 1996; Longley *et al.*, in press b) on the searching success of natural enemies. Predator–prey theory tells us that effects on predation and parasitism rates could be considerable and that this may represent a currently unrecorded indirect impact of pesticides. There is also evidence that natural enemy behaviour can be directly affected by pesticide exposure (e.g. Wiles and Jepson, 1994; Longley and Jepson, 1996) but there has been little effort to investigate the consequences of this for either predator or parasitoid efficiency. Additional knowledge is more likely to relieve the regulatory testing burden rather than expand it, by increasing our understanding of the hazards that pesticides pose and reducing our uncertainty when attempting to predict outcomes.

Recognizing the power and significance of field-based measurements within regulatory test procedures

IPM has guided pesticide ecotoxicologists to recognize the need for farm-scale testing to study pesticide effects on the complex, dynamic and often temporary ecologies of cropping systems. Toxic materials may not be so hazardous when applied because they may be non-persistent or lose toxicity when they contact the soil; beneficial species also may re-populate treated areas rapidly by reproduction, immigration or both. IPM, an ecologically and field-based technology, has developed an implicit recognition that these processes are important but, like pesticide ecotoxicology, has failed to explore larger-scale population dynamics in such a way that pest and natural enemy population trajectories can be predicted with any degree of certainty: ironically, it is now certain aspects of ecotoxicology that are leading the development of a landscape-scale perspective of natural enemy population dynamics (see below). Ecotoxicology and regulatory test procedures traditionally rely on toxicological testing and their associated statistics and implicitly assume that laboratory test data are predictive of field effects. Laboratory test data, however, seldom predict effects in the field, though experience has told us that products with limited effects in the laboratory may be similarly non-hazardous in practice. Among those testing side-effects, this has been incorporated into testing procedures, where toxic effects above threshold trigger values lead to more complex and realistic field-based bioassays. Field testing is important, but often considered to be so specific to each situation that it is only undertaken when the registration process is complete. We re-emphasize the need to retain and develop field-based evaluations, and hope they remain as supplements to pre-registration testing. The balance between pre- and post-registration tests must be struck, and products that show a significant hazard in pre-registration tests should not be registered until safety criteria have been met in semi-field or field tests. Without field testing at early stages, it is unlikely that innovations in spray targeting, doses or formulations will be developed to mitigate side-effects: such refinements are essential to pesticide regimes of the future and are bridges to post-registration tests that compare compounds to refine usage for IPM.

Quantifying pesticide exposure for risk prediction and dose optimization

Predictions of effects from laboratory tests may be improved through the incorporation of exposure data, derived from the field, and these data may promote increased precision in the calculation of dose-rate. To date, however, there is very little direct residue data derived from natural enemies in sprayed systems (Mullie and Everts, 1992). Analyses of spray tracer deposits on invertebrates provide an estimate of direct exposure to liquid insecticide, if not the body burden of active ingredient taken up by the exposed organism, and have been used to derive estimates of optimum dose rates, in combination with dose-response statistics (Everts, 1990; Kjaer and Jepson, 1995; Wiles and Jepson, 1995).

Here, ecotoxicologists have lead IPM specialists: exposure is a key component of risk assessment for wildlife and needs to be considered explicitly in assessing the risk of pesticides of beneficial organisms.

Larger-scale spatial and temporal considerations

Small experimental plots lead to underestimates of the duration of side-effects, because organisms in untreated plots re-invade the treated area over unrealistic time-scales (Everts *et al.*, 1989; Jepson and Thacker, 1990; Thomas *et al.*, 1990; Duffield and Aebischer, 1994). Acknowledging this increasing body of data on natural enemy mobility, research and even some pre-registration field test procedures are now being carried out on a sufficiently large scale (e.g. 1 ha plots in wheat systems) that re-invasion occurs over a more realistic time-scale. Where large plots are impractical, the duration of sampling may be curtailed, recognizing that data become increasingly unrepresentative of the real world as time progresses following treatment. We note that regulatory authorities increasingly apply these rules of thumb to account for spatial and temporal scaling factors, but that recognition of these factors by regulatory agencies is patchy.

Use of models (detail-rich and analytical)

Single-species population models have helped us to understand the processes that underlie the long-term impact of pesticides. They permit us to test our understanding and to extrapolate to scales beyond those that are experimentally tractable. We argue that they have been underused. Such models are not normally intended to be predictive and so great care must be taken in interpretation of their output: the ecological paradigm that there are many possible outcomes from a given set of starting conditions has not penetrated the realm of ecotoxicology. Additionally, it is impractical to encode sufficient detail in any model for realism. For example, we cite impacts of pesticides on ground beetles (Carabidae) (Burn, 1992). Some species became extinct in the heaviest treatment regime; others, quite closely related, responded by becoming very abundant; and others remained unaffected. The underlying mechanisms for these rather profound differences probably lies in differences in the exposure of larvae, in rates and timings of dispersal and in the relative flexibility of phenologies and breeding times between species. Using modelling to predict the distribution of effects (even within this one family of beetles) would have required an excessive effort, but a simplified mechanistic model that explores the underlying features of carabid population biology and ecotoxicology has been developed by Sherratt and Jepson (1993). They explored trade-offs between spraying frequency, field boundary permeability, pesticide toxicity and both dispersal and reproductive rates. Their model has triggered research and increased understanding: its use in the regulatory process, however, must be seen as heuristic rather than absolutely predictive.

Recent research on spiders (Halley *et al.*, 1996) has added sophistication by incorporating cropping practices (other than pesticides) that impact populations, thus enabling more realistic farm landscapes to be simulated. This model is still generalized to the arachnid family Linyphiidae. Future progress in modelling will most likely rise from two sources. The first is spatially explicit models, based on specific farm landscapes, which are overlaid by population models. These may permit pesticide usage tactics to be optimized on a relatively large scale. Evidence that farming system layout and natural enemy refugia affect side-effects needs further modelling and validation in the real world. Secondly, less parameter-rich analytical population models may continue to expose key aspects of biology and interactions with habitat that affect pesticide hazards.

COMMUNITY PERSPECTIVES

Much research has focused on single species, and multi-species effects have not been widely addressed. The extent to which effects can cascade (Kareiva *et al.*, 1996) through communities, destabilize pest populations and cause outbreaks is barely researched. We are increasingly aware of species interactions, and limited data on pest dynamics in sprayed cereal and rice have revealed that outbreaks are more likely and even less predictable (e.g. Cohen *et al.*, 1995). Is there enough evidence now to pursue the community dimension with the same energy as has been expended on test methodologies with single species?

Despite a shortfall in registration procedures and research infrastructure in many countries, there is a positive outlook for limiting the extent of pesticide side-effects through education of end-users. Other examples of pioneering implementation projects are discussed later in this volume.

Developing methodologies for more localized applications (rapid screening, assessment of resurgence risk, IPM compatibility)

The focus of test methodology upon the requirements of industry and registration authorities has tended to distract research emphasis away from the needs of end-users and their immediate advisers. There is an urgent need for rapid, simple and robust bioassays to be established that permit localized risk assessments to be undertaken with field-collected organisms. The farmer field-school philosophy of grower involvement in the development of crop management practices (Chapter 42) has already established that semi-field screening of pesticide impacts upon natural enemies provides important educational messages to growers that have a positive influence on their pest control decisions. There is considerable scope to transfer this technology across the developing world and also into the developed agroecosystems of the west, where decision makers have often been unaware of the role and importance of natural enemies in the farmland that they manage. This will require an interaction between educationalists, scientists and agronomists and a greater degree of community involvement than has traditionally been envisaged.

Exploitation of databases as decision tools

There are now thousands of published articles concerning pesticide side-effects upon natural enemies. Databases permit the organization, synthesis, analysis and rapid retrieval of large bodies of information in such a way that more general patterns and messages could result that are of direct relevance to the better management and use of pesticides. Theiling and Croft (1988) established such a database for pesticide/natural enemy literature, and recent analyses (Croft, 1990; Kovach *et al.*, 1992; Murphy *et al.*, 1992) reveal the potential utility of databases in pesticide management and decision making. The costs of establishing and maintaining such databases are large and currently the critical problem, once such tools have been established, is their updating after conventional funding sources, which tend to be short-term, have dried up.

ADDITIONAL ITEMS OF EMPHASIS

Side-effects testing should place less emphasis on laboratory tests and greater emphasis on semi-field tests and (if needed) field tests, when economically feasible. Pesticide side-effects testing largely began with controlled laboratory tests that enabled scientists to establish standardized methodologies that could be widely exploited (e.g. Hassan, 1983). Research has now progressed to the stage where a number of semi-field-based tests can now be standardized and provide similar levels of accuracy and precision with improved realism (direct exposure to spraying and well as to residues; realistic pesticide–substrate interactions; the presence of refugia; exposure via the food chain, etc.). It is apparent that pesticide industry personnel are now using semi-field tests more widely than laboratory tests in their own screening evaluations because of the increased reliability that this technology provides.

Further development of reliability and validation tests

Reliability and validation tests of sequential decision rules in side-effects testing need to be developed further. A tier-wise procedure for decision making has evolved for evaluation of natural enemy side-effects and is followed by most workers (Samsoe-Petersen, 1990; Hassan *et al.*, 1991). Given the resources and personnel-hours invested in the development of new test methods and regulatory procedures, it is critical that quantitative evaluations of the effectiveness of this process in predicting actual hazards are undertaken (e.g. Bakker and Jacas, 1995). Decisions about the development of new testing methods cannot be made with any reliability until the actual performance of current regulatory test procedures are known. In particular, no test method should elicit excessive numbers of either falsely negative or falsely positive risk predictions. There is mounting evidence that tests on inert substrates create the potential for such unrealistically high rates of exposure to pesticides and formulation additives that some agencies are side-stepping initial screening tests and beginning the testing sequence with tests on leaf or soil substrates. These directions of evolutionary change and development in test methodology could be counter-productive unless they are supported by analyses of pooled data and experience from industry, researchers and regulatory agencies.

Assessing impacts at community level

Methods to assess impacts at the community level may establish a new philosophy of side-effects evaluation. New methods that evaluate changes in the herbivore/carnivore/parasite food web can define hazards using simplified ecological indices that, for example, quantify food chain length (Schoenly *et al.*, 1995; Chapter 41, this volume). Monitoring techniques that explore community structure in well researched systems may provide a new basis for the evaluation of side-effects that incorporate hazards such as pest resurgence risk. These procedures may only operate effectively in certain environments and only with a restricted range of pests, subjected to particularly strong limitation by natural enemies, but they offer a new philosophy that deserves more general attention and they may be more widely applied as food webs are increasingly described in major cropping systems.

Distribution of knowledge

Knowledge about standardized methods, databases of side-effects results, sublethal responses and ecological effects is not widely distributed. Only those with access to the most modern methods of data retrieval and access to research libraries, computer literature searches or the World Wide Web have ready access to data concerning pesticide impacts upon natural enemies. Yet there is a global need for access to this information. Because of poor communications, there is much redundancy in testing of similar materials and natural enemy taxa, and opportunities for synthesis and comparison of data on a world-wide basis are being missed. Global databases of side-effects are available (Theiling and Croft, 1988; Croft 1990; Hassan *et al.*, 1991) and increased effort should be expended in the inventory of current projects and the process of distributing this information globally. Where projects with good communications and access to data are developed in previously under-researched regions, there is good evidence that they may be just as innovative and successful as those undertaken in the West, where the research infrastructure is so well established (e.g. Van der Valk, 1990; Van der Valk and Niassy, 1996).

User (pest manager, grower) feedback in the evaluation of side-effects testing/decision making procedures is not used – or not used early enough. The relative successes and impacts of side-effects testing programmes are not usually quantified or widely discussed amongst the communities of growers and managers that are their intended beneficiaries (Chapter 41). For example, simple rules have been established in rice production areas for integrating the complex of pest control tactics and the ecological assessments necessary to establish more sustainable plant protection programmes

Despite the fact that growers may not have advanced reading skills, sophisticated decision-making strategies have been developed to assist the implementation of complex IPM systems. These philosophies deserve far wider uptake and implementation, because they facilitate integration of grower perceptions concerning adverse impacts with the basic and applied science needed to gain an understanding of systems dynamics and chemical impacts.

There still is a lack of common perspective among those involved in the development, regulation, risk assessment, mitigation (IPM) and costs/benefits of chemical use. Many do not recognize the breadth of perspective or level of basic science needed to understand the process of developing, registering and using pesticides in more selective and effective ways. The International Review Conference on Ecotoxicology demonstrated many times how little we know about what each other does.

CONCLUSIONS AND ACKNOWLEDGEMENTS

The organizers of the conference graciously permitted us to give both pre- and post-meeting perspectives. All the participants benefited from the updates of research and regulatory practices in the major regions. Our perspectives were broadened by being able to address the ways in which prevailing conditions affecting agriculture, natural resource management and chemical use differ between continents and countries, although certain core needs apply to all the situations that we reviewed. We were motivated by this meeting to engage in discussions of how beneficial invertebrate side-effects data can be better used in the USA, where pesticide use is a major component of pest control technology despite efforts to develop less pesticide-intensive IPM in cotton, cereals, legumes and tree fruits (Frisbie and Adkisson, 1986). We were impressed with changes in western Europe and the Asia-Pacific region, where new laws and regulatory practices are impacting pesticide registration and use and are encouraging the exploitation of more selective materials.

Globally, there many other signs of progress, arising particularly from individuals who are applying a broad environmental and ecological perspective in key positions that deal with production, regulation and use of toxins. Change comes with new ideas and our discussions were lively and constructive. We thank the organizers and the Welsh Pest Management Forum for their support of this activity.

REFERENCES

Bakker, F.M. and Jacas, J.A. (1995) Pesticides and phytoseiid mites: strategies for risk assessment. *Ecotoxicology and Environment Safety*, **32**, 58–67.

Burn, A.J. (1992) Interactions between cereal pests and their predators and parasites, in *Pesticides, Cereal Farming and the Environment* (eds P.W. Greig-Smith, G.W. Frampton and A.R. Hardy), HMSO, London, pp. 110–131.

Çilgi, T. and Jepson, P.C. (1992) The use of tracers to estimate the exposure of beneficial insect to direct pesticide spraying in cereals. *Annals of Applied Biology*, **121**, 239–247.

Cohen, J.E., Schoenly, K., Heong, K.L. *et al.* (1995) A food web approach to evaluating the effect of spraying on insect population dynamics in a Philippine irrigated rice ecosystem. *Journal of Applied Ecology*, **31**, 747–763.

Croft, B.A. (1990) *Arthropod Biological Control Agents and Pesticides*, J. Wiley and Sons, New York.

Croft, B.A. and Mullin, C.A. (1990) Physiology and toxicology, in *Arthropod Biological Control Agents and Pesticides* (ed B.A. Croft), J. Wiley and Sons, New York, pp. 127–153.

Duffield, S.J. and Aebischer, N.J. (1994) The effect of spatial scale of treatment with dimethoate on invertebrate population recovery in winter wheat. *Journal of Applied Ecology*, **31**, 263–281.

Duffield, S.J., Jepson, P.C., Wratten, S.D. and Sotherton, N.W. (1996) Spatial changes in invertebrate predation rate in winter wheat following treatment with dimethoate. *Entomologia Experimentalis et Applicata*, **78**, 9–17.

Everts, J.W. (1990) Sensitive indicators of side-effects of pesticides on the epigeal fauna of arable land. PhD thesis, 114 pp.

Everts, J.W., Aukema, B., Hengelveld, R. and Koeman, J.H. (1989) Side-effects of pesticides on ground-dwelling predatory arthropods in arable ecosystems. *Environmental Pollution*, **59**, 203–225.

Everts, J.W., Aukema, B., Mullie, W.C. *et al.* (1991a) Exposure of the ground-dwelling spider *Oedothorax apicatus* (Blackwall) (Erigonidae) to spray and residues of deltamethrin. *Archives of Environmental Contamination and Toxicology*, **20**, 13–19.

Everts, J.W., Willemsen, I., Stulp, M. *et al.* (1991b) The toxic effect of deltamethrin on linyphiid and erigonid spiders in connection with ambient temperature, humidity and predation. *Archives of Environmental Contamiation and Toxicology*, **20**, 20–24.

Frisbie, R.E. and Adkisson, P.L. (1986) *CIPM: Integrated Pest Management on Major Agricultural Crops*, USA Texas Agricultural Experimental Station Publication.

Halley, J.M., Thomas, C.F.G. and Jepson, P.C. (1996) A model for the spatial dynamics of linyphiid spiders in farmland. *Journal of Applied Ecology*, **33**, 471–492.

Hassan, S.A. (1983) Procedures for testing the side-effects of pesticides on beneficial arthropods as being considered by the IOBC International Working Group. *Mitteilungen der Deutschen Gesellschaft für allgemeine und angewandte Entomologie*, **4**, 86–88.

Hassan, S.A., Bigler, F., Bogenschütz, H. *et al.* (1991) Results of the fifth joint pesticide testing programme carried out by the IOBC/WPRS Working Group 'Pesticides and Beneficial Organisms'. *Entomophaga*, **36**, 55–67.

Jagers op Akkerhuis, G. (1993) Walking behaviour and population density of linyphiid spiders in relation to minimising the plot size in short-term pesticide studies with pyrethroid insecticides. *Environmental Pollution*, **80**, 163–171.

Jagers op Akkerhuis, G. and Hamers, T.H.M. (1992) Substrate-dependent bioavailability of deltamethrin for the epigeal spider *Oedothorax apicatus*. *Pesticide Science*, **36**, 59–68.

Jepson, P.C. and Thacker, J.R.M. (1990) Analysis of the spatial component of pesticide side-effects on non-target invertebrate populations and its relevance to hazard analysis. *Functional Ecology*, **4**, 349–355.

Jepson, P.C., Croft, B.A. and Pratt, G.E. (1994) Test systems to determine the ecological risk posed by toxin release from *Bacillus thuringiensis* genes in crop plants. *Journal of Molecular Ecology*, **3**, 81–89.

Kareiva, P., Stark, J.D. and Wennergren, U. (1996) Using demographic theory, community ecology and spatial models to illuminate ecotoxicology, in *Ecological Dimensions* (eds L. Maltby and P. Grieg-Smith), Chapman & Hall, pp. 13–23.

Kjaer, C. and Jepson, P.C. (1995) The toxic effects of direct pesticide exposure for a nontarget weed-dwelling chrysomelid beetle (*Gastrophysa polygoni*) in cereals. *Environmental Toxicology and Chemistry*, **14**(6), 993–999.

Kovach, J., Petzoldt, C., Degni, J. and Tette, J. (1992) A method to measure the environmental impact of pesticides. *New York Food and Life Sciences Bulletin*, **139**, 8.

Longley, M. and Jepson, P.C. (1996) The influence of insecticide residues on primary parasitoid and hyperparasitoid foraging behaviour in the laboratory. *Entomologia Experimentalis et Applicata*, **81**, 259–269.

Longley, M., Çilgi, T., Jepson, P.C. and Sotherton, N.W. (in press a) Measurements of pesticide spray drift deposition into field boundaries and hedgerows: I Summer application. *Environmental Toxicology and Chemistry*.

Longley, M., Jepson, P.C., Izquierdo, J. and Sotherton, J.W. (in press b) Temporal and spatial changes in aphid and parasitoid populations following applications of deltamethrin in winter wheat. *Entomologia Experimentalis et Applicata*.

Mullie, W.C. and Everts, J.W. (1992) Uptake and elimination of 14C deltamethrin by *Oedothorax apicatus* (Archnida: Erigonidae) with respect to bioavailability. *Pesticide Biochemistry and Physiology*, **39**, 27–34.

Murphy, C.F., Jepson, P.C. and Croft, B. (1992) Database analysis of the toxicity of antilocust pesticides to non-target, beneficial invertebrates. *Crop Protection*, **13**, 413–420.

Samsoe-Petersen, L. (1990) Sequences of standard methods to test effects of chemicals on terrestrial arthropods. *Ecotoxicology and Environment Safety*, **19**, 310–319.

Schoenly, K.G., Cohen, J.E., Heong, K.L. *et al.* (1995) Quantifying the impact of insecticides on food web structure of rice–arthropod populations in a Philippine farmers' irrigated field: a case study, in *Food Webs: Integration of Patterns and Dynamics* (eds G.A. Polis and K. Wisenmiller), Chapman & Hall, London, pp. 343–351.

Sherratt, T.N. and Jepson, P.C. (1993) A metapopulation approach to modelling the long-term impact of pesticides on invertebrates. *Journal of Applied Ecology*, **30**, 696–705.

Stark, J.D., Jepson, P.C. and Mayer, D.F. (1995) Limitations to the use of topical toxicity data for predictions of pesticide side-effects in the field. *Journal of Economic Entomology*, **88**, 1081–1088.

Theiling, K.M. and Croft, B.A. (1988) Pesticide side-effects on arthropod natural enemies: a database summary. *Agriculture, Ecosystems and Environment*, **21**, 191–218.

Thomas, C.F.G., Hol, E.H.A. and Everts, J.W. (1990) Modelling the diffusion component of dispersal during the recovery of a population of linyphiid spiders from exposure to an insecticide. *Functional Ecology*, **4**, 357–368.

Van der Valk, H. (1990) Beneficial arthropods, in *Environmental Effects of Chemical Locust and Grasshopper Control – a Pilot Study. Project Report ECLO/SEN/003/NET*, (ed. J.W. Everts), FAO, Rome, pp. 171–224.

Van der Valk, H. and Niassy (1996) Side-effects of locust control on beneficial arthropods – research strategy approaches used by Locustox project in Senegal, in *New Strategies in Locust Control* (eds S. Krall, R. Peveling and B. Daoule Diallo), Birkhauser Verlag, Berlin.

Wiles, J.A. and Jepson, P.C. (1994) Sub-lethal effects of deltamethrin residues on the within-crop behaviour and distribution of *Coccinella septempunctata. Entomologia Experimentalis et Applicata*, **72**, 33–45.

Wiles, J.A. and Jepson, P.C. (1995) Dosage reduction to improve the selectivity of deltamethrin between aphids and coccinellids cereals. *Entomologia Experimentalis et Applicata*, **76**, 83–96.

3

The initiative of the IOBC/WPRS Working Group on Pesticides and Beneficial Organisms

Sherif A. Hassan

The Working Group 'Pesticides and Beneficial Organisms' of the International Organization for Biological Control/West Palaearctic Regional Section (IOBC/WPRS) was established in 1974. The major aim, according to the inaugural meeting held in Zurich in May 1974, was to 'encourage the development of standard methods for testing the side-effects of pesticides on natural enemies that would be acceptable to all cooperating institutions'. It was stressed during the meeting that this should be conducted in cooperation with the pesticide industry as well as international organizations such as European and Mediterranean Plant Protection Organization (EPPO) and Food and Agriculture Organization of the United Nations (FAO). For the IOBC, this Working Group, at that time, was of a 'new type' because all other working groups were concerned with development of integrated control programmes for different agricultural crops and orchards. Pesticides suitable for use in integrated pest management were urgently needed. A further aim of the group was therefore to test the side-effects of commonly used pesticides on the most important natural enemies and to provide an information 'service' to the other IOBC groups.

Now, after 24 years of intensive activities, the Working Group is happy to see that the standard laboratory, semi-field and field methods developed by its members are being used by research workers world-wide and that the testing has become obligatory in several countries. IOBC

Ecotoxicology: Pesticides and beneficial organisms.
Edited by P.T. Haskell and P. McEwen. Published in 1998 by Chapman & Hall, London. ISBN 0 412 81290 8.

methods were adapted by the European Union (EU) for registration purposes. The use of these standard methods allowed the exchange of results from one country to another and saved the cost of repeated testing. Agreement on the choice of beneficial organisms to be tested was difficult and required long debate. Now we have a concept for choice of species based on the role of the natural enemy as a biological control agent as well as on its susceptibility to pesticides.

An international net of laboratories to conduct the tests in 13 countries has been established. The cooperation with EPPO and with the EU to harmonize testing guidelines has given fruitful results. EPPO and EU guidelines based on the IOBC/WPRS test methods have been developed. Guidelines to test the side-effects of pesticides on beneficial organisms were published in an EPPO Bulletin (Hassan *et al.*, 1985) and two IOBC/WPRS Bulletins (Hassan, 1988, 1992). The standard methods developed by the Working Group were accepted for use by the EC for the registration of pesticides. A joint IOBC, Beneficial Arthropods Regulatory Testing Group (BART), Commercial Ecotoxicology Testing Group (COMET) and EPPO initiative for the regulatory ring testing and validation of methods was started in 1995.

ORGANISMS TESTED AND GUIDELINES DEVELOPED

The joint testing of pesticides within organized programmes has become a standard feature of the working group. These testing programmes provide valuable information on the side-effects of pesticides and give the testing members an opportunity to improve testing techniques and develop better guidelines. Joint testing programmes, each including 20 chemicals, have been organized every two years since 1977. Results of testing the side-effects of about 124 pesticides (insecticides, acaricides, fungicides, herbicides and plant growth regulators) on beneficial organisms within six joint pesticides programmes were published in multi-author publications by Franz *et al.* (1980) and Hassan *et al.* (1983, 1987, 1988, 1991, 1994) involving 7, 14, 21, 20, 22 and 24 authors, respectively. The beneficial organisms tested (Table 3.1) included arthropods, entomopathogenic fungi and nematodes. Information on the effect of pesticides on greenhouse natural enemies was published by Hassan and Oomen (1985). Rearing methods for 14 different natural enemies were published by Samsoe-Petersen *et al.* (1989, 16 authors).

Selective pesticides

The joint testing programmes included 20 beneficial arthropods, three entomopathogenic fungi and the nematode *Steinernema feltiae*. All pesticides chosen for the tests were registered in at least one of the IOBC

Table 3.1 Organisms tested

Organism	*Tested by*
Parasitic insects	
Trichogramma cacoeciae	Hassan, Germany
Encarsia formosa	van de Veire, Belgium
Leptomastix dactylopii	Viggiani, Italy
Cales noacki	Vivas, Spain
Aphidius matricariae	Polgar, Hungary
Phygadeuon trichops	Moreth, Germany
Coccygomimus turionellae	Bogenschütz, Germany
Predatory mites	
Phytoseiulus persimilis	Calis and Bakker, Netherlands; Blümel and Stolz, Austria
Amblyseius andersoni	Calis, Netherlands
Amblyseius finlandicus	Sterk, Belgium
Typhlodromus pyri	Calis, Netherlands; Englert, Germany, Boller, Switzerland
Predatory insects	
Chrysoperla carnea	Bigler, Switzerland; Vogt, Germany
A. phidoletes aphidimyza	Helyer, UK
Syrphus vitripennis	Rieckmann, Germany
Semiadalia 11-notata	Brun, France
Coccinella septempunctata	Brun, France
Aleochara blinineata	Samsoe-Petersen, Denmark; Moreth, Germany
Pterostichus melanarius	Lewis, UK
Pterostichus cupreus	Heimbach, Germany
Forficula auricularia	Sauphanor, France
Anthocoris nemoralis	Schaub, Switzerland
Spiders	
Chirancanthium mildei	Mansour, Israel
Fungi	
Verticillim lecanii	Tuset, Spain
Beauveria bassiana	Hokkanen, Finland
Beauveria brongniartii	Coremans-Pelseneer, Belgium
Metarhizium anisopliae	Hokkanen, Finland
Nematodes	
Steinernema feltiae	Vainio, Finland

member countries. This testing programme is not only meant to provide valuable information on the side-effects of pesticides on beneficial organisms: it also gives the testing members an opportunity to improve testing techniques, compare results and exchange experience with colleagues.

Whereas harmlessness of pesticides can easily be shown by laboratory experiments, harmfulness can only be confirmed under practical conditions in the field. Pesticides found to be harmless to a particular beneficial organism in the laboratory test are most likely to be harmless to the same organism in the field and no further testing in semi-field or field experiments is therefore recommended.

Among the 124 pesticides tested to date, the following compounds were found to be harmless to nearly all the beneficial organisms tested, or to have limited persistence:

- **Insecticides and acaricides**: Dipel (*Bacillus thuringiensis*), Applaud (buprofezin), Shell Torque (fenbutatin oxide), Azomate (benzoximate), Dimilin (diflubenzuron), Spruzit-Nova-flüssig (pyrethrum and piperonylbutoxide), Pirimor Granulat (pirimicarb), Cesar S.L. (hexythiazox), Apollo SOSC (clofentezine), Kelthane (dicofol), Tedion V 18 (tetradifon).
- **Fungicides**: Nimrod (bupirimate), Saprol (triforine), Sumisclex (procymidone), Dyrene flüssig (anilazine), Bayfidan (triadimenol), Anvil (hexaconazole), Calixin (tridemorph), Bayleton (triadimefon), Ronilan (vinclozolin), Orthocid 83 (captan), Cercobin-M (thiophanate-methyl), Ortho Difolatan (captafol), Derosal (carbendazim), Daconil 500 (chlorothalonil), Plondrel (ditalimfos), Pomarsol forte (thiram), Dithane Ultra (mancozeb), Baycor (bitertanol), Delan flüssig (dithianon), Vitigran (copper-oxychloride), Impact (flutriafol), Rovral PM (iprodion).
- **Herbicides**: Illoxan (diclofop-methyl), Semeron (desmetryn), Betanal (phenmedipham), Kerb 50 W (propyzamid), Cycocel Extra (chlormequat), Luxan 2,4-D amine (2,4-D aminesalt), Ally (metsulfuron-methyl), Grasp (tralkoxydim), Basagran (bentazone).
- **Plant growth regulators**: Rhodofix (naphthyl acetic acid), Dirigol-M (alphanaphthyl-acetamid).

The Working Group has expanded considerably both in number of members and in its duties. This is largely due to the new EC regulations regarding the obligatory testing of pesticides for their side-effects on beneficial arthropods (Council Directive, 1991). The number of colleagues from the newly developed commercial testing laboratories has increased rapidly in the last few years. The number of participants from the chemical pesticide industry has also increased.

To cope with the increasing workload, three subgroups were established with S. Blümel as subconvenor for method validation, U. Heimbach for the development of new testing methods and G. Sterk for the joint testing of pesticides. At present, all active members (traditional testing members) of the Working Group are actively involved in the work of all three subgroups and regularly hold joint meetings.

The Working Group welcomed the new EC regulations and is happy that the EC has adapted methods and procedures developed by the group. More elements are still under discussion. The Group aims to increase cooperation between governmental research workers (traditional testing members), scientists from the commercial testing laboratories (COMET) and colleagues from the chemical pesticide industry (BART). Its aims are:

- to complete the development of sequential testing procedures (laboratory, extended laboratory, semi-field and field methods) for important natural enemies (subgroup);
- to continue to organize joint pesticide testing programmes and provide information for integrated pest management (IPM) (subgroup);
- to ring-test and validate IOBC standard methods for testing the side-effects of pesticides on beneficial organisms (subgroup);
- to achieve global recognition of IOBC standard test methods;
- to exchange ideas with EPPO, EC and BART experts on the framework and the methods used for testing;
- to discuss the value of extended laboratory methods in the rating of pesticides;
- to compare results of laboratory, extended laboratory, semi-field and field methods;
- to use indicator species to extrapolate the risk on groups of natural enemies and non-target species;
- to standardize pesticide doses for the different types of tests;
- to discuss new ideas for testing on entomopathogenic fungi (Entomophthorales, Deuteromycetes);
- to exchange information and experiences on good laboratory practice (GLP).

REFERENCES

Council Directive (1991) Council Directive 15 July 1991 concerning the placing of plant protection products on the market (91/414/EEC). *Official Journal of the European Communities* L230 pp.1–32, 19 August 1991.

Franz, J.M., Bogenshütz, H., Hassan, S.A. *et al.* (1980) Results of a joint pesticide test programme by the Working Group 'Pesticides and Beneficial Arthropods'. *Entomophaga*, **25**, 231–236.

Hassan, S.A. (1988) Guidelines for testing the effects of pesticides on beneficials. *IOBC/WPRS Bulletin*, **XI**(4), 143.

Hassan, S.A. (1992) Guidelines for testing the effects of pesticides on beneficial organisms. *IOBC/WPRS Bulletin*, **XV**(3), 186.

Hassan, S.A. and Oomen, P.A. (1985) Testing the side effects of pesticides on beneficial organisms by OILB Working Party, in *Biological Pest Control – the Glasshouse Experience* (eds N.W. Hussey and N. Scopes), Blandford Press, Poole, Dorset, p. 240.

Hassan, S.A., Bigler, F., Bogenschütz, H. *et al.* (1983) Results of the second joint pesticide testing programme by the IOBC/WPRS Working Group 'Pesticides and Beneficial Arthropods'. *Z. angew. Entomol.*, **95**, 151–158.

Hassan, S.A., Bigler, F., Blaisinger, P. *et al.* (1985) Standard methods to test the side-effects of pesticides on natural enemies of insects and mites developed by the IOBC/WPRS Working Group 'Pesticides and Beneficial Organisms'. *Bulletin OEPP/EPPO*, **15**, 214–255.

Hassan, S.A., Albert, R., Bigler, F. *et al.* (1987) Results of the third joint pesticide testing programme by the IOBC/WPRS Working Group 'Pesticides and Beneficial Organisms'. *Z. angew. Entomol.*, **103**, 92–107.

Hassan, S.A., Bigler, F., Bogenshütz, H. *et al.* (1988) Results of the fourth joint pesticide testing programme carried out by the IOBC/WPRS Working Group 'Pesticides and Beneficial Organisms'. *Z. angew. Entomol.*, **105**, 321–329.

Hassan, S.A., Bigler, F., Bogenschütz, H., *et al.* (1991) Results of the fifth joint pesticide testing programme carried out by the IOBC/WPRS Working Group 'Pesticides and Beneficial Organisms'. *Entomophaga*, **36**, 55–67.

Hassan, S.A., Bigler, F., Bogenschütz, H. *et al.* (1994) Results of the sixth joint pesticide testing programme of the IOBC/WPRS Working Group 'Pesticides and Beneficial Organisms'. *Entomophaga*, **39**, 107–119.

Heimbach, U., Büchs, W. and Abel, C. (1992) A semi-field method close to field conditions to test effects of pesticides on *Poecilus cupreus* L. (Coleoptera: Carabidae). *IOBC/WPRS Bulletin*, **XV**(3), 159–165.

Samsoe-Petersen, L., Bigler, L., Bogenshütz, H. *et al.* (1989) Laboratory rearing techniques for 16 beneficial arthropod species and their prey/hosts. *Z. Pflanzenkrankh., Pflanzensch*, **96**, 289–316.

4

Reducing insecticide use in plant protection in Israel: policy and programmes

The late David Rosen

INTRODUCTION

Integrated pest management (IPM) is an attempt to achieve a realistic solution to a very serious dilemma. The continuing population explosion has confronted mankind with several major problems, including imminent starvation, depletion of non-renewable energy sources, and the threat of world-wide environmental pollution. It is imperative that whatever we do to alleviate one problem does not aggravate the others. Therefore, in order to feed the ever-increasing population of the world and avoid starvation, we should devise effective means to control the various pests that take such a heavy toll of our agricultural crops – but we should endeavour to do so by employing methods that would tend to conserve energy and reduce, rather than increase, environmental pollution.

Modern chemical pest control, based on the use of synthetic organic pesticides, has provided us with effective and usually quite reliable means for controlling noxious organisms, and has therefore become the prevalent method of pest management in this day and age. However, indiscriminate use of modern pesticides involves some very serious hazards, including the development of resistance in pest populations, toxicity to plants, humans, domestic animals and wildlife, destruction of beneficial natural enemies, contamination of food chains, and general pollution of the environment. With the rising price of these pesticides,

Ecotoxicology: Pesticides and beneficial organisms.
Edited by P.T. Haskell and P. McEwen. Published in 1998 by Chapman & Hall, London. ISBN 0 412 81290 8.

the cost of chemical pest control has often become prohibitive. In view of all these drawbacks, it is no wonder that attention has been increasingly focusing on the development of safer and less expensive alternatives to chemical control.

Of the various alternatives to wholesale use of pesticides that are available to us, biological control by natural enemies has been by far the most successful to date and is the most promising for the foreseeable future. Whether achieved through importation of exotic natural enemies or through conservation and augmentation of locally established species, biological control is usually permanent, quite inexpensive and – despite the recent claims of some well meaning but misguided environmentalists – virtually hazard-free. Unfortunately, it is not always feasible at the present time, because we still do not have effective natural enemies for some serious 'key' pests. Until such enemies are discovered and put to use, chemical pesticides or other means of control will continue to be required.

Integrated pest management provides a reasonable compromise, taking into account the desirability of biological control on the one hand and the indispensability of some form of chemical control on the other. Like diplomacy, IPM is the art of the possible. It represents a holistic approach, harmonizing all available methods to attain optimal pest management and environmental quality. In principle, effective IPM may be achieved through judicious use of relatively selective pesticides, only when absolutely necessary and in the least disruptive modes of application, in combination with a vigorous programme of applied biological control. Other selective alternatives – cultural, mechanical, physical and autocidal controls – should be incorporated into the programme whenever applicable, but biological control by natural enemies should always be a major component. In particular, as the value of naturally occurring biological control agents cannot be overemphasized, their conservation should be the first goal of any IPM programme (Rosen, 1990b; DeBach and Rosen, 1991).

In Israel, reduction of pesticide use in agriculture has become the official policy of the Ministry of Agriculture in the plant protection sciences. Over the years, we have made some progress in the development of biological and integrated control programmes, mainly on citrus (Rosen, 1967a, 1974, 1980; Rivnay, 1968; Harpaz and Rosen, 1971; Rossler and Rosen, 1990), but also on various other crops such as date palms (Kehat *et al.*, 1974) and avocado (Swirski *et al.*, 1988). In 1990, a National Steering Committee for Reducing the Use of Pesticides in Agriculture proposed detailed guidelines for attaining this goal (Rosen, 1990a), and these have set in motion a series of IPM programmes in major cropping systems, with biological and supervised control as their mainstays. Some of these will be briefly reviewed here, beginning with the long-standing programme on citrus.

CITRUS

Citrus, planted to about 30 000 ha, is one of the most important agricultural crops of Israel and a major agricultural export. It is also one of the best-organized agricultural industries. The Citrus Marketing Board of Israel, instituted in 1954, was until recently the sole marketing organization for the entire citrus crop of the country. It maintains its own biological control laboratory, supports research in all aspects of citriculture, and is authorized to carry out certain pest control operations on citrus on a country-wide basis. This high level of organization has been a key factor in the development of our IPM programme on citrus.

The history of biological control on citrus in Israel dates back to 1912, when the cottony-cushion scale, *Icerya purchasi* (Maskell) (Homoptera: Margarodidae), was effectively controlled by the vedalia ladybeetle, *Rodolia cardinalis* (Mulsant) (Coleoptera: Coccinellidae), introduced from Italy. Another outstanding success was brought about in the early 1940s, when *Clausenia purpurea* (Ishii) (Hymenoptera: Encyrtidae) was introduced from Japan against the citriculus mealybug, *Pseudococcus citriculus* (Green) [now known as *P. cryptus* (Hempel)] (Homoptera: Pseudococcidae) (see Rosen, 1967a; Rivnay, 1968). In spite of these early successes, most professional entomologists in Israel accepted Bodenheimer's (1951) view that climatic conditions were the primary factors governing insect populations in a semi-arid subtropical zone like Israel, and that the prospects for biological control in such a region were therefore rather dim. All this would change dramatically in the late 1950s.

In about 1955, the three major pests of citrus in Israel were:

- the Florida red scale, *Chrysomphalus aonidum* (L.) (Homoptera: Diaspididae), against which one or two sprays of mineral oil were applied annually in each infested grove, and organophosphorus insecticides were just being introduced;
- the Mediterranean fruit fly, *Ceratitis capitata* (Wiedemann) (Diptera: Tephritidae), which was controlled with full-coverage sprays of persistent chlorinated hydrocarbon insecticides;
- the citrus rust mite, *Phyllocoptruta oleivora* (Ashmead) (Acari: Eriophyidae), which was controlled with sulphur or other non-selective acaricides.

The credit for changing this situation should go mainly to the late Israel Cohen, then head of the Agrotechnical Division of the Citrus Marketing Board of Israel. In 1956/57, against the advice of leading entomologists, he introduced *Aphytis holoxanthus* (DeBach) (Hymenoptera: Aphelinidae) from Hong Kong, which soon brought about the virtual elimination of the Florida red scale from the main citrus-growing regions of Israel (DeBach, 1960; Cohen, 1969, 1975; DeBach *et al.*, 1971). This spectacular success marked the beginning of a new era in the management of citrus pests in Israel.

It was soon realized that complete biological control of the Florida red scale could not be maintained for long, unless all other pest control practices on citrus were modified to render them less detrimental to the effective natural enemy. Therefore, a programme for selective control of the Mediterranean fruit fly had to be developed. It consists of aerial strip-sprays of small amounts of poison bait, containing protein hydrolysates as a powerful attractant for female flies and malathion as a toxicant. These applications, which are centrally organized and carried out by the Citrus Marketing Board on a country-wide basis, are carefully timed with the aid of a network of traps containing Trimedlure, an effective synthetic attractant for male flies. This method proved to be far superior in effectiveness, and much less disruptive to the citrus ecosystem, than the use of chlorinated hydrocarbons (Avidov *et al.*, 1963; Cohen and Cohen, 1967; Rossler and Rosen, 1990).

Concurrently, all commercial pesticide formulations recommended for use against citrus pests in Israel were evaluated in laboratory tests for their possible effects on *A. holoxanthus* (Rosen, 1967a,b). Mortality of wasps was assayed upon exposure to treated citrus leaves in Munger cells, and their fecundity was tested when ovipositing in treated scale-insect hosts. Effects of pesticides on predatory phytoseiids were also studied in field and laboratory tests (Swirski *et al.*, 1968, 1969). One outcome of these investigations has been the elimination from citrus groves of sulphur preparations (once the prevalent means for controlling mites) and their replacement by more selective acaricides. As a result, the citrus rust mite proceeded to decline through the increased activity of indigenous and some introduced phytoseiid predators.

Biological equilibrium in any ecosystem may, of course, be affected by pest control operations taking place in neighbouring areas. To prevent the adverse effects of pesticidal drift from adjacent fields, appropriate legislation has been enacted prohibiting aerial spraying of cotton and other crops with non-selective pesticides within a distance of 200 m from a citrus grove.

As a result of all these concerted efforts, a much less toxic environment was established in citrus groves and the foundation was laid for an effective programme of IPM. Not only were the three major pests brought under effective control; it was also found that several other potentially injurious species were suppressed by natural enemies. Scale insects were controlled, whenever necessary, with selective mineral oil sprays, and broad-spectrum pesticides were usually applied only as a last resort in problematic groves, where the natural equilibrium had been upset. For a while, the prospects for successful IPM on citrus appeared very promising (Rosen, 1974).

However, an agricultural ecosystem is a complex, dynamic system. Any subtle change in any of its many interdependent components may have far-reaching effects on various other parts. Thus, since the late 1960s, various developments in the citrus ecosystem gradually combined to cause serious setbacks in our IPM programme. These included: the rise of secondary pests such as the Mediterranean black scale, *Saissetia oleae* (Olivier) (Homoptera: Coccidae); the appearance of several notorious invaders such as the spirea aphid, *Aphis spiraecola* (Patch) (Homoptera: Aphididae), the citrus whitefly, *Dialeurodes citri* (Ashmead) and the bayberry whitefly, *Parabemisia myricae* (Kuwana) (both Homoptera: Aleyrodidae); recognition of the citrus flower moth, *Prays citri* (Milliere) (Lepidoptera: Yponomeutidae), as a pest of economic importance that could be controlled only with diazinon; and difficulties with the timing, coverage and phytotoxicity of mineral oil sprays – all of which increasingly induced citrus growers to turn to non-selective carbamate and organophosphorus insecticides. A change to ultra-low volume in the control programme against the Mediterranean fruit fly may have also contributed to the decline of natural enemy populations and a gradual increase of several citrus pests.

Then the opening of new markets in countries like Japan, where stringent quarantine regulations are imposed on imported citrus fruit, caused a lowering of economic thresholds and a general intensification of pest control practices on citrus in Israel – so much so that eventually there were alarming indications that pesticide resistance was developing in certain groves. Inevitably, even such species as the Florida red scale, cottony-cushion scale and citrus rust mite, which had been kept under biological control for so many years, again attained pest proportions in certain localities due to the destruction of their natural enemies.

Efforts to reverse these alarming trends and restore the upset balance in the citrus ecosystem have included:

- introduction of numerous additional natural enemies, which has already resulted in biological control of the Mediterranean black scale and the two whitefly invaders (Blumberg and Swirski, 1977; Argov and Rossler, 1986; Rossler *et al.*, 1986; Swirski *et al.*, 1987);
- development of selective alternatives to conventional chemical control, notably control of the citrus flower moth with pheromone-baited traps (Sternlicht *et al.*, 1978) and in increase in drop size in Mediterranean fruit fly bait sprays;
- refinement of the supervised control system, with field scouts taking detailed bi-weekly counts on representative sampling trees (Rossler and Rosen, 1990);
- post-harvest treatment of citrus fruit by high-pressure rinsing in the packing-house.

Special emphasis has been placed on breeding *Aphytis* spp. for resistance to pesticides (Havron and Rosen, 1994; Havron *et al.*, 1995) – a truly integrative approach to increasing the role and efficacy of natural enemies wherever pesticides continue to be used in an IPM programme.

These endeavours have been accompanied by in-depth studies of the systematics, biology, ecology and population dynamics of the various pests of citrus and their natural enemies. The Citrus Marketing Board, the Volcani Institute of Agricultural Research, the Faculty of Agriculture of the Hebrew University, the Agricultural Extension Service and several other institutions take part in these collaborative efforts, which have already brought about a significant change in the management of citrus pests in Israel. At present, no more than 10–15% of our citrus acreage is still subject to broad-spectrum pesticide applications, and it is hoped that eventually the entire citrus industry of Israel will again be covered by an effective IPM programme.

OTHER CROPS

Similar principles have guided Israel's IPM programmes in other major cropping systems. On avocado, effective biological control of the long-tailed mealybug, *Pseudococcus longispinus* (Targioni Tozzetti) (Homoptera: Pseudococcidae), was followed by introduction of *Bacillus thuringiensis* preparations against the giant looper, *Boarmia selenaria* (Schiffermuller) (Lepidoptera: Geometridae) and other lepidopterous pests. The bayberry whitefly was controlled by imported natural enemies on avocado as well as on citrus, and a biological control project aimed at another recent invader, the pyriform scale, *Protopulvinaria pyriformis* (Cockerell) (Homoptera: Coccidae), is now under way. At present, avocado groves in Israel are not treated with any broad-spectrum pesticides, except for some occasional treatments directed against thrips.

Similarly, on date palms, biological control of the most serious pest – the parlatoria date scale, *Parlatoria blanchardi* (Targioni Tozzetti) (Homoptera: Diaspididae) – was followed by selective chemical control of other pests and provided the basis for an IPM programme. Some pests are excluded from date bunches by the use of fine-mesh nets. Even on such a traditionally heavily treated crop as cotton, an extensive system of supervised control, based on well defined economic thresholds, has resulted in a reduction of pesticidal sprays from an average of 28 to about five per season, and a concomitant conservation of natural enemies.

Efforts are currently being made to reduce the use of pesticides and develop IPM programmes in deciduous fruit orchards, vineyards, greenhouse and field-grown vegetables, cut flowers and spices, as well as area-wide programmes aimed at all crops grown in a given region. Conservation of biological control agents, physical and cultural controls monitoring and control with semiochemicals and extensive pest scouting systems are major components of all these programmes.

ROLE OF EDUCATION AND EXTENSION

Progress in developing biological control and IPM programmes in Israel has largely been made possible by two great advantages of Israeli agriculture: enlightened, well educated farmers and the excellent agricultural extension services. Traditional farming communities throughout the world are usually bastions of conservatism – culturally, politically and professionally. It is often quite difficult to persuade a traditional farmer to change practices that have been used for centuries. But Israel is an immigration country, and is fortunate to have farmers who are certainly not 'traditional' in this respect. They are remarkably dynamic, open to new concepts and ideas, and quick to adopt new methods and technologies.

Even with such outstanding farmers, the success or failure of an IPM programme may largely depend on the availability of competent extension officers, who carry research developments to the farmers and instruct them in sophisticated methods of pest control. It is a well trained government officer, not a pesticide salesman, who provides the Israeli farmer with unbiased advice on all pest control matters. These extension officers, who are held in high esteem by the farmers, are thoroughly versed in the modern concepts of IPM. They maintain close contacts with both farmers and research institutions, take part in field research, and provide an invaluable mechanism of bilateral feedback that may be rightfully regarded as the backbone of Israel's IPM programme (Rosen, 1974).

REFERENCES

Argov, Y. and Rossler, Y. (1986) The introduction of *Encarsia lahorensis* (Howard) (Hymenoptera: Aphelinidae) into Israel for the control of the citrus whitefly, *Dialeurodes citri* (Ashmead) (Homoptera: Aleyrodidae). *Israel Journal of Entomology*, **20**, 1–5.

Avidov, A., Rosen, D. and Gerson, U. (1963) A comparative study on the effects of aerial versus ground spraying of poisoned baits against the Mediterranean fruit fly on the natural enemies of scale insects in citrus groves. *Entomophaga*, **8**, 205–212.

Bedford, E.C.G. (1990) Mechanical control: high-pressure rinsing of fruit, in *Armored Scale Insects: Their Biology, Natural Enemies and Control*, World Crop Pests 4B (ed. D. Rosen), Elsevier Science Publishers, Amsterdam, pp. 507–513.

Blumberg, D. and Swirski, E. (1977) Release and recovery of *Metaphycus* spp. (Hymenoptera: Encrytida) imported for the control of the Mediterranean black scale, *Saissetia oleae* (Olivier) in Israel. *Phytoparasitica*, **5**, 115–118.

Bodenheimer, F.S. (1951) *Citrus Entomology in the Middle East*, W. Junk, The Hague.

Cohen, I. (1969) Biological control of citrus pests in Israel, in *Proceedings of the 1st International Citrus Symposium, Riverside, California (1968)*, pp. 769–772.

Cohen, I. (1975) *Citrus Fruit Technology*, Revivim, Tel Aviv (in Hebrew).

Cohen, I. and Cohen, J. (1967) *Centrally Organized Control of the Mediterranean Fruit Fly in Citrus Groves in Israel*, Agrotechnical Division, Citrus Board of Israel.

DeBach, P. (1960) The importance of taxonomy to biological control as illustrated by the cryptic history of *Aphytis holoxanthus n* sp. (Hymenoptera: Apehlinidae), a parasite of *Chrysomphalus aonidum*, and *Aphytis coheni n* sp., a parasite of *Aonidiella aurantii*. *Annals of the Entomological Society of America*, **53**, 701–705.

DeBach, P. and Rosen, D. (1991) *Biological Control by Natural Enemies*, 2nd edn, Cambridge University Press, Cambridge.

DeBach, P., Rosen, D. and Kennett, C.E. (1971) Biological control of coccids by introduced natural enemies, in *Biological Control* (ed. C.B. Huffaker), Plenum Press, New York, pp. 165–194.

Harpaz, I. and Rosen, D. (1971) Development of integrated control programmes for crop pests in Israel, in *Biological Control* (ed. C.B. Huffaker), Plenum Press, New York, pp. 458–468.

Havron, A. and Rosen, D. (1994) Selection for organophosphorus pesticide resistance in two species of *Aphytis*, in *Advances in the Study of Aphytis (Hymenoptera: Aphelinidae)* (ed. D. Rosen), Intercept, Andover, pp. 209–220.

Havron, A., Rosen, D. and Rubin, A. (1995) Releases of pesticide-resistant *Aphytis* strains in Israeli citrus orchards. *Israel Journal of Entomology*, **29**, 309–313.

Kehat, M., Swirski, E., Blumberg, D. and Greenberg, S. (1974) Integrated control of data palm pests in Israel. *Phytoparasitica*, **2**, 141–149.

Rivnay, E. (1968) Biological control of pests in Israel (a review 1905–1965). *Israel Journal of Entomology*, **3**(2), 1–156.

Rosen, D. (1967a) Biological and integrated control of citrus pests in Israel. *Journal of Economic Entomology*, **60**, 1422–1427.

Rosen, D. (1967b) Effect of commercial pesticides on the fecundity and survival of *Aphytis holoxanthus* (Hymenoptera: Aphelinidae). *Israel Journal of Agricultural Research*, **17**, 47–52.

Rosen, D. (1974) Current status of integrated control of citrus pests in Israel. *EPPO Bulletin* **4**, 363–368.

Rosen, D. (1980) Integrated control of citrus pests in Israel, in *Proceedings International Symposium of IOBC/WPRS on Integrated Control in Agriculture and Forestry (Vienna, 1979)*, pp. 289–292.

Rosen, D. (1990a) *Report of the Steering Committee for Reducing the Use of Pesticides in Agriculture*, Chief Scientist's Office, Israel Ministry of Agriculture (in Hebrew).

Rosen, D. (1990b) IPM: background and general methodology, in *Armored Scale Insects: Their Biology, Natural Enemies and Control*, World Crop Pests 4B (ed. D. Rosen), Elsevier Science Publishers, Amsterdam, pp. 515–517.

Rossler, Y. and Rosen, D. (1990) A case history: IPM on citrus in Israel, in *Armored Scale Insects: Their Biology, Natural Enemies and Control*, World Crop Pests 4B (ed. D. Rosen), Elsevier Science Publishers, Amsterdam, pp. 519–526.

Rossler, Y., Nitzan, Y. and Argov, Y. (1986) The contribution of the Israel Cohen Institute for Biological Control to pest management of citrus. *Alon Ha'notea*, **40**, 654–663 (in Hebrew, with English abstract).

Sternlicht, M., Goldenberg, S., Nesbitt, B.F. *et al.* (1978) Field evaluation of the synthetic female sex pheromone of the citrus flower moth *Prays citri* (Mill.) (Lepidoptera: Yponomeutidae), and related compounds. *Phytoparasitica*, **6**, 101–113.

Swirski, E., Amita, S., Greenberg, S. and Dorzia, N. (1968) Field trials on the toxicity of some carbamates and endosulfan to predaceous mites (Acarina: Phytoseiidae). *Israel Journal of Agricultural Research*, **18**, 41–44.

Swirski, E., Dorzia, N. and Greenberg, S. (1969) Trials on the control of the citrus rust mite (*Phyllocoptruta oleivora* Ashm.) with four pesticides, and on their toxicity to predaceous mites (Acarina: Phytoseiidae). *Israel Journal of Entomology*, **4**, 145–155.

Swirski, E., Blumberg, D., Wysoki, M. and Izhar, Y. (1987) Biological control of Japanese bayberry whitefly, *Parabemisia myricae* (Kuwana) (Homoptera: Aleyrodidae), in Israel. *Israel Journal of Entomology*, **21**, 11–18.

Swirski, E., Wysoki, M. and Izhar, Y. (1988) Integrated pest management in the avocado orchards of Israel. *Applied Agricultural Research*, **3**(1), 1–7.

5

Pesticides and natural enemies: Malaysia perspectives

Zulkifli Ayob and Normah Mustaffa

INTRODUCTION

Agricultural intensification is always associated with increased pest problems, which lead to increased use of pesticides. Being relatively cheap, effective and easy to use, farmers resort to chemical pesticides to overcome their pest problems. However, over-dependence and indiscriminate use of chemical pesticides have generated adverse effects on humans and the environment, causing resistance in pests and the destruction of natural enemies. For example, the indiscriminate use of pesticides resulted in outbreaks of brown planthoppers which destroyed rice worth US$2.4 million in Malaysia in the late 1970s.

Malaysia recorded a substantial usage of pesticides valued at US$110 million in 1994. Although chemical pesticides are well recognized as important inputs, especially at this time when there is a shortage of manpower, there are concerns about the negative impact of pesticides. This has prompted authorities in Malaysia to intensify efforts towards minimizing the risks of pesticide use.

This chapter describes Malaysian experiences in integrated pest management (IPM) with reference to biological control of pests, and the challenges ahead in the country's effort to adopt an environment-friendly approach towards pest management and pesticide regulations.

Ecotoxicology: Pesticides and beneficial organisms.
Edited by P.T. Haskell and P. McEwen. Published in 1998 by Chapman & Hall, London. ISBN 0 412 81290 8.

BIOLOGICAL CONTROL IN IPM

The incorporation of biological control in IPM programmes has been effective in reducing dependence on chemical pesticides, and has contributed to the successful control of important pests of rice, vegetables, cocoa and coconut in Malaysia.

Rice

An IPM programme for rice was initiated for integrated control of rice pests in 1981. Initially, the programme introduced surveillance and monitoring systems for key pests, training of research and extension personnel and strengthening of extension services and farmer organizations.

The effective execution of the IPM programme on rice started with the implementation of the pest surveillance system, which was implemented by the Department of Agriculture (DOA) throughout Peninsular Malaysia to assist farmers in the early detection of pests. Initially, pest monitoring and surveillance were carried out by DOA personnel, but currently this function is also being carried out by the farmers themselves. More recently pest surveillance brigades comprising school children have been established to monitor the pest situation in the field.

Major pests are being monitored regularly, including natural enemies that are important predators of brown planthopper (BPH) – namely *Cyrtorhinus lividipennis* (mirid bugs), *Lycosa pseudoannulata* (spiders), *Coccinella arcuata* (ladybird beetle) (Staphylinid), and *Casnoidea* spp. (Carabidae). Early detection of tungro virus disease is possible with the use of mobile nursery tests and enzyme-linked immunosorbent assay (ELISA) tests. The monitoring and surveillance programme has successfully prevented serious pest outbreaks such as the BPH and tungro virus disease, through guided spraying programmes.

Rodents used to be a serious threat in rice-growing areas and damage by rodents was estimated at 5–15%. The DOA has developed an IPM package for rodents in rice that incorporates biological control and cultural practices. Introduction of barn owls has successfully reduced the rat population in rice areas where nestboxes have been installed. In one project in the state of Selangor, 850 nestboxes were installed in an area covering 17 000 ha, representing a density of one nest/20 ha of rice field. Field records in some areas indicated a 90% reduction in damage (from 8.7% infestation in 1988 to 0.4% in 1993–1995). It was found that farmers were using two applications of rodenticides in 1991–1992, and one application (or none) in 1993–1995, compared with eight applications per planting season prior to the introduction of barn owls (unpublished report).

Of late, rice growers have been encouraged to rear fish (*Clarias batrachus*) in their rice plots as a strategy to discourage use of pesticides. Indirectly, fish rearing discourages farmers from using pesticides unnecessarily and encourages proper water management, resulting in yield increase. In addition, fish rearing provides extra income and a necessary protein source for the farmers.

Vegetables

Diamondback moth (DBM) is the most important pest of cruciferous vegetables in Malaysia. With resurgence recurring every two to three years due to the development of resistance to pesticides, DBM infestations cause heavy crop loss to farmers. The development and implementation of integrated control are necessary in a DBM control programme. In 1975 a biological control programme for DBM was initiated in the Cameron Highlands, Malaysia. *Cotesia plutellae* (Kurdjumov) found locally was discovered in the early 1970s. *Diadegma semiclausum* (Hellen) and *Diadromus collaris* (Gravenhorst) were introduced in the mid 1970s from New Zealand and Australia. In 1987, due to the restrictions imposed on excessive levels of pesticide residues in crucifers by the Singapore government, farmers switched to using *Bacillus thuringiensis* (Ooi, 1990). With the reduced usage of chemical pesticides, the population of parasitoids increased and farmers enjoyed good harvests in spite of using less chemical pesticides. *D. semiclausum* became the dominant parasitoid. Unfortunately, in recent years there has been a reduction in the population levels of parasitoids in the Cameron Highlands attributed mainly to increased usage of chemical pesticides to control other pests, such as leaf miners (*Chromatomyia* sp. and *Liviromyza* spp.).

Cocoa

The cocoa pod borer (CPB) *Conopomorpha cramerella* was first detected in Malaysia in 1986 and has become the major pest of cocoa. *Trichogrammatoidea bactrae fumata* (Tbf), an egg parasitoid, has been shown to be a promising biological agent for the control of cocoa pod borers if released periodically (Lee *et al.*, 1995). The successful use of Tbf depends on the ability to mass-produce the parasitoid economically. Currently, cost-effective mass-rearing techniques are being developed to make it possible to use Tbf as a major component in IPM strategy for CPB.

The black cocoa ant *Dolichoderus thoracicus* is an effective natural suppressant of the mirid *Helopeltis theobromae* in Malaysia. Higher yields were recorded in ant-abundant plots compared with those in an ant-scarce plot during a two-year observation (Ho and Khoo, 1994).

Coconut

Cordia curassavica (Jacq) R. & S. is a shrub that was brought into the country as an ornamental plant and became widespread in Malaysia in the 1970s. It has become a serious weed affecting 2000 ha of coconut. Coconut palms were seriously impaired due to competition with *C. curassavica*, which was successfully suppressed using biological agents *Metrogaleruca obscura* (= *Schematiza cordiae*) and *Eurytoma attiva*. Released in 1977, these natural enemies completely defoliated the cordia bush, which was under control the following year (Ung *et al.*, 1979).

Artona catoxantha (zygaenid moth) is an occasional pest of coconut. Outbreaks occur when there is a reduction in the natural enemy population – in particular, *Ptychomyia remota*. Local experience has shown that in controlling an outbreak of this pest the parasitism level of *P. remota* has first to be determined. If the field parasitism level is found to be above 60%, and if the pest population consists of insects that have passed the third instar, the application of chemical pesticides is not necessary. In this way, *P. remota* will increase in numbers and eventually help to reduce a build-up in the population of *A. catoxantha*.

CURRENT TRENDS IN PESTICIDE CONTROL

Pesticides will remain an essential component in pest control programmes. The present trend in IPM packages is to incorporate the judicious use of chemical pesticides. Application of pesticides based on action threshold levels of pests and natural enemies, surveillance and monitoring systems, use of pheromones, traps, bio-pesticides, and narrow-spectrum pesticides are some of the methods employed to minimize the undesirable side-effects of pesticides.

Another important tool in minimizing risks posed by pesticides is the use of official policies and legislation. In Malaysia, various aspects of pesticides are controlled under the Pesticides Act which was passed by Parliament in 1974. To date the Malaysian Pesticides Board, which is responsible for the enforcement of the Act, regulates:

- the import and manufacture of pesticides;
- premises selling and storing pesticides for sale;
- the advertisement of pesticides;
- the labelling of pesticides;
- the import of unregistered pesticides for research and education purposes;
- the use and handling of certain highly toxic pesticides.

To ensure further that pesticides imported and manufactured do not cause unacceptable adverse effects on the environment, the board has placed greater emphasis on: the evaluation of ecotoxicological and environmental aspects of pesticides; the setting up of a laboratory to analyse pesticides in the environmental samples; and better communications through dialogue and seminars with the pesticide-related industry so as to improve the management of pesticides in the country (Tan, 1996).

To complement the management of pesticides, the DOA has formed a committee on pesticide application technology aimed at improving the application technology and safety aspects of pesticides.

TESTING THE EFFECTS OF PESTICIDES ON NATURAL ENEMIES AND BENEFICIAL ORGANISMS

Risk to natural enemy

The efficient use of biological control involves maintaining a balance in the population levels of beneficial organisms and of pests. It also involves the protection of beneficial organisms. Microbial pesticides such as *B. thuringiensis* (Bt) are harmless to beneficial organisms when sprayed on target pests. It was noted that the population levels of parasitoids increased when Bt was used to control DBM in the Cameron Highlands. However, the populations of these parasitoids have decreased during recent years due to the harmful effects of other pesticides used in controlling other pests.

Future efforts to introduce or integrate biological control in IPM programmes would be futile if no effort was made to maintain the natural enemy population. Assessment of the risk of pesticides to natural enemies is an option that is currently under serious consideration by the Malaysian authorities.

Malaysian Working Group: Effects of Pesticides on Natural Enemies and Beneficial Organisms

Testing the side-effects of pesticides on beneficial arthropods is gaining more attention in many countries in Europe. Pesticide testing programmes have been initiated by the International Organization for Biological Control of Noxious Animals and Plants (IOBC) to provide information about the side-effects of pesticides on beneficial organisms. Methods that have common characteristics and common evaluation techniques were developed so that results can be compared (Hassan, 1988). Currently there are efforts to introduce testing programmes for evaluating pesticide impact on natural enemies in tropical developing countries in Asia.

IOBC/IIBC International Training Workshop

In March 1995, the IOBC's Working Group on Pesticides and Beneficial Organisms organized a training course in Malaysia designed to provide researchers from Asian countries with hands-on skills in measuring pesticidal effects on natural enemies at the laboratory, semi field and field levels (Anon., 1995). In total 16 participants from 11 countries attended the course. The effort made by the IOBC and the International Institute for Biological Control (IIBC) was timely as it provided valuable information on the risks of pesticides to natural enemies and gave an insight into current works in Europe. The workshop made several recommendations, one of which was a proposal to the participating countries to set up national and regional working groups that could coordinate the development and use of appropriate methods for testing side-effects on natural enemies.

Formation of the Malaysian Working Group

Following the recommendations of the workshop, and recognizing the risks of pesticides to beneficial organisms in the crop ecosystem, the Malaysian Pesticides Board formed a Malaysian Working Group on 'Effects of Pesticides on Natural Enemies and Beneficial Organisms' in June 1995. The working group is made up of 11 members from the various research organizations, the Agriculture University of Malaysia and extension agencies. The first meeting took place at the Department of Agriculture, Kuala Lumpur, on 18 September 1995. The main agenda discussed in the meeting was the setting of objectives and the terms of reference of the working group. Members agreed to the following objectives:

- to collect information of work done on testing of pesticides on natural enemies focusing on countries of the Southeast Asia region;
- to standardize testing procedures and if necessary develop new procedures on the effects of pesticides on natural enemies and beneficial organisms;
- to identify and set priorities on the types of pesticides for further research;
- to make recommendations to the Pesticide Board.

Progress of the working group

Database

The group has made a head start gathering much information about the existing tests on the effects of pesticides on natural enemies – specifically those done in Southeast Asia. A database containing compilations of tests on the effects of pesticides of beneficial organisms has been established. Initial findings indicated that limited work has been done in this field of study. There were variations in the test methods employed by these research workers, and differences in categorizing the effects of pesticides on the test species. However, information gathered could provide a basis for a registration agency to develop procedures such as pesticide labelling to restrict the use of products that pose a hazard to key natural enemies.

Selection of test species

On the selection of natural enemies for testing, there is a need to gather and examine more information before appropriate test species could be selected. However, the basis of selection will be based on the relevance of the test species within the agriculture systems, and on its abundance – similar to selection principles recommended by the European Standard Characteristics of Beneficial Regulatory Testing (ESCORT) (Barrett *et al.*, 1994). As it is not possible to test all species, selection would be limited in numbers and confined initially to important target pests and crops such as rice and vegetables.

Testing methods

Presently the group is still in the process of gathering information on testing methods. Standardization of testing methods could be adapted from those developed by the IOBC or European and Mediterranean Plant Protection Organization (EPPO) guidelines wherever possible. The assessment of the potential risks to natural enemies could be a scheme similar to the European one, which involves simple laboratory tests, field trials and decision-making schemes.

PROPOSALS

Joint tests – regional working group for the Asian region

Formation of a regional working group for Asia would be beneficial as it could provide networking among member countries. Also the group could be involved in the harmonization of tests and decision-making schemes, and could coordinate joint tests if need be. It is proposed that formation of this group be initiated and led by IOBC/IIBC through special funding.

Training programme

International training workshops similar to those organized by IOBC/IIBC in Malaysia should be continued on a regular basis and extended to research workers and those involved in the pesticide regulatory agencies from Asian countries. Similarly, locally organized training and workshops should be initiated by the respective countries to create awareness, and also coordinated test projects. Availability of funds would be required to conduct such training programmes.

Communication networking – Asian Networking on Natural Enemies and Pesticides (ANNEP)

ANNEP is an informal network set up during IOBC/IIBC training in Malaysia coordinated from IIBC-UK. Members are participants and resource persons of the training course. This network could be used effectively as a platform to provide information on current tests and establishment of database facilities. It is proposed that ANNEP be formalized and its scope of activities be expanded.

CONCLUSION

Malaysia recognizes the importance of the need to minimize the risk posed by hazardous chemicals, including pesticides, and is making efforts in that direction. Use of biological control in IPM programmes is a positive option although there is concern about the risk of pesticides to natural enemies and beneficial organisms. Testing the effects of pesticides on selected natural enemies and beneficial organisms may be necessary in the future as part of a registration process for pesticides in Malaysia.

The International Review Conference on Ecotoxicology: Pesticides and Beneficial Organisms was an extremely important platform from which current information on pesticides and beneficial organisms could be tapped. Malaysia was honoured to be able to share her experiences – in particular, the setting up of the National Working Group on the Effects of Pesticides on Natural Enemies and Beneficial Organisms – and hopes that, through the conference, we have built a foundation for further cooperation in this area of studies.

ACKNOWLEDGEMENTS

The authors would like to express their gratitude to the Welsh Pest Management Forum for extending the invitation to Mr Zulkifli Ayob (Director, Pest Control Branch) to attend the International Review Conference on Ecotoxicology: Pesticides and Beneficial Organisms. Sincere appreciation is directed to the Director General of Agriculture, Malaysia, for his support and permission to attend this conference and present this paper.

REFERENCES

Anon. (1995) *Executive Report: IOBC/IIBC International Training Workshop on Evaluation of Pesticides Effects on Natural Enemies and Its Implications for Pesticide Registration*, IIBC.

Barrett, K.L., Grandy, N., Harrison, E.G. *et al.* (1994) *Guidance Document on Regulatory Testing Procedures for Pesticides with Non Target Arthropods. From the ESCORT workshop, Wageningen, Netherlands, March 1994*, SETAC-Europe, Brussels.

Hassan, S.A. (1988) Guidelines for testing the effects of pesticides on beneficials. IOBC/WPRS Bulletin, XI(4)m 143.

Ho, C.T. and Khoo, K.C. (1994) Some factors influencing sustenance of high activity of black cocoa ant *Dolichoderus thoracicus* (Hymenoptera: Formicidae) abundance and cocoa borer *Conopomorpha cramerella* (Lepidoptera: Gracillariidae) in cocoa estates, in *Proceedings Fourth International Conference on Plant Protection in the Tropics*, pp. 218–220.

Lee, C.T., Tay, E.B., Lee Ming Teng, Sulaiman Hashim (1995) Mass rearing of egg parasitoid TBF. Recent advances in the management of Cocoa Pod Borer. Malaysia Cocoa Board.

Ooi, P.A.C. (1990) Role of parasites in managing diamondback moth in the Cameron Highlands, Malaysia, in *Proceedings of the 2nd International Workshop, Diamondback Moth and Other Crucifer Pests*, pp. 225–262.

Tan, S.H. (1996) Current Trends in Pesticide Control. Paper presented in Seminar on Pesticides in the Agroenvironment: Fate and Impact, University of Malaya, pp. 7–8

Ung, S.H., Yunus, A. and Chin, W.H. (1979) Biological control of *Cordia curassavica* (Jacq) R & S in Malaysia by *Schematiza cordiae* (Barb.) (Coleop: Galerucidae). *Malayan Agricultural Journal*, **52**, 154–165.

6

Can we achieve harmonization of regulatory requirements in Europe?

Katie Barrett

INTRODUCTION

This chapter summarizes the position with respect to regulatory testing requirements for pesticides, the risk assessment and management proposals and the assessment of products for use in integrated pest management (IPM) systems. This may sound relatively straightforward, but of all the areas of ecotoxicology and pesticide registration this particular topic has generated more debate than any other, for it seems that almost every national authority, every expert entomologist, and each agrochemical company has a slightly different opinion and position.

For other areas of ecotoxicology – for example, testing with aquatic and avian species – there are well established guideline methodologies from OECD (Organization for Economic Cooperation and Development), EPA (Environmental Protection Agency) or the EC (European Community), which can be used to determine toxicity; this data is then used in subsequent risk assessment schemes and management proposals are made. The aims of the testing are well defined.

Until the recent harmonization of regulatory requirements for pesticides, with the ratification of EC 91/414 (Council Directive, 1991), the requirement for testing with beneficial and non-target species varied between national authorities. The importance placed upon testing with

Ecotoxicology: Pesticides and beneficial organisms.
Edited by P.T. Haskell and P. McEwen. Published in 1998 by Chapman & Hall, London. ISBN 0 412 81290 8.

this group of organisms depended to some extent upon the agronomic practice of the country and the extent to which IPM was practised. When discussing IPM it is important to clarify and define what is included. In some regions, the use of pesticides only when the presence of infestation or disease is identified is considered IPM; for others it involves the active introduction of beneficial species, and there are various degrees of IPM between these two extremes. In some countries it is claimed that IPM accounts for as much as 70% of the cropped and managed area; in other countries IPM is almost exclusively confined to greenhouse use. For some the issue is not so much the impact on beneficial species, but the impact on non-target organisms, which may be found in both within-crop and off-crop situations. This is of particular concern in areas managed for game birds, where these beneficial species form an important part of the chicks' diet (Sotherton and Moreby, 1992).

INFLUENTIAL GROUPS

A number of key organizations have contributed to the quest for European harmonization of beneficials testing for pesticide regulatory purposes. With the drafting of EC 91/414 a number of expert groups were set up under the auspices of EPPO (European and Mediterranean Plant Protection Organization) in conjunction with the Council of Europe (CoE), to prepare proposals and a sequential framework which would form the basis of the regulatory requirements in different areas. Such a group was set up to address the area of testing with arthropod natural enemies. The individuals participating in the group have changed over the years but it is made up of members of regulatory authorities, industry and independent experts, and is currently chaired by Dr Peter Oomen, of the Plant Protection Service, The Netherlands.

The EPPO scheme, as currently published (EPPO/CoE, 1994), has provided the initial skeleton around which subsequent work and discussions have taken place. During its conception, the group consulted widely, with regulators from other countries and expert groups working in this area, including IOBC (International Organization for Biological and Integrated Control of Noxious Animals and Plants) and BART (Beneficial Arthropod Regulatory Testing group).

The IOBC has been responsible for the development of the majority of the laboratory-based testing methods used today. It is the largest single group of experts providing data on a range of pesticide products, and has enabled valuable recommendations to be made on the use of these products in IPM and ICM (integrated crop management) systems.

The BART group, formed in 1989, was initially an industry group set up primarily to address the issue of regulatory testing and beneficial arthropods. At the time there was concern within the agrochemical industry that the regulatory testing requirements for this group of species would exceed that for any other area of ecotoxicology. They had independently formulated their own proposals for regulatory testing requirements, which was based on the principles applied to other key areas: a sequential testing scheme generating a limited, quality data set with four key indicator species using proven reproducible methods that had been validated through extensive ring testing (Barrett, 1992). In its quest to develop the methods, the membership of the group widened over time to include testing experts from a number of organizations.

As a result of discussions between these groups, it became apparent that a number of issues needed to be resolved. The only way to achieve this was to bring all the relevant parties together to form a workshop.

THE ESCORT WORKSHOP

The ESCORT (European Standard Characteristics of Beneficial Regulatory Testing) workshop took place in March 1994 in Holland. The workshop, organised by BART, EPPO and IOBC, was funded by the EC and run under the auspices of SETAC (Society for Environmental Toxicology and Chemistry). Thirty-five people participated, representing different European national authorities, industry and experts working in the field of entomology. The objective of the workshop was to develop a guidance document for the testing of the effects of pesticides on non-target arthropods for regulatory purposes, particularly with respect to EC Directive 91/414, which could be used in conjunction with the risk assessment scheme developed by the EPPO/CoE working group.

The participants at the workshop were split into five working groups to cover specific topics, including:

- selection of test species;
- laboratory and extended laboratory testing;
- semi-field and field testing;
- principles of testing;
- interpretation of data: risk assessment and management.

Each group debated their topic initially in private, and then presented their recommendations back to the whole group. Based on the comments received here they went back and refined their recommendations. The workshop was chaired by Dr Nicky Grandy from the OECD. The OECD is also keen to add testing methods on arthropod species to its guidelines – hence their interest and involvement.

In determining which species should be tested, it first had to be decided for what purpose the data was intended. Was it to provide a minimal data set to indicate the toxicity of these products to arthropod species, or to determine the suitability of the products for IPM use? It was agreed finally that the species tested for regulatory purposes would be selected from arthropod natural enemies, or beneficial species, since information on the potential use of the product in IPM programmes could be obtained. However, it was recognized that registration testing could not provide all the information necessary to cover all potential IPM situations, and that work such as that carried out by the IOBC would still be particularly valuable.

Finally 13 species were identified as the preferred test species (Table 6.1). For all products it was agreed that two standard species – a parasitoid and a predatory mite – would be tested to provide a basic data set on all products which could be used for comparative and ranking purposes. Two additional species should also be tested for each product, depending on the proposed use. The species are divided into four taxonomic groups, and are further divided by crop type. If appropriate, the additional two species chosen would come from the ground-dwelling predator and foliage-dwelling predator groups.

A sequential testing procedure was advocated, with tests being conducted initially in the laboratory under carefully controlled conditions. These include use of an artificial substrate (glass or sand) for initial exposure to the pesticide residues, under closely controlled environmental conditions. Where harmful effects were determined at this level of testing, a more realistic route of exposure could be used in an extended laboratory test, for which it was recommended that the product would be applied at a single rate equivalent to the highest recommended for

Table 6.1 ESCORT recommendations for test species

Crop type	*Parasitoids*	*Predatory mites*	*Ground-dwelling predators*	*Foliage-dwelling predators*
Orchard (greenhouse, forest and vineyard)[a]	*Aphidius rhopalosiphi* *Trichogramma cacoeciae* *Leptomastix dactylopii* *Drino* sp.	*Typhlodromus pyri* *Amblyseius* sp.	*Pardosa* sp. *Poecilus cupreus*	*Orius* sp. *Episyrphus balteatus* *Chrysoperla carnea* *Coccinella septempunctata*
Arable[b]	*Aphidius rhopalosiphi* *Trichogramma cacoeciae*		*Poecilus cupreus* *Pardosa* sp. *Aleochara bilineata*	*Episyrphus balteatus* *Chrysoperla carnea* *Coccinella septempunctata*

[a] The resemblance in species occurring in glasshouses, vineyards and orchards is such that these outlets can be included in one category for preliminary assessments
[b] The arable category includes all vegetables, cereal and foliage crops.

field application using a representative lead formulation. For products to be applied two or three times per season, it was recommended that they should be applied at twice the recommended application rate initially; and for products applied more frequently, further testing was recommended that would indicate the effects of repeat application and potential residue build-up.

For products still indicating harmful effects, the next step in the sequential testing scheme is the semi-field test, where the study is conducted under more realistic use conditions in the field, in an enclosure or cage.

The ultimate step in the evaluation procedure would be a field study conducted under realistic conditions of use with natural populations of beneficial and non-target species. These field studies form part of the sequential side-effects testing procedure and are separate from those conducted to assess efficacy. For all steps in this sequential testing procedure it is recommended that a representative formulation is used, and that at the field study level the method of application should also reflect the method that would ultimately be recommended, as this could affect the outcome of the study.

The recommendations made at the ESCORT workshop were published as a SETAC guidance document (Barrett *et al.*, 1994) and form the text cited in the current version of Annexes II and III of EC 91/414.

RESEARCH ACTIVITIES AND ISSUES OF CONCERNS

In addition to the recommendations outlined, a number of actions and areas requiring research were identified. In recommending this sequential testing approach, it was recognized that the methods used needed to be ring tested and validated, in order to determine which parameters gave greatest variability and to identify ways of improving the methods so as to reduce this inherent variability in the results that they produced. This would ensure that regulators could have a degree of confidence in the data supplied from different sources. It was also agreed that the blanket trigger values applied to all species within the EPPO scheme should be revised to be species specific. To achieve this, the joint testing initiative of IOBC, BART, COMET and EPPO was set up, within which working groups for each of the preferred test species are actively developing, improving and evaluating the methodologies available.

Since the publication of the ESCORT workshop recommendations a number of issues have continued to be debated. For example:

- The merit of dose-response/multi-concentration testing versus single concentration testing. Some authorities and companies advocate this approach, which generates additional information – allowing a degree of extrapolation to different usage patterns and agricultural situations.

- Assessing the effects of multi-application products. This is not proving to be practical using the first tier laboratory tests with glass or an inert substrate; however, method development at the extended laboratory testing level, applying multiple applications to plants, is under development for a number of species.
- Extrapolation to non-target species. Many people working in this field feel that it should be possible to extrapolate from data generated on beneficial species to non-target species, but we appear to lack the scientific evidence to support this.
- Use of the data in risk assessment and mitigation recommendations.

This last point is of particular concern at present. Within the UK a proposal has now been ratified that provides the regulating authority with a scheme by which they can evaluate the data and make label recommendations (Chapter 24). Through discussions in a number of fora it is apparent that whilst there is a lot of common ground and agreement between interested parties, there are also some issues about which it is difficult, at this time, to see how an acceptable compromise can be achieved. The agronomic practices are very different in some member states, and whilst it is an admirable objective to divorce IPM assessment from registration, it is proving more difficult in reality for some authorities. The situation has been further complicated by the different approaches taken by the national authorities in reviewing data. In some organizations the same experts are regulating not only on the suitability of the compound for use in the agricultural environment, but also for the classification given to the product with respect to its use in IPM systems. In other authorities these roles are separate, and two reviews are conducted. Due to the different agronomic practices within Europe, the proposals for risk mitigation have also differed significantly. In those countries where IPM is practised widely, it has been proposed that more information is provided to the farmer on the label, giving details for the species tested, those affected and those unaffected. In some countries buffer zones have been recommended as a viable risk mitigation practice. In countries with smaller fields, or where a strip cultivation approach is used, this is not practical. Thus a number of issues are still to be resolved.

Not only is IPM becoming the focus of increasing attention for regulatory authorities: there is also an increased awareness at the consumer level. In response to public concern over the use of chemical crop protection products, the large supermarkets are providing information to the public about the advantages of IPM, using beneficial species to control crop pests instead of chemicals. They are also increasing the pressure on their growers and suppliers to adopt ICM and IPM methods.

At the other end of the scale, there is increased awareness within Brussels that there is no 'European' policy on IPM and they are currently reviewing agricultural practices in the member countries and assessing the viability of different systems, although for the reasons given previously it is unlikely that a single policy could hope to address all the different agricultural systems and agronomic practices.

The European Crop Protection Association (ECPA) is involved in a similar mission. Overall there is motivation towards increased use of IPM, and cooperation at the national and international level.

CONCLUSION

What of the future? A revised EPPO scheme is in preparation, which takes into account the recommendations of ESCORT, but also some of the issues currently under debate.

With so much activity again focused on this area of ecotoxicology it has been recommended that the time is right for a second SETAC workshop. This suggestion has been discussed with the EC authorities and plans are in progress for a meeting in 1998.

To achieve true 'harmonization' of procedures and recommendations in all areas may be an unrealistic objective and it may be better to accept that there needs to be a degree of flexibility allowing national authorities to adapt recommendations to suit their individual needs and agricultural environments.

REFERENCES

Barrett, K.L. (1992) Intepretation of pesticide effects on beneficial arthropods. The BART group approach to regulatory testing. *Aspects of Applied Biology*, **31**, 165–170.

Barrett, K.L., Grandy, N., Harrison, E.G. *et al.* (1994) *Guidance Document on Regulatory Testing Procedures for Pesticides with Non Target Arthropods. From the ESCORT workshop, Wageningen, Netherlands, March 1994*, SETAC-Europe, Brussels.

Council Directive (1991) Council Directive 15 July 1991 concerning the placing of plant protection products on the market (91/414/EEC). *Official Journal of the European Communities*, **L230**, 1–32, 19 August 1991

EPPO/CoE (1994) Decision making scheme for the environmental risk assessment of plant protection products – arthropod natural enemies, *EPPO Bulletin* **24**, 17–35.

Sotherton, N.W. and Moreby, S.J. (1992) Interpretation of pesticide effects on beneficial arthropods. The importance of beneficial arthropods other than natural enemies in cereal fields. *Aspects of Applied Biology*, **31**, 11–18.

Part Two

Defining the Problem

7

Introduction

Sherif A. Hassan

Parasites and predators of agricultural pests reduce the population of their prey or host and help to limit damage caused by the pest. Modern plant protection therefore recommends reduction in the use of chemical pesticides to a minimum. If the use of pesticides is indispensable, selective pesticides should be chosen. One of the major aims of the international working group 'Pesticides and Beneficial Organisms' of the International Organization for Biological Control of Noxious Animals and Plants, West Palaearctic Regional Section (IOBC/WPRS), which was established in 1974, was therefore to coordinate international activities to develop standard methods to test the side-effects of pesticides on the most important natural enemies and to choose selective pesticides suitable for use in integrated control programmes. Review of the literature published before the IOBC initiative showed that diverse methods, modes of exposure, monitoring techniques and rating systems were used. This made it extremely difficult to summarize and compare results on the side-effects of compounds, but by using standard methods the results of the IOBC testing programmes can easily be presented in clear and simple tables.

Standard laboratory, semi-field and field methods to test the side-effects of pesticides on beneficial organisms were developed by the Working Group according to common characteristics. Guidelines to test the side effects of pesticides on natural enemies were published in multi-author publications (Hassan *et al.*, 1985; Hassan, 1988, 1992). Joint IOBC testing programmes provide valuable information to users of integrated control and give the organizations carrying out the tests an opportunity to improve their testing techniques and develop better guidelines.

Ecotoxicology: Pesticides and beneficial organisms.
Edited by P.T. Haskell and P. McEwen. Published in 1998 by Chapman & Hall, London. ISBN 0 412 81290 8.

Whereas harmlessness of pesticides can easily be shown by laboratory experiments, harmfulness can only be confirmed under practical conditions in the field. Pesticides found to be harmless to a particular beneficial organism in the laboratory test are most likely to be harmless to the same organism in the field and no further testing in semi-field or field experiments is therefore recommended.

Testing the side-effects of pesticides on beneficial organisms has become obligatory in several countries and this made the development of internationally approved guidelines even more important and urgent. The use of standard methods will allow the exchange of results between countries and save the cost of repeated testing. The harmonization of testing guidelines with international organizations such as the European and Mediterranean Plant Protection Organization (EPPO) and the European Commission (EC) is important.

Information on rearing and testing methods for the following natural enemies were published: *Encarsia formosa* (Hoogcarspel and Jobsen, 1984; Oomen, 1985); *Trichogramma cacoeciae*, initial toxicity (Hassan, 1974; 1977), persistence (Hassan, 1980); *Cales noacki*; *Leptomastix dactylopii* (Viggiani and Tranfaglia, 1978); *Phygadeuon trichops* (Plattner and Naton, 1975; Naton, 1983); *Coccygomimus turionellae* (Bogenschütz, 1975, 1984; Bogenschütz *et al.*, 1986); *Pales pavida* (Huang, 1981); *Chrysoperla carnea* (Suter, 1978; Rumpf *et al.*, 1992; Vogt, 1992); *Aphidoletes aphidimyza* (Helyer, 1991); *Amblyseius potentillae* (Overmeer and van Zon, 1982); *Phytoseiulus persimilis* (Samsoe-Petersen, 1983; Stolz, 1990); *Typhlodromus pyri* (Boness *et al.*, 1982; Duso *et al.*, 1992); different phytoseiid mites (Bakker and Calis, 1989); Carabidae (Heimbach, 1988); spiders (Mansour and Nentwig, 1988; Wehling and Heimbach, 1991; Mansour *et al.*, 1992); *Verticillium lecanii* (Tuset, 1975, 1988); *Coelotes terrestris* (Albert and Bogenschütz, 1984); *Anthocoris nemoralis* (Stäubli *et al.*, 1984); *Aleochara bilineata* (Samsoe-Petersen, 1987); effects on soil-dwelling invertebrates (Samsoe-Petersen *et al.*, 1992); pathogenic fungus (Keller and Schweizer, 1991); naturally occurring field organisms (Gendrier and Reboulet, 1992). Methods of rearing 16 different beneficial insect and mite species were discussed in working group meetings and published by Samsoe-Petersen *et al.* (1989).

STANDARD CHARACTERISTICS OF TEST METHODS

Recognizing that no single test method would provide sufficient information to assess the side-effects of pesticides on a beneficial organism, a combination of tests that include laboratory, semi-field and field methods to be carried out in a particular sequence is recommended. The methods were developed according to the following standards.

Laboratory tests

Test (a): Susceptible life stage (e.g. adults of parasites, developmental stages of mites, larvae of predatory insects)

1. Exposure of organisms to fresh pesticide deposit applied on glass plate, leaf, sand, sandy soil.
2. Exposure of beneficial fungus, nematodes and collembola in contaminated standard medium (e.g. based on broth, agar or soil).
3. Even film of pesticide, standard amount of 1.5–2 mg fluid/cm^2 on glass or leaf and 2–6 mg fluid/cm^2 on sand are used.
4. Highest recommended application rate of pesticide.
5. Laboratory-reared or field-collected organisms of uniform age.
6. Adequate exposure period before evaluation.
7. Adequate ventilation.
8. Water-treated control in each experiment; toxic standard in at least one experiment per year.
9. Assessment of the reduction in beneficial capacity (egg laying, parasitism) as well as mortality.
10. Four evaluation categories:
    1 = harmless (< 30%);
    2 = slightly harmful (30–79%);
    3 = moderately harmful (80–99%);
    4 = harmful (> 99%).

Test (b): Less susceptible life stage (e.g. parasites within their hosts, adults of mites, adults of predatory insects)

1. Direct spraying of organisms and substratum. Points (3) to (10) of Test (a) are applicable.

Test (c): Duration of harmful activity (persistence)

1. Exposure to pesticide residues applied on plants or soil at intervals after treatment.
2. Weathering in the field under rain cover with periodical exposure to direct sunshine or under simulated field conditions (summer day).
3. Pesticide application according to good agricultural practice (GAP).
4. Experiments and assessment of toxicity as in Test (a), points (4) to (10).
5. Repeating of test at intervals until there is loss of toxicity (category 1 result), or up to one month after treatment.
6. Four evaluation categories:
   A = short-lived (< 5 days);
   B = slightly persistent (5–15 days);

C = moderately persistent (16–30 days);
D = persistent (> 30 days).

Test (d): Extended laboratory

1. Experiments are carried out in the laboratory under defined experimental conditions.
2. A susceptible life stage of organism is used.
3. Adequate ventilation and air exchange to prevent the accumulation of pesticide fumes. Points (4) to (12) of the semi-field test (below) are applicable.

Semi-field test

1. Experiments are carried out in the field with climatic factors to be left unaffected as much as possible, but rain cover can be used where necessary.
2. Appropriate time, crop and season for the chemical, but choosing conditions to represent the worst case.
3. Experiments to be repeated under different weather conditions when considered necessary.
4. Beneficial organisms (possibly a susceptible life stage) to be present on the crop during spraying – if practical – or to be released as soon as possible after spraying.
5. Laboratory-reared or field-collected organisms of uniform age.
6. Highest recommended application rate of pesticide.
7. Application according to GAP.
8. Adequate exposure period before evaluation.
9. Water-treated control and toxic standard in each experiment.
10. Assessment of the reduction in beneficial capacity (egg laying, parasitism, prey intake, population changes) as well as mortality.
11. Four evaluation categories:
    1 = harmless (< 25%);
    2 = slightly harmful (25–50%);
    3 = moderately harmful (51–75%);
    4 = harmful (> 75%).

Field

Test (a): Naturally occurring organisms

1. Crops or soil inhabited by naturally occurring beneficials are directly sprayed.
2. Experiment to be repeated at different locations.

3. No release of beneficial organisms in the same year as the experiment.
4. Sampling is carried out at intervals before and after treatment(s).
5. Highest recommended dose rates and number of treatments following GAP.
6. Experiments are carried out at the appropriate time and season for the chemical.
7. Adequate exposure period before evaluation.
8. Water-treated control and toxic standard in each experiment.
9. Mortality, survival and population changes may be monitored.
10. Plot design and number of individuals to exceed a certain limit to allow statistical analysis.
11. Four evaluation categories:
    1 = harmless (< 25%);
    2 = slightly harmful (25–50%);
    3 = moderately harmful (51–75%);
    4 = harmful (> 75%).

Test (b): Released organisms

1. Laboratory-reared or field-collected beneficial organisms of uniform age are released in field plots and are directly sprayed. Points 4 to 11 of the field test are applicable. Categories as under Field Test (a).

Functions of the different types of tests

(a) Laboratory: susceptible life stage

To prove the harmlessness of pesticides, and to screen out harmless or low toxicity preparations (harmless pesticides are not tested any further).

(b) Laboratory: less susceptible life stage

To help to estimate hazard and differentiate between toxic preparations.

(c) Laboratory: duration of harmful activity (persistence)

To help to estimate hazard. The impact of a pesticide in the field is greatly affected by its persistence. Short-lived pesticides can often be successfully used in integrated control programmes.

(d) Laboratory: extended laboratory

To help to estimate hazard under simulated field conditions.

(e) Semi-field and field

To assess the hazard of pesticides, and to provide information relevant to practice.

Selection of test species

As it is impossible to test all species that may be exposed to pesticides, it is recognized that a limited number of species must be selected. It is important that the beneficial organisms chosen should be relevant to the crops on which the particular pesticide is to be used. Registration testing cannot provide the detailed information necessary for integrated pest management (IPM), and expert groups like the IOBC Working Group 'Pesticides and Beneficial Organisms' should continue their work to assess the suitability of products for specific IPM programmes. It is possible that products which cannot be classified as harmless in the first laboratory tests may nevertheless be considered for use in IPM programmes and identified through the work of research scientists.

The process of selection of test species should be based principally on sensitivity to pesticides, relevance to the field environment and amenability.

Sensitivity

The use of sensitive species is recommended but it is accepted that there is no 'most sensitive' species. The sensitivity of the test species, in combination with the appropriate laboratory methods, should be sufficient to predict unacceptable effects on these species under realistic conditions of use, with a level of certainty. It is also desirable that the tests do not produce harmful effects where there would be no risk, i.e. they should be effective at differentiating between 'harmful' and 'harmless' products. Tables 7.1 to 7.4 show the relative sensitivity of a number of natural enemies as tested by the IOBC Working Group. It can be seen from the tables that *Trichogramma cacoeciae* is more sensitive than eight other Hymenoptera. The susceptibility of this species compared with *Encarsia formosa* was equal in 73% of the preparations tested, with 19% higher and only 4% lower. The differences from the other parasitic Hymenoptera tested (Table 7.1) were even larger. *Trichogramma* is therefore considered to be a good indicator species for this order of insects. *Typhlodromus pyri* (Table 7.2) was also shown to be a good indicator species for seven other strains of predatory mites including susceptible and resistant strains of *Amblyseius* and *Phytoseiulus* species. The data in Table 7.3 show that *Phytoseiulus persimilis* is less susceptible than the other predatory mites. Table 7.4 shows that *Chrysoperla carnea* is more tolerant to

Table 7.1 *Trichogramma cacoeciae* as an indicator for other parasites: susceptibility compared with several related species

Species	*Number of experiments*	*Susceptibility compared with* Trichogramma		
		Equal	*Higher*	*Lower*
Encarsia formosa	86	63 (73.3%)	19 (22.1%)	4 (4.6%)
Aphidius matricaria	58	46 (79.3%)	10 (17.2%)	2 (3.5%)
Phygadeuon trichops	113	75 (66.4%)	33 (29.2%)	5 (4.4%)
Coccygomimus turionellae	122	73 (59.8%)	47 (38.5%)	2 (1.7%)
Leptomastix dactylopii	118	74 (62.7%)	42 (35.6%)	2 (1.7%)
Drino inconspicua	25	18 (72%)	7 (28%)	0
Opius sp.	20	9 (45%)	11 (55%)	0
Pales pavida	20	13 (65%)	7 (35%)	0

Table 7.2 *Typhlodromus pyri* as an indicator for predatory mites: susceptibility compared with several related species

Species	*Number of experiments*	*Susceptibility compared with* Typhlodromus pyri		
		Equal	*Higher*	*Lower*
Amblyseius spp.	81	69 (85.2%)	10 (12.3%)	2 (2.5%)
Amblyseius spp. R	18	14 (77.9%)	4 (22.2%)	0
Phytoseiulus persimilis	22	9 (40.9%)	13 (59.1%)	0
Phytoseiulus persimilis R	59	34 (57.6%)	23 (39%)	2 (3.4%)
Typhlodromus pyri R	59	46 (78%)	12 (20.3%)	1 (1.7%)
Typhlodromus pyri Feld	16	11 (68.7%)	4 (25%)	1 (6.3%)
Amblysius finlandicus Feld	61	42 (68.8%)	15 (24.6%)	4 (6.6%)

pesticides and therefore not suitable as an indicator species for several aphid predators.

Relevance

The test species should be relevant within agricultural systems, relatively abundant and widely distributed geographically. The species should also have demonstrated beneficial capacity, and have functional or taxonomic similarity to the groups they represent.

Table 7.3 Susceptibility of *Phytoseiulus persimilis*, resistant strain, compared with several related species.

Species	*Number of experiments*	*Susceptibility compared with* Phytoseiulus persimilis		
		Equal	*Higher*	*Lower*
Amblyseius potentillae	101	61 (60.4%)	2 (2%)	38 (37.6%)
Typhlodromus pyri	81	43 (53.1%)	2 (2.5%)	36 (44.4%)
Typhlodromus pyri R	59	39 (66.1%)	3 (5.1%)	17 (28.8%)
Chiracanthium (Spinne)	39	21 (53.8%)	10 (25.6%)	8 (20.6%)
Coelotes (Spinne)	21	12 (57.1%)	4 (19.1%)	5 (23.8%)

Table 7.4 Susceptibility of *Chrysoperla carnea* compared with several related species

Species	*Number of experiments*	*Susceptibility compared with* Chrysoperla carnea		
		Equal	*Higher*	*Lower*
Syrphus corollae	85	41 (48.1%)	6 (7.1%)	38
Coccinella septempunctata	22	13 (59.1%)	0	9
Harmonia axyridis	40	27 (67.5%)	7 (17.5%)	6
Semiadalia 11-notata	57	27 (47.4%)	5 (8.8%)	25
Anthocoris nemorum	69	43 (62.3%)	17 (24.6%)	9

Amenability

Although amenability is not the most important factor in the selection process, it is important that the species are relatively easy to maintain and handle so that the tests may be successfully carried out in different laboratories. It is advantageous if the species can be cultured in the laboratory, to ensure availability and also, for example, homogeneity in age.

Choice of species for regulatory testing

As a first stage in identifying suitable species for use in tests for pesticide registration, crop environments that contain different non-target arthropods were considered. Two main crop categories were identified: arable and orchards. The resemblance in species occurring in greenhouses, vineyards and orchards is such that these were included in one category. The arable category includes all vegetable, cereal and forage

Table 7.5 Selection of relevant test species for regulatory testing: species should be selected according to the intended use of the product (here the species belonging to different groups are categorized into two main field environments – orchards and arable crops – in which they are particularly relevant)

Crop type	*Parasitoids*	*Predatory mites*	*Ground-dwelling predators*	*Foliage-dwelling predators*
Orchard (greenhouse, forest and vineyard)	*Aphidius rhopalosiphi* *Trichogramma cacoeciae* *Leptomastix dactilopii* *Drino* sp.	*Typhlodromus pyri* *Amblyseius* sp.	*Pardosa* sp. *Poecilus cupreus*	*Orius* sp. *Episyrphus balteatus* *Chrysoperla carnea* *Coccinella septempunctata*
Arable crops	*A. rhopalosiphi* *T. cacoeciae*		*P. cupreus* *Pardosa* sp. *Aleochara bilineata*	*E. balteatus* *C. carnea* *C. septempunctata*

crops. Some crops requiring unique species were also identified, e.g. forestry and citrus.

On the basis of these crop groupings, a table of preferred test species was prepared (Table 7.5). Species were categorized into one of four functional groups: parasitoids, predatory mites, ground-dwelling predators and foliage-dwelling predators.

It is recommended that two sensitive standard species and two species relevant to the intended use of the product are tested. The recommended standard species are *Aphidius rhopalosiphi* and *Typhlodroums pyri*. *Trichogramma* can be used instead of *Aphidius*. *Amblyseius* could replace *Typhlodromus* if tests show this species to be more sensitive and amenable. To fulfil requirements, the test substance to be used should be the lead formulation rather than the active ingredient.

Choice of species for IPM testing

More elaborate testing is recommended for IPM. More relevant natural enemies should be included in the testing programmes. Tables 7.6 and 7.7 show lists of natural enemies that would be tested for IPM purposes. These lists indicate the relevance of beneficial organisms in 16 different crops or groups of crops.

Table 7.6 Choice of beneficial for IPM testing: one organism from each of the four groups of natural enemies is chosen (the testing of at least four natural enemies per pesticide is recommended)

Parasitic insects	*Predatory insects*		*Predatory mites and spiders*
	Soil-living	*Foliage-living*	
Trichogramma	*Aleochara*	*Chrysoperla*	
Typhlodromus			
Coccygomimus	*Bembidion*	*Anthocoris*	*Erigone*
Encarsia	*Pterostichus*	*Coccinella*	
Phytoseiulus			
Aphidius		*Metasyrphus*	
Chiracanthium			
Phygadeuon		*Semiadalia*	*Ambylseius*
		Orius	*Coelotes*
		Aphidoletes	

Table 7.7 Choice of natural enemies relevant to the crop on which the particular pesticide is to be used (the testing of at least four natural enemies per pesticide is recommended)

Crop	*Parasitic insects*	*Predatory insects*		*Predatory spiders and mites*
		Soil-living	*Foliage-living*	
Cereals	*Ahidius*	*Bembidion*	*Coccinella*	*Erigone*
		Pterostichus	*Semiadalia*	
Corn	*Trichogramma*	*Bembidion*	*Coccinella*	*Erigone*
	Aphidius		*Chrysoperla*	
Sugar beet	*Phygadeuon*	*Aleochara*	*Metasyrphus*	*Erigone*
	Trichogramma		*Semiadalia*	
Potato	*Aphidius*	*Bembidion*	*Chrysopherla*	*Erigone*
Forage, rape	*Trichogramma*	*Bembidion*	*Chrysoperla*	
Fruit orchard	*Trichogramma*		*Anthocoris*	*Typhlodromus*
	Aphidius		*Chrysoperla*	*Amblyseius*
Vineyards	*Trichogramma*		*Chrysoperla*	*Typhlodromus*
			Anthocoris	
Cabbage	*Trichogramma*	*Aleochara*	*Metasyrphus*	
	Phygadeuon	*Bembidion*	*Aphidoletes*	
Peas	*Trichogramma*	*Bembidion*	*Orius*	*Erigone*
Beans	*Aphidius*	*Aleochara*	*Coccinella*	*Phytoseiulus*
	Phygadeuon		*Chrysoperla*	
Onions	*Phygadeuon*	*Aleochara*	*Orius*	
GH – Cucumber	*Encarsia*		*Orius*	*Phytoseiulus*
	Aphidius		*Chrysoperla*	

Table 7.7 *continued*

Crop	Parasitic insects	Predatory insects		Predatory spiders and mites
		Soil-living	Foliage-living	
GH - Tomato	*Encarsia*		*Aphidoletes*	*Phytoseiulus*
	Aphidius		*Chrysoperla*	
GH - Pepper	*Aphidius*		*Chrysoperla*	*Phytoseiulus*
	Encarsia		*Apidoletes*	
GH - Ornament	*Encarsia*		*Aphidoletes*	*Phytoseiulus*
	Aphidius		*Chrysoperla*	
Forest	*Coccygomimus*	*Bembidion*	*Coccinella*	*Coelotes*
	Trichogramma	*Pterostichus*	*Chrysoperla*	*Chiracanthium*

EXTRAPOLATION

The beneficial species selected for testing should belong to different taxonomic groups so that information on the potential hazard to other non-target organisms ('neutral species') may also be provided. Results from insecticide primary screening tests and honeybee tests, if available, may provide additional useful information about the potential risk to arthropod species.

Extrapolation from one species to another has to be made with caution. However, sensitive species are used here, and the tests are carried out under 'worst-case' conditions. If these laboratory tests indicate no toxicity, the likelihood that significant effects will occur in the field is very small.

Extrapolation to non-target 'neutral' species

Effects on beneficial organisms may give an indication about the potential effects of a product on non-target (neutral) species of the same taxonomic group and the same trophic level.

Extrapolation to functional and ecological groups

Effects on beneficials may give an indication about the potential effects on species of the same functional and ecological group. Based on present knowledge, the extrapolation from results obtained in tests with sensitive indicator species to species from the same taxonomic order appears to be feasible, at least in some groups. Results from experiments with hymenopterous species and predatory mites are given in Tables 7.1 and 7.2.

Extrapolation of laboratory and semi-field data to field situations

Whereas harmlessness of pesticides can easily be shown by laboratory experiments, harmfulness can only be confirmed under practical conditions in the field. Pesticides found to be harmless to a particular beneficial organism in a laboratory test are most likely to be harmless to the same organism in the field and no further testing in semi-field or field experiments is therefore recommended. Because behaviour and ecology are restricted in controlled laboratory tests, extrapolation between species is more justified at this stage of testing. Work to compare results of the different types of standard tests (laboratory, semi-field and field) was recently published in an IOBC/WPRS Bulletin (Vogt, 1994). Information included in the bulletin should help research workers to interpret the results of the different types of tests.

In semi-field conditions, extrapolation between species will be more difficult than in the laboratory because behaviour becomes more relevant. Even similar species can behave quite differently, which may result in different exposure and thus different susceptibility.

In the majority of cases, harmlessness – as demonstrated in laboratory assays – can be extrapolated to field conditions for the test species. However, several factors can influence this prediction, such as body size, ecology, behaviour, sensitivity of the strain tested, the test method (overspray vs. residual method), the temperature at which the test is conducted, the presence of food, and the life stage tested in the laboratory test, particularly in the case of insect growth regulator (IGR) products. All these may affect the degree of exposure. It was recommended that a review of existing data, especially from field trials, would allow evaluation of the validity of extrapolation of results for one organism to predict likely effects on another organism.

REFERENCES

Albert, R. and Bogenschütz, H. (1984) Prüfung der Wirkung von Pflanzenschutzmitteln auf die Nutzarthropode *Coelotes terrestris* Wider (Araneida, Agelenidae) mit Hilfe eines Glasplattentest. *Anz. Schadlingskde., Pflanzenschutz, Umweltschutz*, **57**, 111–117.

Bakker, F.M. and Calis, J.N.M. (1989) A laboratory method for testing side effects of pesticides on Phytoseiid mites, based on a ventilated glass box: the coffin box. *Mededelingen Faculteit Landbouwwetenschappen Rijksuniversiteit Gent*, **54/3a**, 845–851.

Bogenschütz, H. (1975) Prüfung des Einflusses von Pflanzenschutz mitteln auf Nutzinsekten. *Z. angew. Entomol.*, **77**, 438–444.

Bogenschütz, H. (1984) Über die Wirkung von Pflanzenbehandlungsmitteln auf die Parasitierungsleistung der Schlupfwespe *Coccygomimus turionella*. *Nachrichtenbl. Deut. Pflanzenschutzd., Braunschweig*, **36**, 65–67.

Bogenschütz, H., Albert, R., Hradetzky, J. and Kublin, E. (1986) Ein Beitrag zur Prüfung der unerwünschten Wirkung von Pflanzenbehandlungsmitteln auf

Nutzarthropoden im Laboratorium. *Agrar- und Umweltforschung in Baden-Württemberg*, **11**, 5–25.

Boness, M., Englert, W.D., Haub, G. *et al.* (1982) *Richtline für die Prüfung der Auswirkung von Pflanzenbehandlungsmitteln auf Raubmilben im Weinbau*, BBA, Braunschweig.

Duso, C., Camporese, P. and van der Geest, L.P.S. (1992) Toxicity of a number of pesticides to strains of *Typhlodromus pyri* and *Amblysius andersoni* (Acari: Phytoseiidae). *Entomophaga*, **37**, 363–372.

Gendrier, J.-P. and Reboulet, J.-N. (1992) Choix de produits phytosanitaires en vergers. *Phytoma – La Défense des végétaux*, **438**, 26–30.

Hassan, S.A. (1974) Eine Methode zur Prüfung der Einwirkung von Pflanzenschutzmitteln auf Einparasiten der Gattung *Trichogramma* (Hymenoptera, Trichogrammatidae). Ergebnisse einer Versuchsreihe mit Fungiziden. *Z. angew. Entomol.*, **76**, 120–134.

Hassan, S.A. (1977) Standardized techniques for testing side-effects of pesticides on beneficial arthropods in the laboratory. *Journal of Plant Diseases and Protection*, **83**(3), 158–163.

Hassan, S.A. (1980) Reproduzierbare Laborverfahren zur Pr fung der Schadwirkungsdauer von Pflanzenschutzmitteln auf Eiparasiten der Gattung *Trichogramma* (Hymenoptera, Trichogrammatidae). *Z. angew. Entomol.*, **89**, 281–289.

Hassan, S.A. (1988) Guidelines for testing the effects of pesticides on beneficials. *IOBC/WPRS Bulletin*, **XI**(4), 143.

Hassan, S.A. (1992) Guidelines for testing the effects of pesticides on beneficial organisms. *IOBC/WPRS Bulletin*, **XV**(3), 186.

Hassan, S.A., Bigler, F., Blaisinger, P. *et al.* (1985) Standard methods to test the side-effects of pesticides on natural enemies of insects and mites developed by the IOBC/WPRS Working Group 'Pesticides and Beneficial Organsisms'. *Bulletin OEPP/EPPO*, **15**, 214–255.

Heimbach, U. (1988) Nebenwirkungen einiger Fungizide auf Insekten. *Nachrichtenblatt des Deutschen Pflanzenschutzdienstes*, **40**(12), 180–183.

Helyer, N. (1991) Laboratory pesticide screening method for the aphis predatory midge *Aphidoletes aphidimyza* (Rondani) (Diptera: Cecidomyiidae). *Biocontrol Science and Technology*, **1**, 53-58.

Hoogcarspel, A.P. and Jobsen, J.A. (1984) Laboratory method for testing the side effects of pesticides on *Encarsia formosa* (Hymenoptera: Aphelinidae). Results with pesticides used on tomato in glasshouses in the Netherlands. *Z. angew. Entomol.*, **97**, 268–278.

Huang, P. (1981) Zue Laborzucht von *Pales pavida* Meig. (Dipt., Tachinidae) am Ersatzwirt *Galleria mellonella* L. (Lep., Galleriidae). *Z. Pflanzenkrankh., Pflanzensch*, **88**, 177–188.

Keller, S. and Schweizer, C. (1991) Die Wirkung von Herbiziden auf das Sporulierungsvermögen des blattlauspathogenen Pilzes *Erynia neoaphidis*. *Anz. Schadlingskde., Pflanzenschutz, Umweltschutz*, **64**, 134–136.

Mansour, F. and Nentwig, W. (1988) Effects of agrochemical residues on four spider taxa: laboratory methods for pesticide tests with web-building spiders. *Phytoparasitica*, **16**(4), 317–325.

Mansour, F., Heimbach, U. and Wehling, A. (1992) Effects of pesticide residues on ground-dwelling lycosid and micryphantid spiders in laboratory tests. *Phytoparasitica*, **20**(3), 195–202.

Naton, E. (1983) Testing the side-effects of pesticides on *Phygadeuon trichops*. *Anz. Schadlingskde., Pflanzenschutz, Umweltschutz*, **56**, 82–91.

Oomen, P. (1985) Guidelines for the evaluation of side-effects of pesticides. *Encarsia formosa. Bulletin OEPP/EPPO*, **15**, 257–265.

Overmeer, W.P.J. and van Zon, A.Q. (1982) A standardized method for testing the side-effects of pesticides on the predacious mite *Amblyseius andersoni* (Acarina: Phytoseiidae). *Entomophaga*, **27**, 357–364.

Plattner, H.C. and Naton, E. (1975) Zur Prüfung der Auswirkung von Pflanzenschutzmitteln auf Nutzarthropoden. *Bay. Landw. Jb.*, 143–147.

Rumpf, S., Storch, V., Vogt, H. and Hassan, S.A. (1992) Effects of juvenoids on larvae of *Chrysoperla carnea* Steph. (Chrysopidae). *Acta Phytopatholica et Entomologica Hungarica*, **27**(1-4), 557–563.

Samsoe-Petersen, L. (1983) Laboratory method for testing side effects of pesticides on juvenile stages of the predatory mite, *Phytoseiulus persmilis* (Acarina, Phytoseiidae) based on detached bean leaves. *Entomophaga*, **28**, 167–178.

Samsoe-Petersen, L. (1987) Laboratory method for testing side-effects of pesticides on the rove beetle *Aleochara bilineata* adults. *Entomophaga*, **32**(1), 78–81.

Samsoe-Petersen, L., Bigler, L., Bogenshütz, H. *et al.* (1989) Laboratory rearing techniques for 16 beneficial arthropod species and their prey/hosts. *Z. Pflanzenkrankh., Pflanzensch*, **96**, 289–316.

Samsoe-Petersen, L., Biere, M. and Büchs, W. (1992) Interpretation of laboratory measured effects of slug pellets on soil dwelling invertebrates. *Aspects of Applied Biology*, **31**, 87–96.

Stäubli, A., Hächler, M., Antonin, P. and Mittaz, C. (1984) Tests de nocivité de divers pesticides envers les ennemis naturels des principaux ravageurs des vergers de poiriers en Suisse romande. *Revue suisse Vitic. Arboric. Hortic.*, **16**, 279–286.

Stolz, M. (1990) Testing side effects of various pesticides on the predatory mite *Phytoseiulus persimilis* Athias-Henriot (Acarnia: Phytoseiidae) in laboratory. *Pflanzenschutzber*, **51**(3), 127–138.

Suter, H. (1978) Prüfung der Einwirkung von Pflanzenschutzmitteln auf die Nutzarthropodenart *Chrysopa carnea* Steph. (Neuroptera: Chrysopidae). Methodik und Ergenbnisse. *Schweiz. landw. Forschung*, **17**, 37–44.

Tuset, J.J. (1975) Effets d'inhibition du développment du champignon *Cephalosporium lecanii* Zimm. 'in vitro' causés par des produits antiparasitaires, in *VIIth International Plant Protection Congress, Section V. Biological and Genetic Control*, Moscow, pp. 201–208.

Tuset, J.J. (1988) *Verticillium lecanii*, hongo entomopatógeno que combate en los agrios al coccido 'caparreta' (*Saissetia oleae*). *Phytoma Espana*, **4**, 31–35.

Viggiani, G. and Tranfaglia, A. (1978) A method for laboratory test of side-effects of pesticides on *Leptomastix dactylopii* (How.) (Hym., Encyrtidae). *Boll. Lab. Ent. Agr. Portici*, **35**, 8–15.

Vogt, H. (1992) Untersuchungen zu Nebenwirkungen von Insektiziden und Akariziden auf *Chrysoperla carnea. Med. Fac. Landbouww. Univ. Gent*, **57**, 2b.

Vogt, H. (1994) Side effects of pesticides on beneficial organisms: comparison of laboratory, semi field and field results. *IOBC/WPRS Bulletin*, **17**(10), 178.

Wehling, A. and Heimbach, U. (1991) Untersuchungen zur Wirkung von Pflanzenschutzmitteln auf Spinnen (Araneae) am Beispiel einiger Insektizide. *Nachrichtenblatt des Deutschen Pflanzenschutzdienstes*, **43**(2), 24–30.

Part Two A

Laboratory Testing

8

Standard laboratory methods to test the side-effects of pesticides (initial and persistent) on *Trichogramma cacoeciae* Marchal (Hym., Trichogrammatidae)

Sherif A. Hassan

Parasites of the genus *Trichogramma* are distributed world-wide and play an important role as natural enemies of lepidopterous pests on a wide range of agricultural crops. The use of *Trichogramma* in biological control has gained widespread interest in many countries. At present, about 18 species of *Trichogramma* are being mass-reared to control pests on corn, sugarcane, tomato, rice, cotton, sugar beet, apple, plum, vineyard, pasture, cabbage, chestnut, sweet pepper, pomegranate, paddy and forests in at least 23 countries.

Regarding complicity in assessing the side-effects of pesticides on beneficial arthropods, the Working Group 'Pesticides and Beneficial Organisms' of the International Organization for Biological Control, West Palaearctic Regional Section (IOBC/WPRS) recommends a combination of several laboratory test methods together with semi-field and field trials. The choice of methods was made to screen out harmless preparations by using a worst-case method at the beginning of the sequential testing; these preparations are not tested any further. This is followed by several laboratory tests to differentiate between chemicals with high initial toxicity on the natural enemy.

Ecotoxicology: Pesticides and beneficial organisms.
Edited by P.T. Haskell and P. McEwen. Published in 1998 by Chapman & Hall, London. ISBN 0 412 81290 8.

The sequential procedure includes the following.

(a) 'First screening', laboratory: initial worst-case toxicity test at the highest recommended concentration of the product on the most susceptible life stage – the adult parasitoid, which is exposed to a fresh, dry pesticide film applied to glass plates. The method was published by Hassan (1974, 1977, 1992) and a guideline was drafted for use by German registration authorities (Hassan, 1988). This worst-case test allows the classification of products as harmless, but further testing may be necessary if the test product is harmful. Because field testing is time-consuming and costly, the use of three additional laboratory methods, (b)–(d) below, may be recommended, depending on the intended use of the chemical.
(b) Less susceptible life stage, laboratory: spraying of product on parasite pupae within the host eggs.
(c) Duration of harmful activity: exposure of adults to pesticide residues applied on plants at intervals after treatment (Hassan 1980).
(d) Extended laboratory method: exposure of adults to a pesticide film applied to plant leaves under controlled laboratory conditions.
(e) Semi-field: similar test but carried out under true field conditions at the appropriate time of the year for the chemical.
(f) Field test: relevant crops are directly sprayed with the product after laboratory-reared organisms are released. Experiments are conducted at the appropriate season for the chemical and repeated at different locations.

In the present study, methods (a), (b) and (c) were carried out to conduct the experiments. This combination has the advantage of including two different developmental stages of the natural enemy (adult and pupa) that differ greatly in their susceptibility and vulnerability to pesticides. The persistent test (c) adds information towards the overall impact of the chemical on the natural enemy.

MATERIALS AND METHODS

The experiments were carried out using standard methods of the IOBC/WPRS Working Group (Hassan, 1977, 1992). The preparations and the concentrations tested are given in Tables 8.1–8.3.

(a) Susceptible life stage (adults of parasites)

The initial toxicity was tested by exposing the adult parasites to a fresh dry pesticide film applied to glass plates at the recommended concentration. The exposure cage consisted of two square glass plates and an aluminium frame (13 cm long, 1.5 cm high and 1 cm wide). Each of three

Table 8.1 Insecticides/acaricides: results of three different laboratory tests on *Trichogramma cacoeciae*: initial toxicity on adults (susceptible life stage), toxicity on pupae within *Sitotroga* eggs (less susceptible life stage) and persistence toxicity on adults (duration of harmful activity)

Preparation	*Concentration tested (%)*	*Adult*	*Pupa within host egg*	*Persistence*
Dipel (*Bacillus thuringiensis*)	0.10	1	–	–
Torque (fenbutatin-oxid)	0.05	1	–	–
Dimilin (diflubenzuron)	0.05	1	–	–
Apollo SOSC (clofentezine)	0.04	1	–	–
Cesar (hexythiazox)	0.025	1	–	–
Insegar (fenoxycarb)	0.06	1	–	–
Applaud (buprofezin)	0.03	1	–	–
Dimilin (diflubenzuron)	0.05	1	–	–
Trigard (cyromazine)	0.067	1	–	–
Neudosan (Kali-Seife)	2.0	1	–	–
Delfin WG (*Bacillus thuring.*)	0.10	1	–	–
Novodor FC (*Bac. thuring. tenebr.*)	0.50	1	–	–
Micro Germin (*Verticillium lec.*)	0.20	1	–	–
Nomolt (teflubenzuron)	0.10	1	1	–
AAzomate (benzoximate)	0.15	2	–	–
Match (Lufenuron)	0.20	2	–	–
Admiral (Pyriproxifen)	0.25	2	–	–
Kelthane (dicofol)	0.15	3	1	2
Evisect S (thiocyclam)	0.03	3	3	2
Cropotex (flubenzimine)	0.10	3	1	4
Pirimor-Granulat (pirimicarb)	0.10	4	1	1
Croneton (ethiophencarb)	0.10	4	1	2
Tedion V 18 (tetradifon)	0.20	4	1	2
Asepta Lindane (lindane)	0.10	4	3	2
Dimecron 20 (phosphamidon)	0.25	4	3	2
Spruzit-Nova-flüssig (pyrethrum+)	0.10	4	4	2
Unden (propoxur)	0.15	4	4	2
Basudine vloeibar (dilzinon)	0.21	4	4	2
Phosdrine W 10 (mevilphos)	0.58	4	4	2
Telmion (Rapsöl)	4.0	4	2	2
Vertimec (Abamectin)	0.375	4	3	2
Dipterex WP 80 (trichlorphon)	0.10	4	2	3
Thiodan 35 Spritzp. (Endosulfan)	0.10	4	3	3
Hostaquick (heptenophos)	0.10	4	4	3
Peropal (azocyclotin)	0.10	4	1	3
Imidan (phosmet)	0.25	4	2	–
Zolone Flow (Phosalon)	0.24	4	1	4
Polo (Difenthiuron)	0.10	4	1	4
Plictran 25 W (cyhexatin)	0.10	4	1	4
Rubitox Sritzp. (Phosalone)	0.20	4	1	4

Table 8.1 *continued*

Preparation	*Concentration tested (%)*	*Adult*	*Pupa within host egg*	*Persistence*
Ambush (permethrin)	0.02	4	1	4
Orthen (acephate)	0.15	4	2	4
Maitac (amitraz)	0.30	4	2	4
Decis (deltamethrin)	0.06	4	2	4
Gusathion (azinphos-methyl)	0.20	4	3	4
Kilval (vamidothion)	0.125	4	1	4
Vydate L (oxamyl)	0.15	4	1	4
Rody (fenpropathrin)	0.05	4	1	4
Klartan (fluvalinate)	0.06	4	1	4
Baythroid 50 (cyfluthrin)	0.05	4	2	4
Karate (lambda-cyhalothrin)	0.075	4	2	4
Tamaron (methamidophos)	0.15	4	4	4
Torak E (dialiphos)	0.25	4	3	4
Lannate (methomyl)	0.10	4	4	4
Sumicidin (fenvalerate)	0.075	4	4	4
Actellic 50 (pirimiphos-methyl)	0.20	4	4	4
Ultracid (methidathion)	0.075	4	4	4
Folithion (fenitrothion)	0.10	4	4	4
Hostaquick (heptenophos)	0.10	4	4	4
Ekamet (etrimfos)	0.20	4	4	4
Asepta Nexion (bromophos)	0.27	4	4	4
Birlane EC 40 (chlorfenvinphos)	0.33	4	4	4
Dursban Spritzp. (chlorpyrifos)	0.25	4	4	4
Ambush C (cypermethrin)	0.04	4	4	4
Perfekthion (dimethoate)	0.21	4	4	4
Hostathion (triazophos)	0.24	4	4	4

Initial toxicity: 1 = harmless (< 30%); 2 = slightly harmful (30–79%); 3 = moderately harmful (80–99%); 4 = harmful (> 99%)
Persistence: 1 = short-lived; 2 = slightly persistent; 3 = moderately persistent; 4 = persistent.

Table 8.2 Fungicides: results of three different laboratory tests on *Trichogramma cacoeciae*: initial toxicity on adults (susceptible life stage), toxicity on pupae within *Sitotroga* eggs (less susceptible life stage) and persistence toxicity on adults (duration of harmful activity)

Preparation	*Concentration tested (%)*	*Adult*	*Pupa within host egg*	*Persistence*
Nimrod (bupirimate)	0.04	1	–	–
Cercobin-M (thiophanat-methyl)	0.10	1	–	–
Ortho Difolatan (captafol)	0.20	1	–	–
Orthocid 83 (captan)	0.25	1	–	–
Bayleton (triadimefon)	0.20	1	–	–

Table 8.2 *continued*

Preparation	*Concentration tested (%)*	*Adult*	*Pupa within host egg*	*Persistence*
Ronilan (vinclozolin)	0.05	1	–	–
Derosal (carbendazim)	0.05	1	–	–
Daconil 500 (chlorothalonil)	0.30	1	–	–
Milgo-E (ethirimol)	0.18	1	–	–
Ortho-Phaltan 50 (folpet)	0.33	1	–	–
Topas (penconazole)	0.04	1	–	–
Baycor (bitertanol)	0.37	1	–	–
Delan flüssig (dithianon)	0.20	1	–	–
Vitigran (copper-oxychlorid)	1.00	1	–	–
Impact (flutriafol)	0.16	1	–	–
Rovral PM (iprodion)	0.15	1	–	–
Saprol (triforine)	0.15	1	–	–
Sumisclex (procymidone)	0.15	1	–	–
Dyrene flüssig (anizaline)	0.40	1	–	–
Bayfidan (triadimenol)	0.05	1	–	–
Anvil (hexaconazole)	0.03	1	–	–
Calixin (tridemorph)	0.075	1	–	–
Alto 100 SL (cyproconazol)	0.08	1	–	–
Score EC 250 (difenoconazol)	0.05	1	–	–
BioBlatt Mehltaumittel (lecithin)	0.15	1	–	–
Topsin M (Thiophanat-methyl)	0.114	1	–	–
Bavistin (Carbendazim)	0.40	2	–	–
Aliette (Fosetyl)	1.375	2	–	–
Captan 83 W (Captan)	0.82	2	–	–
Dithane Ultra (mancozeb)	0.10	2	–	–
Pomarsol forte (thiram)	0.20	2	–	–
Rubigan Vloeibaar (fenarimol)	0.12	2	–	–
Antracol (propineb)	0.20	2	–	–
Omnex WP 10 (penconazol)	0.025	2	–	–
Tilt (propiconazole)	0.08	3	1	1
Scala (Pyrimethanil)	0.50	3	1	2
Dithane Ultra (mancozeb)	0.20	3	1	2
Trimidal EC (nuarimol)	0.08	3	1	2
Plondrel (ditalimfos)	0.075	3	1	3
Netzschwefel Bayer (sulphur)	0.40	3	1	4
Corbel (fenpropimorph)	0.17	4	1	3
Euparen (dichlofluanid)	0.20	4	1	3
Sportak (prochloraz)	0.187	4	1	3
Euparen (dichlofluanid)	0.15	4	1	3
Nevikén (lime-sulphur)	3.00	4	2	3
Polyram-Combi (metiram)	0.42	4	1	4
Afugan WP 30 (pyrazophos)	0.05	4	1	4
Thiovit (sulphur)	0.40	4	1	4

Table 8.2 *continued*

Preparation	*Concentration tested (%)*	*Adult*	*Pupa within host egg*	*Persistence*
Morestan (chinomethionate)	0.10	4	1	4
Dithane M 22 (maneb)	0.50	4	1	4
Euparen M (Tolylfluanid	0.625	4	1	4
Dithane M 45 (Mancozeb)	0.90	4	1	4
Thiram (Thiram)	0.80	4	2	4

Initial toxicity: 1 = harmless (< 30%); 2 = slightly harmful (30–79%); 3 = moderately harmful (80–99%); 4 = harmful (> 99%)
Persistence: 1 = short-lived; 2 = slightly persistent; 3 = moderately persistent; 4 = persistent.

Table 8.3 Herbicides/plant growth regulators: results of three different laboratory tests on *Trichogramma cacoeciae*: initial toxicity on adults (susceptible life stage), toxicity on pupae within *Sitotroga* eggs (less susceptible life stage) and persistence toxicity on adults (duration of harmful activity)

Preparation	*Concentration tested (%)*	*Adult*	*Pupa within host egg*	*Persistence*
Betanal (phenmedipham)	2.25	1	–	–
Hyvar X (bromacil)	0.20	1	–	–
Gesatop 50 (simazin)	0.375	1	–	–
Fusilade (fluazifop-butyl)	0.25	1	–	–
Luxan 2,4-D amine (aminesalt)	0.432	1	–	–
Tribunil (metabenzthiazuron)	0.67	1	–	–
Ally (metsulfuron-methyl)	0.076	1	–	–
Dirigol-N (alphanaphthyl-acetamid)	0.02	1	–	–
Exp.30004 A (ioxynil)	0.24	1	–	–
Lontrel 100 (clopyralid)	0.12	1	–	–
Targa (quizalofop-ethyl)	0.30	1	–	–
Grasp (tralkoxydim)	0.50	1	–	–
Basagran (bentazone)	0.40	1	–	–
Tramat 500 (ethofumesat)	1.00	1	–	–
Starane 180 (fluroxpyr)	0.50	1	–	–
Arelon flüssig (isoproturon)	0.75	1	–	–
Goltix 70 WG (metamitron)	2.50	1	–	–
Pyramin (Chloridazon)	1.50	1	–	–
Butisan S (Metazachlor)	1.50	1	–	–
Banvel 70WP (Dicamba)	0.35	1	–	–
Duplosan KV (mecoprop-p)	1.00	2	–	–
Focus (Cycloxydim)	1.50	2	–	–
Illoxan (diclofop-methyl)	0.75	2	–	–
Ustinex PA (amitrol+diuron)	1.00	2	–	–
Gesaprim 50 (atrazin)	0.67	2	–	–

Table 8.3 *continued*

Preparation	*Concentration tested (%)*	*Adult*	*Pupa within host egg*	*Persistence*
Basta (glufosinate-ammonium)	0.50	2	–	–
Roundup (glyphosate)	1.00	2	–	–
Faneron (bromofenoxim)	1.70	3	1	–
Gallant Super (haloxyfop)	0.50	3	1	1
Cycocel Extra (chlormequat)	0.70	3	1	3
Kerb 50 W (propyzamid)	0.75	3	1	4
Fervinal Plus (sethoxydim)	0.79	4	1	1
Semeron (desmetryne)	0.25	4	1	2
Avenge (difenzoquat)	1.00	4	1	2
Rhodofix (1-naphthyl-acetic acid)	0.15	4	1	2
Certrol B (bromoxynil)	0.33	4	1	2
Ramrod (propachlor)	1.00	4	3	2
Kumulus (Netzschwefel)	2.00	4	2	4
Aretit flüssig (dinoseb)	1.25	4	4	4
Prosevor (carbaryl)	0.125	4	4	4
Touchdown (Glyphosattrimesium)	2.50	4	1	
Aresin (monolinuron)	0.75	4	2	–

Initial toxicity: 1 = harmless (< 30%); 2 = slightly harmful (30–79%); 3 = moderately harmful (80–99%); 4 = harmful (> 99%)
Persistence: 1 = short-lived; 2 = slightly persistent; 3 = moderately persistent; 4 = persistent.

sides of the frame contained six ventilation holes (1 cm diameter), covered with tightly woven black material. Two portable openings on the fourth side of the frame were used to introduce the *Trichogramma*, host eggs and food. The cage was held together with two clamps. The glass plates were sprayed with the pesticide at the recommended concentration as indicated in Table 8.1. The experiment started with a 24-hour period of forced exposure. At the end of this period the parasites, if still alive, were given host eggs so that their parasitization capacity could be measured. Eggs of the angoumois grain moth *Sitotroga cerealella* (Oliv.) were offered on days 2, 3 and 5 of the experiment. The capacity of parasitism per *Trichogramma* adult female and the reduction in capacity compared with the control (treated with water) were used to measure the effect of the chemical. Three replicates were used for each treatment. The pesticides were then classified into four categories as shown in Table 8.1.

(b) Less susceptible life stage (parasites within their hosts)

Seven-day-old *T. cacoeciae* pupae within *Sitotroga* eggs were directly sprayed and the emerging parasites, if any, were tested for their capacity to parasitize host eggs. The same experimental cage (but with untreated glass plates) and the method of assessment of parasitism described above were used. Three replicates, each with about 300 eggs, were used for each treatment.

(c) Duration of harmful activity (persistence toxicity on adults)

The technique used to test the persistence of pesticide residues involves the spraying of potted vine plants, maintaining them under a field or field-simulated environment, and exposing adult *Trichogramma* to samples of the treated leaves, taken at different time intervals after application. Exposure tests were carried out at 3, 10, 17, 24 and 31 days after the treatment of the vine plants. Three replicates were used for each treatment. The same experimental cage as in (a), but with untreated glass, was used. The sampled leaves were spread inside the cage to cover the entire lower surface. The reduction in parasitism compared with control was plotted on a probit scale against time. The persistence is the time required for the pesticide residue to lose effectiveness so that a reduction in parasitism of less than 30%, compared with the control, is reached.

RESULTS OF TESTING ON *TRICHOGRAMMA*

The results of testing the side-effects of 160 pesticides on *T. cacoeciae* using three different types of test methods were compared. The results of testing 66 insecticides/acaricides (Table 8.1), 53 fungicides (Table 8.2) and 42 herbicides/plant growth regulators (Table 8.3) on *Trichogramma* using the three different laboratory methods showed that the chemicals differed markedly in their initial as well as their residual toxicity. In each table, the preparations were listed according to their increasing toxicity in the initial contact test. In each evaluation category, the duration of harmful activity (persistence) was considered to be more important than the effect of the preparation on the parasite within its host (less susceptible life stage).

The results showed that the preparations differ greatly in their initial toxicity as well as in their persistence. Seventeen insecticides/acaricides, 34 fungicides and 27 herbicides/plant growth regulators were harmless. Twenty three insecticides/acaricides, 19 fungicides, seven herbicides/plant growth regulators were harmful or moderately harmful to the adult parasite but harmless or slightly harmful to the *Trichogramma* pupa within the host eggs. Thirteen insecticides/acaricides, six fungicides and seven herbicides/plant growth regulators were harmful in the initial toxicity test but were short-lived to moderately persistent and therefore are much more useful for integrated control. Short-lived preparations are likely to have much less total impact on the natural enemy than persistent ones and are much more suitable for use in modern plant protection. Persistent chemicals affect natural enemies for longer periods of time and are therefore likely to have a much greater impact on the natural enemy in the field.

REFERENCES

Hassan, S.A. (1974) Eine Methode zur Prüfung der Einwirkung von Pflanzenschutzmitteln auf Eiparasiten der Gattung Trichogramma (Hymenoptera: Trichogrammatidae). Ergebnisse einer Versuchsreihe mit Fungiziden. *Z. angew. Entomol.* **76**, 120–134.

Hassan, S.A. (1977) Standardized techniques for testing side-effects of pesticides on beneficial arthropods in the laboratory. *Journal of Plant Diseases and Protection*, **83**(3), 158–163.

Hassan, S.A. (1980) Reproduzierbare Laborverfahren zur Pr fung der Schadwirkungsdauer von Pflanzenschutzmitteln auf Eiparasiten der Gattung *Trichogramma* (Hymenoptera, Trichogrammatidae). *Z. angew. Entomol.*, **89**, 281–289.

Hassan, S.A. (1988) Guidelines for testing the effects of pesticides on beneficials. *IOBC/WPRS Bulletin*, **XI**(4), 143.

Hassan, S.A. (1992) Guidelines for the evaluation of side-effects of plant protection product on *Trichogramma cacoeciae*. *IOBC/WPRS Bulletin 1992/XV/3*, pp. 18–39.

9

Development and ring-testing of a standardized laboratory test for parasitic wasps, using the aphid-specific parasitoid *Aphidius rhopalosiphi*

Michael Mead-Briggs, K. Brown, M. Candolfi, M. Coulson, S. Klepka, Ch. Kühner, M. Longley, S. Maise, E. McIndoe, M. Miles, Ch. Neumann and A. Ufer

INTRODUCTION

Parasitic wasps are an important part of the non-target arthropod fauna within most crop systems and it is therefore necessary to understand the impact of plant protection products upon them. The largest group of primary parasitoids of aphids are the Aphidiinae, a subfamily of the Braconidae, with more than 400 species currently known. One species from this group, *Aphidius rhopalosiphi* (DeStefani-Perez), has previously been chosen as an indicator for determining the effects of products on parasitic Hymenoptera (European Standard Characteristics of Beneficial Regulatory Testing – ESCORT meeting; Barrett *et al.*, 1994). Subsequently, through a joint initiative of Beneficial Arthropod Registration Testing (BART), European and Mediterranean Plant Protection Organization (EPPO), Commercial Ecotoxicology Testing Group (COMET) and

Ecotoxicology: Pesticides and beneficial organisms.
Edited by P.T. Haskell and P. McEwen. Published in 1998 by Chapman & Hall, London. ISBN 0 412 81290 8.

International Organization for Biological Control (IOBC), an *Aphidius* Experts Group was established to study the methods available for assessing the effects of plant protection products on such wasps. Over the past two years this group has evaluated various aspects of the methods being used for both laboratory and 'extended' laboratory studies. It has established a database of results obtained from a large number of such studies and this has helped to improve our understanding of both the sensitivity and the natural variability of the test system.

DESIGN AND EVALUATION OF THE STANDARD LABORATORY TEST

Selection of the test species

The biology and behaviour of several species of Aphidiinae have been widely studied (Hågvar and Hofsvang, 1991), but a single species had to be chosen to allow comparisons between studies. Several criteria were considered, including the relevance of the species in terms of the crops in which they were active, the ease of culture of both the wasp and its aphid host, the ease of handling, and suitability for use in both laboratory and field-based experiments. Since selecting *A. rhopalosiphi* as the standard test species, there has been further debate as to whether other closely related species, such as *A. colemani* Viereck or *A. matricariae* Hal., might prove equally suitable. Both of these wasps are reared commercially for the biocontrol of aphids in glasshouses and are therefore more widely available than *A. rhopalosiphi.* A recent comparison of the sensitivity of these three wasps to the insecticide dimethoate showed that differences between species were small (Maise *et al.*, 1997). This would suggest that, if a change in the standard test species were to be recommended, this would not invalidate the results of previous studies. One important advantage of *A. colemani* over *A. rhopalosiphi* is that it can be used in extended laboratory and semi-field experiments with hosts from more than one type of cropping system – for example, *Rhopalosiphum padi* (L.) for cereals and *Myzus persicae* (Sulz.) or *Aphis gossypii* Glover for glasshouse and horticultural crops. The current thinking within the Experts Group is that *A. colemani* could replace *A. rhopalosiphi* as the standard test species in due course. However, further comparative studies on the biology of these two species are required before such changes are made. For the present, the development and ring-testing of methods will continue using both species.

Brief description of methods

Currently, the laboratory bioassay in commonest use for *A. rhopalosiphi* is developed from that of Mead-Briggs (1992), which itself drew heavily on an earlier IOBC test design for *A. matricariae* (Polgar, 1988). The test procedures, which are in line with the recommendations in the SETAC Guidance Document (Barrett *et al.*, 1994), involve the exposure of adult wasps to fresh product residues on a glass surface. The wasps, of uniform age and sex ratio, are confined over the treated glass plates for 48 h within shallow-walled, ventilated arenas. After 48 h, the surviving female wasps are taken for fecundity assessments to confirm that their reproductive performance has not been adversely affected. For this, the wasps are individually confined for 24 h over untreated pots of aphid-infested seedling barley. They are then removed and the parasitized aphids are left to develop for approximately 10 days, after which the number of mummies produced is recorded. In the past, it has been recommended that at least three replicate arenas, each holding 10 wasps, are prepared for each treatment for exposure to product residues. Of the surviving females, 10 from each treatment are then evaluated for their fecundity. However, this level of replication is now thought to be insufficient for statistical comparisons and this issue will be discussed further below.

Investigations into the test design

In designing and improving the bioassay, a number of specific parameters have been evaluated to determine whether they have a significant impact on the outcome of tests. Since there is a tendency for individual laboratories to adapt test guidelines to suit their own facilities and practices, it was considered important to determine which aspects of the experimental design needed to be followed rigorously and which could be taken to be more flexible.

Source and age of the test insects

It is accepted that it would be impractical for all testing laboratories to have to work with wasps derived from a single source. As a consequence, there is a risk that there will be differences in the susceptibility of insects taken from different cultures. However, initial data from studies with both *A. rhopalosiphi* and *A. colemani* have suggested that such differences are likely to be relatively minor and may in fact be less than the small variations seen between individual cohorts of insects taken from the same culture (M. Longley, unpublished data; Maise *et al.*, 1997).

The method of producing cohorts of wasps for bioassays will vary between laboratories, depending on the facilities available. The intent should be to produce insects of a uniform size and age for tests. The importance of size on performance, in terms of both survival and fecundity, is still in the process of being evaluated, but in the future it may be possible to set quality control criteria on the basis of the mean body weight of a sample of insects taken from those being tested. A concern would be that if the culture of host aphids became overcrowded, the aphids and any parasitoids derived from them would be smaller and this might influence their susceptibility.

The age of the test insects is also considered important, particularly with respect to their fecundity (Shirota *et al.*, 1983). This factor is certainly easier to control, as a cohort of mummies harvested from a culture can be placed in an emergence chamber for wasps to be collected daily. It has been decided that the test insects should be less than 48 h old. Since the mean longevity of wasps held at 18°C and 75% relative humidity is approximately 13 days (Shirota *et al.*, 1983), natural mortality levels should remain low – ideally below 10% if the ambient test conditions are suitable.

Design of the test arenas

The wasps are exposed to fresh, dry product residues on glass plates. These plates form the floor and ceiling of shallow arenas, the walls of which are made from an untreated metal frame (Mead-Briggs, 1992). To allow ventilation, the frame has mesh-covered holes on each side. There is no evidence that the dimensions of the arenas (either square or round, and nominally 5–12 cm in diameter and 1–2 cm deep) would significantly affect the outcome of studies, but it was considered that they should be as shallow as possible, to reduce the unsprayed area within each unit. The Experts Group had only recorded a few instances of wasps being actively repelled on to the walls of the arenas and this was typically attributed to the physical properties of the product residues (e.g. oily deposits), rather than their biochemical properties.

Test conditions

Ambient test conditions of 18–22°C are currently being recommended since it is believed that these will not unduly stress the test insects. It has been shown that the relative humidity within the test room can influence survival, with humidities of < 50% significantly increasing wasp mortality in all treatments including the control. A relative humidity of 60–90% has therefore been recommended.

The importance of light – its intensity and photoperiod – during the initial exposure phase of the bioassay has been investigated. Regimes of

480 lux and 8300 lux and of 16 h and 24 h photoperiods were evaluated, but were not found to have a significant effect on the outcome of tests where deltamethrin was applied at a rate equivalent to its LC_{50}. Observations have shown that the wasps generally remain active in the arenas while illumination is provided and are less active in the dark, resting on either the glass plates or the walls of the arena. The opinion of the Experts Group is that a reduction of the photoperiod used in the bioassay from 24 h (Mead-Briggs, 1992) to 16 h would not adversely influence the outcome of tests. It would also allow the same photoperiod to be used for both culturing and testing.

Fecundity assessments

One aspect of the test that has continued to be a cause of variability has been the assessment of sublethal treatment effects on wasp fecundity. This has generally involved taking wasps from the treated test arenas after 48 h and confining them over pots of untreated seedling wheat or barley on which host aphids were present. Both *Rhopalosiphum padi* and *Metopolophium dirhodum* (Wlk.) have been used as hosts but as yet no comparison has been made of the relative merits of each. Since it is already known that there is considerable variability in the reproductive performance of individual wasps, further research in this part of the test is currently in progress; e.g. studies into the importance of standardizing the age of the host aphid.

ANALYSIS OF TEST DATA

Influence of product formulation

Data were collated from laboratory studies with 50 product formulations in which wasps had been exposed to fresh residues on glass (Figure 9.1). These comprised a random selection of fungicides, herbicides and insecticides that had been evaluated by three separate laboratories. Harmful treatments were detected in 59% of the tests with the fungicide and herbicide formulations (i.e. > 30% mortality when treatment data were corrected for any control mortality). Of these products, 23 were subsequently evaluated in extended laboratory tests, with wasps being confined over freshly treated plants. Only three (13%) of the compounds still resulted in harmful effects, suggesting that there had been a high percentage of 'false positive' results in the glass plate tests. Since many of these products are thought to contain inherently non-insecticidal active ingredients, it seemed likely that chemicals in the product formulations could be responsible for the high proportion of harmful effects detected. The bioavailability of these compounds – often a mixture of

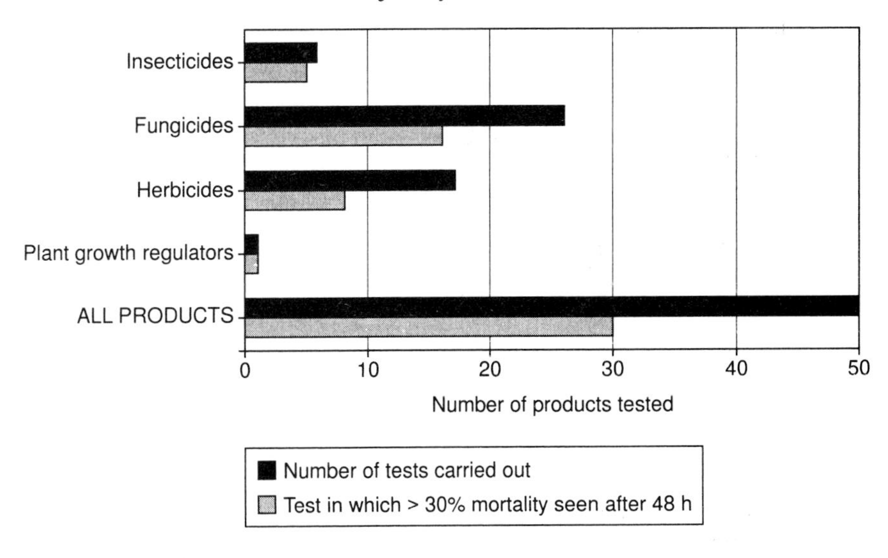

Figure 9.1 Results of laboratory tests in which wasps were confined over freshly treated glass plates. Data drawn from tests with 50 product formulations.

solvents, surfactants and oils – is far greater on a glass surface than on a treated plant. One would therefore expect a reduction in the number of positive effects seen at the next level of testing. A comparison of the types of formulation tested indicated that wettable granule (WG) formulations were considerably less likely to result in harmful effects than emulsifiable concentrate (EC) formulations (Figure 9.2).

These findings raise the question as to whether the glass plate test is too sensitive for this particular indicator species. Can we reliably predict harmful treatment effects in the field from this form of bioassay or will there be too many false positive effects? The latter would limit the value of this test system for product comparisons. Two possible solutions have been proposed: one is to move the thresholds for categorizing harmful or non-harmful effects; the other is to alter the test so as to make it closer to reality – either excised leaves or whole plants are used for the treated exposure surface, rather than glass plates.

Setting trigger values for categorizing harmful effects

At present, a 30% 'effect', whether it be 30% wasp mortality or a 30% reduction in their fecundity, or a combination of the two, is normally taken as the threshold for triggering higher tiers of testing with this species. For the 45 studies with fungicide and herbicide products (Figure 9.1), 25 tests (i.e. 56%) resulted in > 30% mortality of wasps within 48 h. Under current regulations these products would be labelled as being

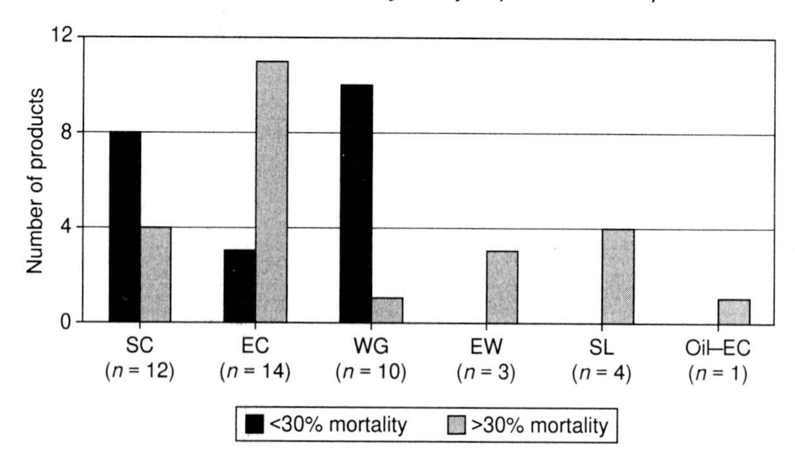

Figure 9.2 Influence of formulation type on numbers of positive treatment effects with 45 fungicide and herbicide formulations detailed in Figure 9.1. **SC**, suspension concentrate; **EC**, emulsifiable concentrate; **WG**, wettable granule; **EW**, oil in water emulsion; **SL** soluble concentrate; **Oil-EC**, oil-based emulsifiable concentrate.

harmful to parasitoids unless additional studies were carried out to prove otherwise. To reduce the numbers of products initially judged to be harmful, one could either move the threshold (e.g. from 30% mortality to 40%, or even 50%), or shorten the exposure period (e.g. from 48 h to 24 h). One argument in support of the latter is that the test environment potentially stresses the test insects and so increases the risk of erroneous harmful effects being detected over time. If these particular tests had been stopped after 24 h, the number of 'harmful' products would have been reduced to 21 (i.e. 47%). If the trigger threshold was moved to 50%, then the number of 'harmful' products would have been reduced to 19 (i.e. 38%). In both cases, none of those products that were subsequently found to be harmful in extended laboratory tests would have been missed by these changes. On the basis that the glass plate test represents very much a 'worst-case' situation, the Experts Group is currently recommending that the duration of exposure to residues in this bioassay is reduced from 48 h to 24 h. Although the possibility of also moving the 30% trigger value has been discussed, the Experts Group thought that further data would be required before this could be justified.

Having accumulated a considerable amount of data on the performance of *A. rhopalosiphi* in fecundity assessments, the authors believe that the use of the Overmeer–van Zon formula, as proposed by Polgar (1988) and Mead-Briggs (1992), is inappropriate for *Aphidius* spp. This is primarily due to the high natural variability of the reproductive performance of these wasps. The problems associated with the high variability being recorded in fecundity assessments with this and other test species was highlighted by Schmuck *et al.* (1996). In a subsequent analysis, the standard deviation (σ^{n-1}) and the mean ($\bar{x}$) number of mummies per wasp (n = 10 per test) were calculated for 45 laboratory tests carried out with *A. rhopalosiphi* under uniform conditions. The coefficient of variation for the control treatment data in each test (i.e. $100 \times [\sigma^{n-1}/\bar{x}]$) was found to be between 29% and 126% (median = 51%). In these circumstances, detecting a 30% reduction in the fecundity of treated insects with sufficient confidence would require an impractical level of replication (Lipsey, 1990). It would therefore seem logical that a greater emphasis be placed on the use of mortality data for assigning 'harmful' or 'harmless' labels to test products and that the fecundity data should merely be used as a means of confirming that the treated test insects are still capable of reproducing.

CONCLUSIONS

Through the activities of the *Aphidius* Experts Group, and particularly through their sharing of test data, there has been a considerable advance in our understanding of the complexities of laboratory studies with these parasitoids. The results of studies have demonstrated the sensitivity of this test system and the influence of product formulation, and also the potential sources of variability between individual tests. It is intended that the test methods currently in use will continue to be refined following further research. Key issues currently being addressed include the choice of test species and the setting of trigger values.

REFERENCES

Barrett, K.L., Grandy, N., Harrison, E.G. *et al.* (1994) *Guidance Document on Regulatory Testing Procedures for Pesticides with Non Target Arthropods. From the ESCORT workshop, Wageningen, Netherlands, March 1994*, SETAC-Europe, Brussels.

Hågvar, E.B. and Hofsvang, T. (1991) Aphid parasitoids (Hymenoptera: Aphididae): biology, host selection and use in biological control. *Biocontrol News and Information*, **12**, 13–41.

Lipsey, M.W. (1990) *Design Sensitivity: statistical power for experimental research*, Sage, Newbury Park, California.

Maise, S., Candolfi, M.P., Neumann, C. *et al.* (1997) A species comparative study: sensitivity of *Aphidius rhopalosiphi*, *A. matricariae* and *A. colemani* (Hymenoptera: Aphididae) to Dimethoate 40 EC under worst-case laboratory conditions, in *New Studies in Ecotoxicology* (papers resulting from posters given at the Welsh Pest Management Forum Conference: Ecotoxicology, Pesticides and Beneficial Organisms, Cardiff, UK, 14–16 October 1996), pp. 45–49.

Mead-Briggs, M.A. (1992) A laboratory method for evaluating the side-effects of pesticides on the cereal aphid parasitoid *Aphidius rhopalosiphi* (DeStefani-Perez). *Aspects of Applied Biology*, **31**, 179–189.

Polgar, L. (1988) Guideline for testing the effect of pesticides on *Aphidius matricariae* Hal. (Hym., Aphidiidae). *Bulletin IOBC/WPRS Working Group on Pesticides and Beneficial Organsisms*, **1988/XI**, 4.

Schmuck, R., Mager, H., Künast, C. *et al.* (1996) Variability in the reproductive performance of beneficial insects in standard laboratory toxicity assays – implications for hazard classification of pesticides. *Annals of Applied Biology*, **128**, 437–451.

Shirota, Y., Carter, N., Rabbinge, R. and Ankersmit, G.W. (1983) Biology of *Aphidius rhopalosiphi*, a parasitoid of cereal aphids. *Entomologia, Experimentalis et Applicata*, **34**, 27–34.

10

Current status of a ring-tested method to determine pesticide effects on the predatory mite *Typhlodromus pyri* (Scheuten) (Acarina: Phytoseiidae) in the laboratory

Sylvia Blümel, B. Baier, F. Bakker, U. Bienert, M. Candolfi, S. Hennig-Gizewski, Ch. Kühner, B. Jäckel, F. Louis, A. Ufer and A. Waltersdorfer

INTRODUCTION

For the authorization of plant protection products in the European Union under Directive 91/414/EEC, data about the effect of pesticides on non-target arthropods are required. As a consequence of the recommendations of the Workshop on 'European Standard Characteristics of Beneficials Regulatory Testing' in 1994 (Barrett *et al.*, 1994) the *Typhlodromus pyri* test group (consisting of 11 laboratories) was established as one of the 13 testing groups, or rings, to develop standardized methods for the evaluation of pesticide effects on different non-target arthropod key species.

The aim of the 11 laboratories in the group is the development of a validated method for laboratory testing of the effects of plant protection products on *T. pyri*, in order to identify compounds that can be classified

Ecotoxicology: Pesticides and beneficial organisms.
Edited by P.T. Haskell and P. McEwen. Published in 1998 by Chapman & Hall, London. ISBN 0 412 81290 8.

as harmless to the test organism. The requirements formulated in the SETAC Guidance Document (Barrett *et al.*, 1994) propose development of a residual contact test under so-called worst-case exposure conditions with the lead formulation of a plant protection product. Two existing test methods – the 'open' method described by Louis and Ufer (1995) and the 'coffin-cell' method described by Bakker and Calis (1989) – were adapted and compared concerning their suitability as standard test methods. Results of the last ring test are presented in this chapter and achievements and critical issues relating to the development of the method are described and discussed.

METHODS AND MATERIALS

Selection of test species

T. pyri was chosen as the test species because it was found to be one of the most important predatory mites in Europe and other geographical areas as regards abundance, beneficial capacity, amenability and pesticide sensitivity (Kostiainen and Hoy, 1996).

Test principles

The 'open' method of Louis and Ufer (1995) uses uncovered glass slides as test arenas, whereas the 'coffin-cell' method (Bakker and Calis, 1989) is carried out in closed but ventilated glass cells. For the open method (Figure 10.1), protonymphs of *T. pyri* are left to develop during continuous exposure to the pesticide residue in the same test arenas. In the improved coffin-cell method (Figure 10.2), all cells are connected to a central tube through which air is sucked, with the air flow kept constant during the trial; after 7 days the cages are opened and surviving individuals are transferred to a second series of test units (which are the open test arenas shown in Figure 10.1) for determination of effects on reproduction.

In both methods, predator protonymphs (the most susceptible developmental stage) are exposed to dried pesticide residue on the prepared test units. Pollen is supplied as a food source. The test animals are laboratory cultured under the same standardized conditions as used during the trial: 25 ± 0.5°C, 75 ± 15% relative humidity, and a photoperiod with 16 h light (Bakker *et al.*, 1992).

The endpoints of the study are the assessment of the pre-imaginal mortality (a direct lethal effect), the escape rate (an indirect effect caused by repellency) and the reproduction rate per female (an indirect, sublethal effect). Mortality assessment is conducted during the first 7 days of the trial period and reproduction assessment from day 7 to day 14.

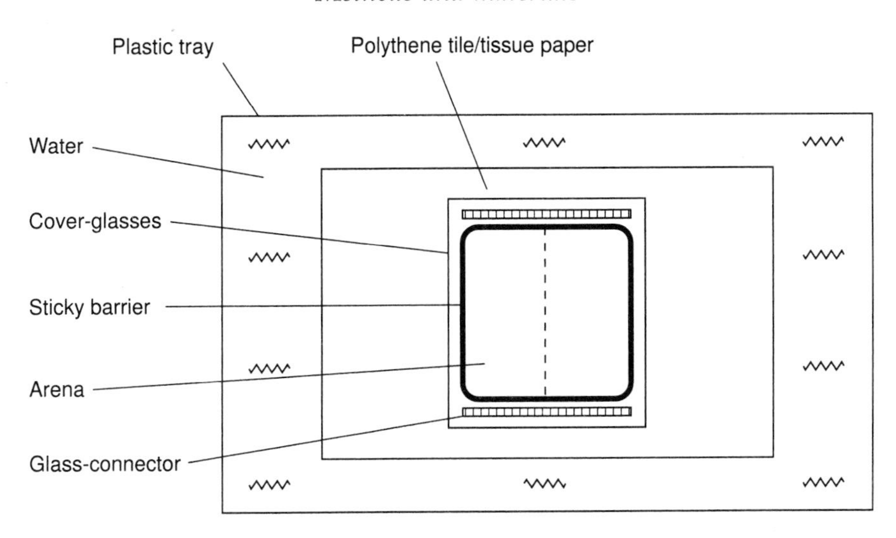

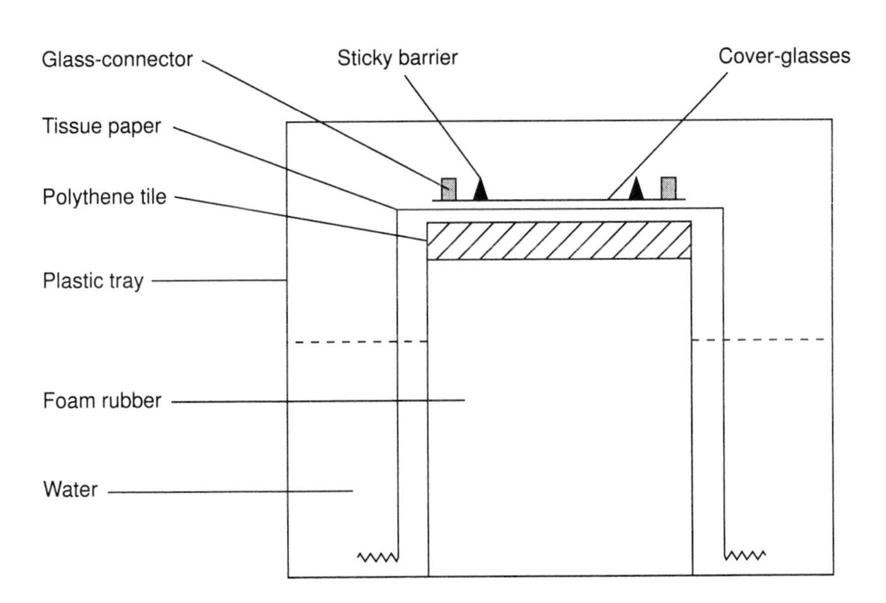

Figure 10.1 Design of the open cell method. Glass cover slides are fixed on glass bars and placed on wet filter paper. The thin gap between the glass plates is filled with drinking water for the predators by capillary forces from the underlying filter paper. The test area is limited by a glue barrier.

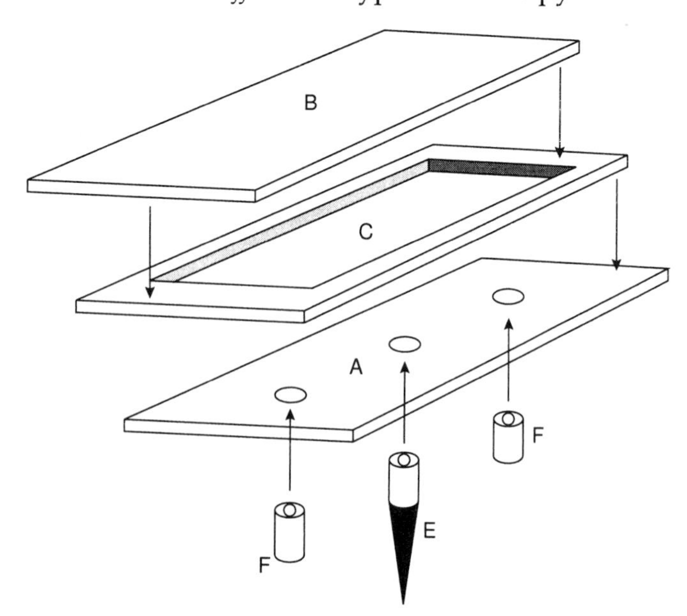

Figure 10.2 Design of the coffin-cell method. The improved version consists of a bottom glass plate (A) with three holes, a top glass plate (B), and a middle frame (C) of white Teflon® containing a central cavity with slanting edges. These three plates are fixed to the cell with clamps on each side of the test unit. The cells are provided with two tubes (F) for air inlet and outlet, and another tube (E) containing a cotton taper sucking up drinking water for the test animals.

The application of the test compound is carried out with calibrated spraying equipment at the maximum recommended field rate of the test product, calculated as the predicted initial environmental concentration (PIEC, μg a.i./cm^2) (Barrett *et al.*, 1994).

During the ring test, Parathion-ethyl 20 EC and Pirimicarb 500 DG were used as test substances, both derived from defined sources. Every test included water-treated control samples. Five separate replicates per treatment, each using 20 predator protonymphs, were carried out.

Achievements and critical issues of both methods

A number of parameters were standardized during the three completed ring-test series:

- culturing (rearing) of the test animals both for maintenance and for the trial;
- principle of the test;
- test system;

- test design;
- calibration of spraying equipment and application techniques;
- food choice and preparation;
- conduct of the trial, including a detailed time schedule and the mode of assessment.
- choice of the validity criteria
- calculations and statistical analysis (with regard to the SETAC Guidance Document (Chapman *et al.*, 1995)

The open method is easy to handle and does not require transfer of the test mites to new test arenas for the reproduction assessment. A big problem is the escape of the test animals, especially when compounds causing repellent effects are tested.

In the closed test cells, the number of mites escaping from the test units is reduced but the high expenditure of time and equipment for a standard routine test are a disadvantage. Also, before the coffin-cell method was combined with the open method for reproduction assessment, the evaluation of the number of eggs laid per female was difficult. Growth of fungi on deteriorated pollen posed problems of evaluation.

Validity criteria of the test

The mortality rate, escape rate, sex ratio and reproduction rate per female were chosen as validation criteria and were characterized by preliminary threshold values during the ring test. These threshold values were selected with regard to the statistical reliability of the trial data and the inherent characteristics of the test animal. If one of the validation criteria is not fulfilled, the test must be repeated with the method that is more appropriate to the agreed threshold value.

- The mortality rate combines the number of dead mites and the number of escaped individuals on day 7 and should not exceed 20% of the mean value of all replicates in the control samples.
- 'Escapers' are those predatory mites that cannot be found on the test unit or that are stuck on the glue barrier. The escape rate is calculated on day 7 in relation to original number of mites. If the escape rate in the treated samples exceeds 20% (mean value of a treated group adjusted to the control; Abbott, 1925) on day 7, the laboratory test should be repeated with the combined coffin-cell method.
- The sex ratio on day 7 should be a proportion of three females to one male as a minimum. If necessary, males originating from another replicate of the same treatment should be added until the appropriate sex ratio is reached.
- The mean number of eggs per female should exceed four during the reproduction period (7 days) in the control samples.

RESULTS AND DISCUSSION

The results are presented in Figures 10.3 to 10.6. Percentage data in the figures are expressed as transformed values with 20% equivalent to 26.57% in the angular transformed scale. The symbols on the confidence limits are those of the different laboratories carrying out the tests.

During trials with the open method, the combined mortality/escape rate in the control samples remained below 20% in the results from eight

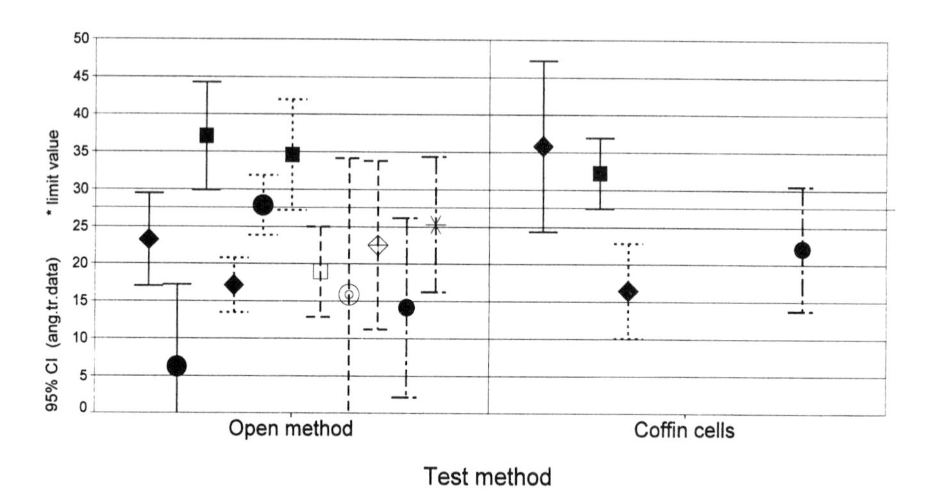

Figure 10.3 *Typhlodromus pyri* ring test: combined mortality/escape rate in control samples (results from different test laboratories).

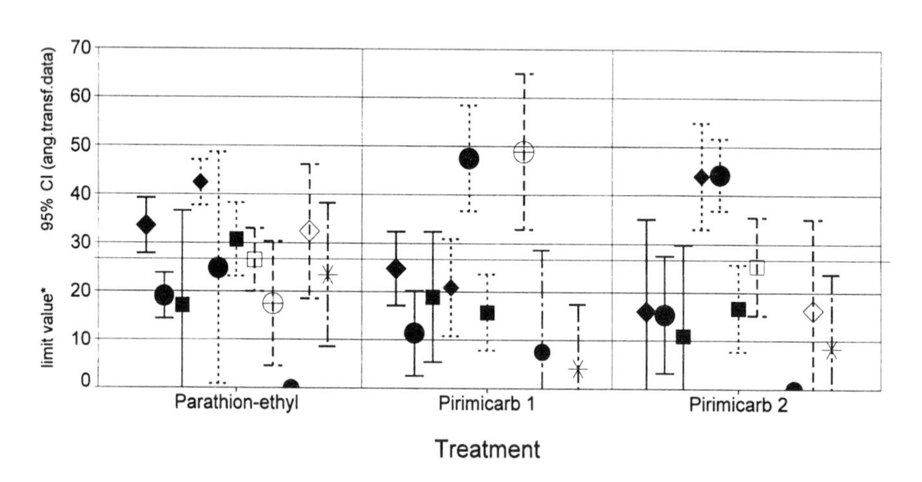

Figure 10.4 *Typhlodromus pyri* ring test: escape rate in treated samples using the open method (Abbott adjusted for the control; results from 11 test laboratories).

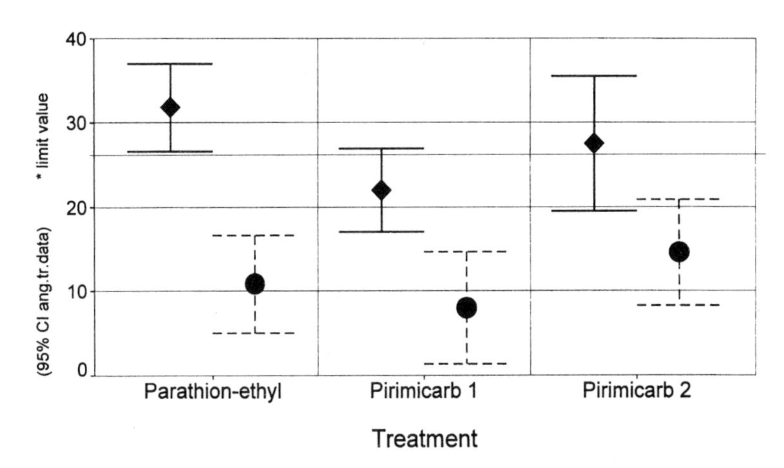

Figure 10.5 *Typhlodromus pyri* ring test: a comparison of the escape rate between the open (◆) and coffin cell (●) methods between four laboratories (Abbott adjusted for the control).

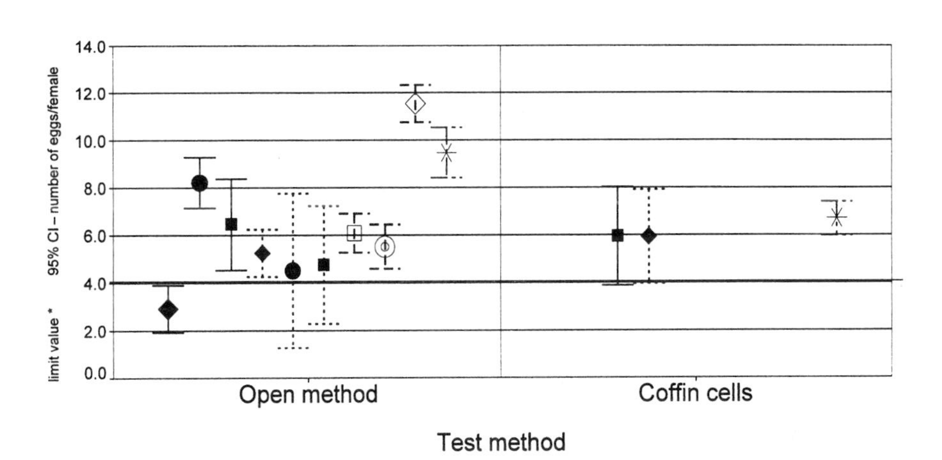

Figure 10.6 *Typhlodromus pyri* ring test: a comparison of reproduction in control samples between the open and coffin cell methods (results from different laboratories).

of the 11 test laboratories, with a mean value of 15.7% (Figure 10.3). The escape rate in the treated samples was influenced by the test compound. In the arenas treated with 1 μg/cm^2 and 2 μg/cm^2 Pirimicarb 500 DG (referred to as Pirimicarb 1 and Pirimicarb 2 in Figures 10.4 and 10.5), the escape rate adjusted for the control mean values reached 14.8% and 15.2%, while the 20% limit was exceeded in two test laboratories. In contrast, Parathion-ethyl 20 EC applied at laboratory-specific LD_{50} dose rates caused a non-compliance of the 20% limit in twice as many test laboratories (Figure 10.4), resulting in a mean value of 20.6%.

Only four of the ring-test laboratories carried out trials with the combined coffin-cell method, resulting in a mortality/escape rate of 21% on average in the control samples – which is higher than in the open method (Figure 10.3) based on the results of two test laboratories.

An improvement was achieved in the escape rate in the treated samples with 8% for Pirimicarb (500 DG 1 $\mu g/cm^2$) and 2% for Pirimicarb (500 DG 2 $\mu g/cm^2$), but especially with Parathion-ethyl 20 EC treatment with 8.7% escapers on average (Figure 10.5). The sex ratio could be adjusted in all trials with regard to the validation requirements. The minimum production of four eggs per female in the control samples was achieved in all but one of the test laboratories (Figure 10.6). No significant influence of the *T. pyri* strains used in the tests was observed (data not presented).

CONCLUSIONS

The results of the ring test to date showed that, depending on the test compound, both methods meet the requirements of the validation criteria – but to a different extent.

Both methods seem to meet the worst-case requirements for laboratory Tier I testing; however, the open method is not suitable for compounds with high repellency effect because of high escape rates.

Statistical analysis revealed that most of the significant differences were caused by the handling regimes of the different test laboratories and this therefore seems to be the most important variability factor.

There is a need for further data from the coffin-cell method to provide a more detailed comparison. Dose-response tests will be carried out to compare the effect of methodology on the test results and to characterize better the test strain used. Other topics still to be investigated are the development of tests for products with other formulations, other routes of exposure and other modes of action, such as IGRs.

REFERENCES

Abbott, W.S. (1925) A method of computing the effectiveness of an insecticide. *Journal of Economic Entomology*, **18**, 265–267.

Bakker, F.M. and Calis, J.N.M. (1989) A laboratory method for testing side effects of pesticides on Phytoseiid mites, based on a ventilated glass box: the coffin box. *Mededelingen Faculteit Landbouwwetenschappen Rijksuniversiteit Gent*, **54/3a**, 845–851.

Bakker, F.M., Grove, J.A., Blümel, S. *et al.* (1992) Side-effect tests for phytoseiids and their rearing methods. *IOBC/WPRS Bulletin*, **15**(3), 61–81.

Barrett, K.L., Grandy, N., Harrison, E.G. *et al.* (1994) *Guidance Document on Regulatory Testing Procedures for Pesticides with Non Target Arthropods. From the ESCORT workshop, Wageningen, Netherlands, March 1994*, SETAC-Europe, Brussels.

Chapman, P.F., Crane, M., Wiles, J.A. *et al.* (1995) *Asking the Right Question: Ecotoxicology and Statistics*, Report of the Workshop in Egham, Surrey, UK, 26–27 April 1995, SETAC-Europe, Brussels.

Kostiainen, T. and Hoy, M. (1996) *The Phytoseiidae*, Monograph 17 of the University of Florida, Gainsville.

Louis, F. and Ufer, A. (1995) Methodical improvements of standard laboratory tests for determining the side-effects of agrochemicals on predatory mites (Acari: Phytoseiidae). *Anz. Schadlingskde., Pflanzenschutz, Umweltschutz*, **68**, 153–154.

11

Testing side-effects of pesticides on carabid beetles: a standardized method for testing ground-dwelling predators in the laboratory for registration purposes

Gerhard Peter Dohmen

INTRODUCTION

In the last decade there has been increasing interest in the possible side-effects of pesticides, first on beneficial arthropods and later more generally on non-target organisms. Accordingly, there was a need for appropriate test methodologies. The BART group (Beneficial Arthropod Regulatory Testing group) was therefore founded as an industry initiative. Its different working groups for specific methods involved contributions from the respective experts from academia and governmental organizations. Most of the work could be based on existing methods developed by the IOBC (International Organization for Biological and Integrated Control of Noxious Animals and Plants). In the case of carabid beetles tests have been described by Edwards *et al.* (1984), Chiverton (1988) and particularly Heimbach (1992). For the specific purposes of registration testing, BART introduced a set of quality criteria:

Ecotoxicology: Pesticides and beneficial organisms.
Edited by P.T. Haskell and P. McEwen. Published in 1998 by Chapman & Hall, London. ISBN 0 412 81290 8.

- A method should give reliable and reproducible results.
- The method should have been validated by more than one institute and give comparable results.
- For registration purposes the studies should be designed to comply with the principles of good laboratory practice.
- The test substance application technique should be quantifiable and reproducible.
- The duration of the study and number of assessments should be adequate to assess the potential safety.
- The number of test organisms and replicates should be appropriate to assess critical levels.
- The test design should take account of randomization or systematic error.
- An 'untreated' (water-treated) control and a toxic standard should be included.
- Environmental conditions and other observations that may influence test results should be monitored.

The principle of the method is based on a tiered testing scheme (Hassan *et al.*, 1985). This means that the first tier has to be harsh enough to detect with a high level of safety those substances that will present a substantial risk to beneficial populations in the field. At the same time, an unnecessary amount of false positive results should be prevented. Hence, the test design should reflect worst-case conditions. The test species and endpoints should be sufficiently sensitive, but also bear ecological relevance.

METHOD DEVELOPMENT

Bearing the above-mentioned points in mind, the existing methods were scrutinized for their suitability for registration testing. *Pterostichus melanarius*, the test species proposed in the IOBC method (Hassan *et al.*, 1985), is frequently encountered in agricultural areas and thus has ecological relevance. However, this species was found to be very robust and accordingly would not be a protective indicator for other ground-dwelling beetles. In a series of tests several different carabid species have been investigated and Figure 11.1 gives an example of a comparison between several species in a laboratory test using dimethoate as test substance.

It has been suggested that *Bembidion* might be a better choice but we did not find consistent significant differences. Depending on the type of experiment, the life stage of the organism or the application rate, either *Bembidion* or *Poecilus* may react more sensitively. In this case *Bembidion* sp. proved to be sufficiently sensitive at the high application rate, but not at the lower one (Figure 11.1). The handling of *Poecilus* is much

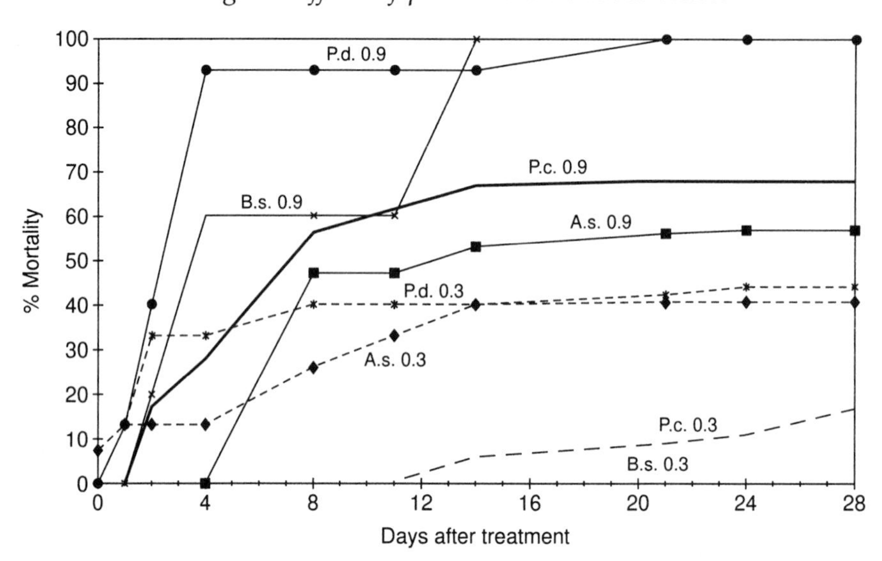

Figure 11.1 Effect of dimethoate at two rates (0.3 and 0.9 l/ha) on the mortality of several carabid beetle species: **A.s.**, *Amara* sp.; **B.s.**, *Bembidion* sp.; **P.c.**, *Poecilus cupreus*; **P.d.**, *Platynus dorsalis*.

easier, particularly in semi-field situations; which in the end allows better and reproducible results. Summarizing all the relevant information, *Poecilus cupreus*, a medium-sized carabid beetle frequently encountered in agricultural sites, was found to be the most suitable test species with respect to sensitivity, ecological relevance and amenability. It can be laboratory bred (Heimbach, 1989) and is now also commercially available. In addition, it is suitable to be tested in the next tier under semi-field conditions.

Therefore we recommend *P. cupreus* as a standard test species. In principle, other carabid beetles may be tested using the method described here, but their suitability should be demonstrated by a sufficiently low control mortality and high sensitivity as shown by the results with the toxic reference substance.

As mentioned above, the first tier in testing has to be conducted under worst-case conditions. This includes maximizing exposure of the animals to the test substance. A glass plate, which is frequntly used with other test species in the first tier, is for various reasons not suited for maintaining carabid beetles for a prolonged period. In order to find a test substrate that shows sufficient effect while at the same time not stressing the beetles in an abnormal way, a quartz sand, a natural soil and a standard artificial soil (described in OECE Guideline 207) were tested in parallel in a ring test. Figure 11.2 shows clearly that the effects were

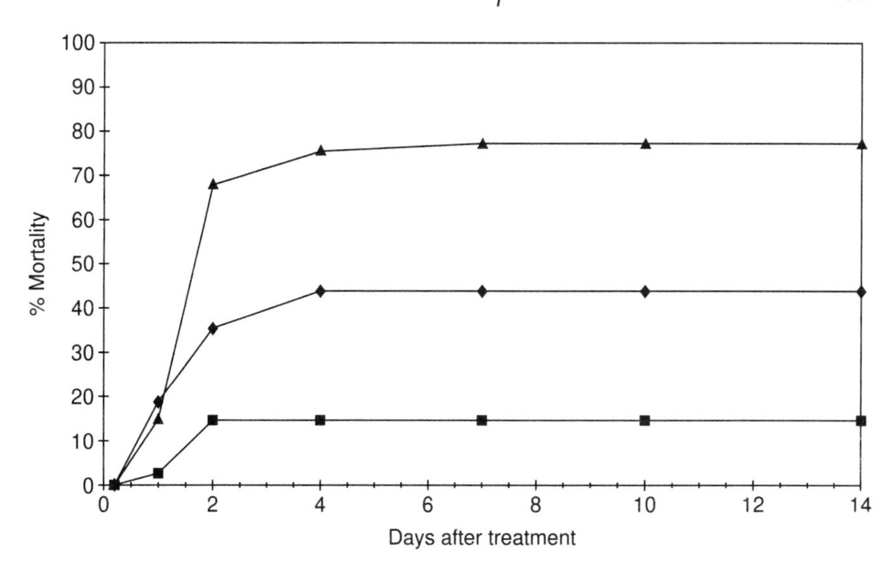

Figure 11.2 Effect of the type of substrate on mortality of Afugan-treated (1 l/ha) *Poecilus cupreus*: mean of a ring test with five laboratories. Substrate: ◆ natural soil; ■ artificial soil; ▲ quartz sand.

most pronounced on quartz sand, while control mortality was negligible on all soil types.

The test design implies that five replicate boxes are used with six beetles in each. Statistical reasons would speak in favour of keeping the beetles separate to increase the number of true replicates. Thereby, however, one might lose information on the interactions between the beetles, and this was seen to be more important – particularly since the test design has been shown to be sufficient to detect statistically significant differences in the required range.

The purpose of the toxic reference substance is to provide an indication for the suitability of the test animals and test conditions. Therefore it should give a reliable, distinct and easily observable effect of a certain degree which should on average be less than 100% (otherwise it cannot sufficiently be demonstrated that test conditions were according to the protocol, unless the time in which this effect occurs can be used as indicator). Afugan (active ingredient pyrazophos) was found to give reliable and consistent results. When applied at a rate of 1 l/ha in the laboratory test, an average mortality of 70% is reached. Figure 11.3 shows the results of a series of individual experiments in one laboratory with respect to Afugan-induced mortality.

The following method is based on previous proposals by IOBC and Heimbach and on the findings of the method development and validation

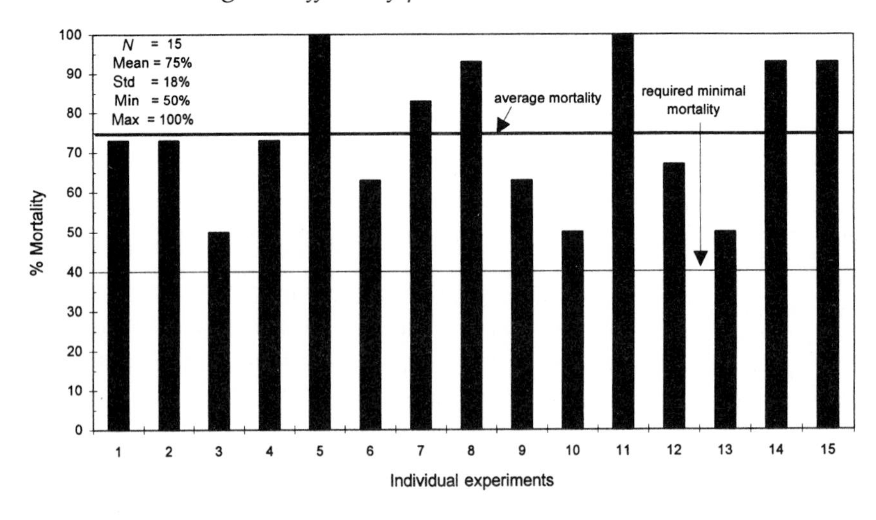

Figure 11.3 Mortality of *Poecilus cupreus* in Afugan treatments (1 l/ha) in a series of 15 separate experiments.

process of the carabid group, which consisted of the following members: K. Barrett, AgrEvo (formerly Schering); K. Brown, Ecotox; G.P. Dohmen, BASF; L. Harrison, Shell; U. Heimbach, BBA; R. Kleiner, Biochem; S. Klepka, IBACON; R. König and D. Heimann, Noack; C. Kühner, GAB; G. Levis and E. Pilling, Zeneca; J. Römbke, ECT (formerly Battelle); R. Schmuck, Bayer.

TEST METHOD

Principle of the test

Thirty beetles in five replicates are exposed to the test substance (at its maximum application rate), a toxic reference substance and water (control) in boxes with quartz sand as a ground substrate, together with their initial feed. Assessments are made during a 14-day period on mortality, behaviour, other sublethal parameters and feed consumption. If increasing effects are observed in the second week, the test is prolonged for one week.

Equipment

Plastic boxes (e.g. Bellaplast) with dimensions of about 18 × 13.5 × 6 cm (length, width, height) have been shown to be suitable test containers (to be used only once). They have a plastic lid which is cut out up to 1 cm to the edges and covered again by a net (mesh size about 2 mm).

The upper part of the boxes may be painted with fluon to impede the beetles from escaping.

The bottom of the boxes is covered by a layer of about 1 cm quartz sand (250 g dry sand on a surface of about 175 cm^2). The quartz sand (> 99% silicium dioxide) should have a particle size of > 80% within the range 0.1–0.4 mm. For the test, the sand has to be moistened to about 70% of its maximum water-holding capacity; this corresponds to about 45 g water in 250 g quartz sand.

The application of the test substance should be carried out using appropriate equipment such that nozzle type, pressure and speed ensure a realistic spray deposition and droplet size. The application rate should be verified, e.g. by weighing the quantity of spray on an exposed target (same type as used in the test) ensuring that only the deposition on the actual target area is taken into account (e.g. using covers for the walls which are removed after spraying). The treatments should be carried out in a windless environment and at temperatures between 10 and 25°C.

The beetles are kept in a ventilated area during the experimental period at 20°C (± 2°C) and a light/dark regime of 16/8 h (light intensity preferably > 200, < 2000 lux). A high relative humidity (> 70%) is advisable; otherwise the substrate has to be moistened more frequently.

Test organism

As a test species, *Poecilus cupreus* L. (Coleoptera, Carabidae) has been selected. It is encountered frequently in agricultural sites and can be bred easily for many generations (Heimbach, 1989, 1992). *P. cupreus* is a middle-sized predatory ground beetle; it is diurnal and univoltine, like most carabid beetles, and needs to hibernate as an adult in order to reach maturity. The beetles can be kept in boxes with moist natural soil or peat; fly pupae are a suitable feed (those with a thick cuticle should be cut). Females and males can easily be distinguished by the tarsi of the forelegs: male tarsi are significantly wider than those of the females.

The age of the beetles used for the test should be between 2 and 10 weeks after hatching from the pupae. They have to be kept without feeding for a minimum of 3 days before the test begins. For at least this period the beetles should also be acclimatized to the test conditions (temperature and light). It is practical to have already sorted the beetles into the appropriate boxes at this stage in groups of three males plus three females, with the necessary number of replicates, and to have spare beetles prepared in the same way.

Test design

The beetles are exposed to the maximum recommended application rate of the test substance. (In addition, a higher rate may be tested for a substance that is applied repeatedly, in which case the highest likely residues should be tested. If significant effects are observed at the highest rate, lower rates may also be applied to include plant cover or make assessments for off-crop situations.) The beetles are kept in groups of six per box (three males plus three females) with five replicates per treatment. The performance of the beetles in the test substance treatment is compared with a water-sprayed control and a toxic reference substance, both also with 30 beetles in five replicates. The standard duration of the experiment is 14 days with regular assessments; if increasing effects are observed during the second week of the experiment, the test is prolonged by one week.

Reference substance

Several ring tests have demonstrated the suitability of pyrazophos as the toxic reference substance. If the formulated product Afugan (containing nominally 294 g pyrazophos/l) is used at a rate equivalent to 1 l product/ha, an average mortality of 70% is reached.

A different reference substance may also be employed but it has to be shown (by a sufficiently large number of tests) that the applied rate of this new reference substance causes an average mortality of less than 100% and more than 40%. The minimum acceptable mortality should not be lower than 30% less than the average mortality for this compound.

Validity criteria

- Beetle mortality in the controls should not exceed 10% during the 14-day experimental period (or 15% in 21 days).
- The mortality in the toxic reference treatment (Afugan at an application rate of 1 l/ha) should be at least 40%.

Performance of the test

For the last three days before test substance application, the beetles are kept without food. Shortly before spraying the beetles are checked (missing or injured individuals are replaced) and the sand is moistened again to 70% of its water-holding capacity. Any cracks in the sand immediately before the application should be avoided. One *Musca* pupa (the integuments of the pupae are punctured with a needle) per beetle is placed on the sand. The walls of the boxes must be protected during spraying and this can be done by inserting the same type of box but with its base removed. The beetles are placed in the boxes and sprayed immediately afterwards; they should not have time to dig themselves into the sand. Thus a maximum exposure of the beetles, their food and the sand surface is achieved. Spraying should be carried out at the maximum recommended rate of the test substance with a standard water volume equivalent to 400 l/ha.

After the application the wall protection is removed and the boxes are covered with the lid and kept for 2 hours in a well ventilated place. Until the end of the testing period the boxes are kept in a ventilated room (20°C ± 2°C, light/dark ratio of 16/8 h, preferably 200–2000 lux). A high air humidity (> 70%) is favourable. After 4, 7 and 10 days the water content of the sand has to be checked and water added to 70% of the maximum water-holding capacity (if air humidity is low and ventilation is strong, this process has to be carried out more frequently).

The first check is carried out 1–3 hours after the treatment; the condition of the beetles (mortalities, damaged beetles or behavioural anomalies) is checked without searching for beetles in the sand. (If effects are observed at this stage it is recommended to carry out a further assessment on the same day 5–7 hours after treatment). The following assessments are carried out at 24 hours and 2, 4, 7, 10 (or 11) and 14 days after treatment. At each check, dead beetles are removed; knocked-down beetles (lying on their backs) are marked (e.g. with Tipp-Ex) and placed in a distinct corner of the box. If they do not move from this place (still lying on their backs) until the next evaluation, they are removed from the box and recorded as dead. Any symptoms occurring (knock-down, behavioural changes, etc.) and the number of dead or damaged beetles are recorded.

Food consumption is assessed on days 2, 4, 7, 10 and 14 after treatment. The fly pupae are collected and recorded as consumed (partly or completely eaten) or untouched. Missing pupae are recorded as consumed. One fresh pupa per surviving beetle is added to the boxes on each of these days, except day 14.

The assessments are made without searching and mixing the sand during the experiment. Only at the last assessment is the sand also searched for missing beetles and for remaining untouched food.

If more than one beetle dies during the second week of the experiment and in addition the mortality is equal to or higher than in the first week within the test substance treatment, the experiment has to be prolonged for one week with two further assessments.

Data and reporting

The number of dead males and females, of sublethally damaged beetles (e.g. abnormal behaviour, uncoordinated movements, lying on the back) and the average food consumption per living beetle should be listed per box for each treatment and assessment date. A graph showing the cumulative food consumption per surviving beetle during the course of the experiment and, where appropriate, cumulative mortality versus time should be included in the report. Any deviations from the standard protocol should be recorded; where necessary, an appropriate statistical analysis of the observed sublethal effects has to be carried out (e.g. analysis of variance).

ACKNOWLEDGEMENTS

The method development and validation process via a series of ring tests was achieved through considerable input of time and resources from the following company laboratories: BASF-AG, Bayer, BBA, BioChem, Ecotox, ECT, GAB, IBACON, Noack and Zeneca.

REFERENCES

Chiverton, P. (1988) Laboratory method for testing initial toxicities of pesticides to the carabid beetle *Bembidion lampros*. *SROP/OILB*, **11**(4), 107–110.

Edwards, P.J., Wilkinson, W. and Coulson, M. (1984) A laboratory toxicity test for carabid beetles, in *British Crop Protection Conference, Pests and Diseases* 4A-21, BCPC, Brighton, pp. 359–362.

Hassan, S.A., Bigler, F., Bl *et al.* (1985) Standard methods to test the side-effects of pesticides on natural enemies of insects and mites developed by the IOBC/WPRS Working Group 'Pesticides and Beneficial Organisms'. *Bulletin OEPP/EPPO*, **15**, 214–255.

Heimbach, U. (1989) Massenzucht von *Poecilus cupreus* (Col., Carabidae). *Verhandlungen der Gesellschaft fur Okologie, Osnabruck*, **XIX/I**, 228–229.

Heimbach, U. (1992) Laboratory method to test effects of pesticides on *Poecilus cupreus* (Coleoptera: Carabidae). *IOBC/WPRS Bulletin*, **XV**(3), 103–109.

12

Testing side effects of pesticides on spiders (*Pardosa* spp.) in the laboratory

Udo Heimbach, A. Wehling, M.P. Candolfi, M. Coulson, M. Mead-Briggs, J. Römbke, S. Schmitzer, R. Schmuck, A. Ufer, J.A. Wiles and H. Wilhelmy

INTRODUCTION

Spiders occur in high densities and with a large variety of species in agricultural crops (Sunderland, 1991). Their polyphagous nature and relative abundance in many crops makes them an important group of beneficial arthropods. Therefore spiders (*Pardosa* spp.) were selected in a SETAC guidance document as test organisms for registration purposes (Barrett *et al.*, 1994). A draft of a method was developed for testing ground-dwelling spiders of the genus *Pardosa* by Wehling and Heimbach (1994) and Wehling (1995). Because test methods have to produce reproducible results in order to gain international acceptance, at the moment this method is in the stage of validation at several testing laboratories in Europe. Results presented in this chapter were elaborated by the following laboratories: Federal Biological Research Centre for Agriculture and Forestry, Braunschweig; Sprinborn Laboratories, Horn; Zeneca Agrochemicals, Jealott's Hill; AEU, Southampton; ECT Ökotoxikologie, Flörsheim; IBACON, Rossdorf; Bayer, Monheim; BASF, Limburgerhof; Huntingdon Life Sciences, Huntingdon; Dr U. Noack Laboratorium, Sarstedt.

Ecotoxicology: Pesticides and beneficial organisms.
Edited by P.T. Haskell and P. McEwen. Published in 1998 by Chapman & Hall, London. ISBN 0 412 81290 8.

MATERIALS AND METHODS

The method used for testing spiders of the genus *Pardosa* was described by Wehling and Heimbach (1994). Spiders used in the tests are collected in agricultural fields in autumn or spring, depending on the normal timing of application of the pesticide being tested.

The spiders have to be acclimatized for at least 5 days in the laboratory before being used in tests. They are kept individually and sprayed in small containers of about 100 cm^2, with humid sand as the substrate. The containers are closed with gauze. Usually 20 replicates (10 males and 10 females) are set up for each of the three following variants: the testing substance, the water-treated control and a toxic standard, which should ideally result in an average of 65% (± 35%) mortality. At regular intervals flies such as *Drosophila* spp. are supplied as a food source and any water that has evaporated from the containers has to be replaced. After application the test units are kept in conditions of constant 20 ± 2°C and about 80% air humidity at 16/8 h light/dark regime for a period of 14 days. During this period behavioural and mortality assessments are made and the number of flies consumed is counted. Spiders that are severely knocked-down for more than 18 hours during the test are considered to be dead. Some important points of the method are summarized in Table 12.1.

RESULTS AND DISCUSSION

To validate the method, most tests were carried out using the EC formulation of the pyrethroid Karate (50 g a.i./l λ-cyhalothrin). In all except two of the experiments presented in Table 12.2, control mortality did not exceed 10%. Most of the experiments met the criteria for the toxic reference product of achieving at least 40% mortality. Species used in the test (*Pardosa agrestis, P. amentata, P. lugubris, P. palustris, P. pullata*) varied between laboratories and seasons. Body weight of spiders was not determined for the first experiments. When immature spiders were used, only the genus could be determined, not the species, and this was quite often the case for autumn-collected spiders.

Autumn versus spring tests using λ-cyhalothrin

The reactions of autumn and spring-collected spiders following treatment with the pyrethroid Karate are given in Table 12.2. Autumn-collected spiders were less sensitive than spring-collected ones. At a rate of 2 g a.i./ha applied to spring-collected spiders, a higher toxicity was usually observed than at 3 g a.i./ha on autumn-collected ones. Different sensitivity of *Pardosa* spp. depending on the season of collection in the field has

Table 12.1 Short description of the *Pardosa* spp. test in the laboratory

Factor	*Points*
Test species	*Pardosa* spp. collected in the field, depending on the use of product, subadult in autumn or subadult/adult in spring
Test organisms	10 males and 10 females, subadult or adult (one replicate) kept for at least 5 days under laboratory conditions
Food	5 *Drosophila* per spider directly after application and after 1, 2, 3, 4, 7, 9 and 11 days (also other appropriate fly species can be used)
Test substrate	Quartz sand, deionized water added to 70% of the maximum water-holding capacity
Test containers	Plastic containers (11.5 × 11.5 × 6 cm)
Application	Calibrated spraying equipment
Volume rate	400 1 per hectare (equvalent to 4 mg/cm^2)
Application rate	The highest recommended field rate of the product
Test duration	14 days; assessments after 2, 4 and 6 hours; 1, 2, 3, 4, 7, 9, 11 and 14 days after application (if more than 3 spiders die or if the feeding rate is 50% lower than in the control between day 7 and day 14, the test has to be prolonged for one week)
Validity of the test	Control mortality not higher than 10% Toxic standard with 65 ± 35% effect
Test parameters	Mortality, behaviour and feeding rate (number of flies eaten per spider)

also been reported by Wehling (1995) and Hof (1993). The same was observed for carabid beetles (Heimbach *et al.*, 1997), where insects were more sensitive when they had overwintered.

Spider characteristics: sex and body weight

In the spring experiments, males were more sensitive than females in most cases, whereas in the autumn usually the females were more sensitive (Table 12.2). This was also observed by Suhm (1996) for *Pardosa* spp. Higher male sensitivity in spring might be due to higher natural mortality and shorter life span of males at this time of the year. Because of the different sensitivity of males and females, equal numbers of males and females should be used in all tests. However, often it is not possible to distinguish the sex when spiders are too young, and this can be a problem with autumn-collected spiders. In most cases the body weight of male spiders was lower than that of females (Table 12.2). Higher sensitivity of males compared with females was also shown for linyphiid spiders (Everts, 1990; Dinter, 1995).

Table 12.2 Results obtained by different laboratories with different rates of λ-cyhalothrin using spring or autumn collected *Pardosa* spp.

Season and laboratory	*Rate (g a.i./ha)*	*Week 1 mortality (%)*[a]	*Week 2 mortality (%)*[a]	*Week 1 % food of control*	*Week 2 % food of control*	*Weight of spiders (mg)*[b]
Spring 94						
1	2	40/40	60/90	8	30	–
2	2	60/100	60/100	10	140	–
3	2	50/50	50/50	50	97	–
4	2	40/80	40/80	18	134	–
Spring 95						
5	2	100/100	–	–	–	–
6	2	0/50	0/60	–	–	37/16
7	2	100/91	100/91	0	216	–
1	2	78/30	78/40	60	99	13.4/13.5
3	2	70/90	70/90	50	140	35.3/23.4
Spring 96						
1	1.5	70/80	80/100	14	57	15.4/14.8
1	1.5	50/100	60/100	7	35	25.9/24.3
1	2	80/100	90/100	10	16	25.5/24.1
8	2.5	90/80	90/90	3	42	19.0/20.1
8	3.5	40/70	90/90	8	28	26.8/25.9
3	2.5	60/80	60/80	54	86	32.6/21.5
3[b]	2.5	60/80	60/80	62	88	65.8/33.1
9	2	90/100	90/100	21	157	42.1/34.2
9	2.5	90/100	90/100	0	157	36.6/28.6
Autumn 94						
6	3	40/20	40/20	–	–	25/17
7	2	10/0	10/0	58	140	–
7	3	25/13	25/13	64	107	–
1	2	70/60	80/60	6	33	6–10
4	3	77/20	77/20	8	90	22–27
Autumn 95						
1	3	60/60	70/60	26	93	17.3/14.3
3	3	10/40	20/50	51	119	35.5/29.6
8	3	20/50	20/50	36	142	15.1/16.3
10	2	30	30	79	101	26.2/28.8
10	3	60	70	68	103	26.2/28.8

[a] female/male
[b] (with cocoons)

Spiders of smaller size were typically more sensitive than larger ones (Table 12.2). This was also demonstrated by a test in which different rates of λ-cyhalothrin (1.8–4.2 g a.i./ha) were sprayed on autumn-collected spiders. The weights of the spiders varied between 5 mg and about 30 mg, with the majority of them being between 7 and 15 mg. Spiders of different weights were randomly distributed between the different rates that were sprayed. The range of toxicity for the different rates of the pesticide did not differ greatly (Table 12.3) but smaller spiders were more sensitive than larger ones. The two smallest surviving spiders had body weights of 8.5 mg at an application rate of 1.8 and 2.4 g a.i./ha; the smallest ones surviving a rate of 3 g a.i./ha or more were at least 21.0 mg in weight. A narrower range of toxicity of the pesticide should be expected with less weight variance of the test spiders. This shows the need to use spiders of standardized body size to obtain reproducible results. Suhm (1996) could also show an influence of body size and physiological stage on the sensitivity of *P. lugubris*. The variability in the sensitivity of different *Pardosa* spp. tested needs further studies. Wehling (1995) could not find important differences in the reaction of two *Pardosa* species. Also the data in Table 12.2 do not indicate large differences in species sensitivity.

Mortality assessment

The pyrethroid used produced a distinct knock-down effect on spiders in all experiments but, depending on the experiment, more or fewer of them were able to recover within a few days. The majority of mortality appeared during the first days after application but even in the second week some spiders died, especially during the moulting process. Spiders

Table 12.3 Mortality (%) of autumn collected *Pardosa* spp. of different weight two weeks after application of λ-cyhalthrin at rates between 1.8 and 4.2 g a.i./ha. (10 replicates per concentration for each weight group of the spiders)

λ-Cyhalothrin rates (g a.i./ha)	*% Mortality*		
	All spiders *∅ = 12.2 mg* *(n = 20)*	*Large spiders* *∅ = 16.1 mg* *(n = 10)*	*Small spiders* *∅ = 8.3 mg* *(n = 10)*
0	0	0	0
1.8	65	50	80
2.4	70	50	90
3	85	70	100
3.6	90	80	100
4.2	90	80	100

seem to be very sensitive during this phase. In many cases no sublethal effects were detected in the period immediately prior to death. High sensitivity during the moult was observed in several other studies.

Feeding assessment

The food uptake of spiders exposed to λ-cyhalothrin was in all cases reduced during the first week after application (Table 12.2). However, in many cases spider consumption rates during the second week of the test were very similar to, and in some cases exceeded, those in the control. The reduction in food consumption was usually in parallel with the observation of sublethal effects on the spiders during the tests. Further studies are necessary for a better risk assessment of reduced food uptake by spiders exposed to pesticides.

Effects of endosulfan

Experiments by the different laboratories using endosulfan (Thiodan 35 flüssig) resulted in an average mortality of 30% for autumn-collected spiders (seven experiments) at a rate of 40 g a.i./ha compared with 70% for spring-collected ones (six experiments) at a rate of 30 g a.i./ha. The variability between results of different laboratories was higher for endosulfan when compared with λ-cyhalothrin. This might be due to the volatility of this compound, which makes standardization of the testing method more complicated.

Comparing the laboratory data of the two pesticides mentioned in this chapter with effects in the field, spiders reacted more sensitively in the laboratory experiment, thus fulfilling a worst-case situation. The differences in toxicity of the two products observed between laboratory and field situations were much greater for endosulfan than for λ-cyhalothrin (Wehling, 1995).

REFERENCES

Barrett, K.L., Grandy, N., Harrison, E.G. *et al.* (1994) *Guidance Document on Regulatory Testing Procedures for Pesticides with Non Target Arthropods. From the ESCORT workshop, Wageningen, Netherlands, March 1994*, SETAC-Europe, Brussels.

Dinter, A. (1995) Untersuchungen zur Populationsdynamik von Spinnen (Arachnida: Aranea) in Winterweizen und deren Beeinflussung durch insektizide Wirkstoffe. PhD thesis, 383 pp.

Everts, J.W. (1990) Sensitive indicators of side-effects of pesticides on the epigeal fauna of arable land. PhD thesis, 114 pp.

Heimbach, U., Metge, K., Abdelgader, H. and Hoffmann, U. (1997) Vergleichende Untersuchungen zur Sensitivität von verschiedenen Käferarten und Wolfsspinnen gegenüber Insektiziden. *Mitt. Biol. Bundesanst. Land- und Forstwirtsch. Berlin Dahlem H*, **333**, 52–66.

Hof, A. (1993) Weiterentwicklung einer Testmethode zur Prüfung der Auswirkungen von Pflanzenschutzmitteln auf Wolfsspinnend der Gattung *Pardosa*. Diplomarbeit.

Suhm, M. (1996) Spinnen (Aranea) im Labortest zur Ermittlung der Toxizität von Pflanzenschutzmitteln Untersuchungen zur Eignung eines Labor-Testsystems. PhD thesis, 196 pp.

Sunderland, K.D. (1991) The ecology of spiders in cereals, in *Proceedings 6th International Symposium on Pests and Diseases*, Halle/Salle, Germany, pp. 269–280.

Wehling, A. (1995) Zur Prüfung der Auswirkungen von Pflanzenschutzmittteln auf Spinnen

Wehling, A. and Heimbach, U. (1994) *Proposed Guideline for Testing the Effects of Plant Protection Products on Spiders of the Genus* Pardosa *(Araneae: Lycosida) in the Laboratory*, BBA Guideline (draft, not yet published).

13

Two-step test system using the plant-dwelling non-target insect *Coccinella septempunctata* to generate data for registration of pesticides

Richard Schmuck, M. Candolfi, R. Kleiner, P. Klepka, S. Nengel and A. Waltersdorfer

INTRODUCTION

Data on the impact of pesticides on non-target arthropods are required for registration of pesticides in the European Union (EU Directive 91/414/EEC). The polyaphidophagous ladybird beetle *Coccinella septempunctata* (L.) (Coleoptera: Coccinellidae) was selected as an indicator species for plant-dwelling insects on the basis of its known sensitivity to pesticides and its widespread distribution in orchards and arable crops throughout Europe. Although test guidelines for this non-target insect already exist, there is a need to standardize and validate these to insure that they comply with international testing requirements (Barrett *et al.*, 1994). The objective of this work was to evaluate a two-step test system to determine its reliability and to determine the comparability of test results. Six research facilities participated in this work: M. Candolfi (Springborn Lab.), R. Kleiner (BioChem), A. Waltersdorfer (AgrEvo), S. Nengel (GAB/IFU), P. Klepka (IBACON) and S. Hennig-Pusch (Bayer AG).

Ecotoxicology: Pesticides and beneficial organisms.
Edited by P.T. Haskell and P. McEwen. Published in 1998 by Chapman & Hall, London. ISBN 0 412 81290 8.

MATERIALS AND METHODS

First step: glass plate test

The first step is a residual contact test based on the method of Pinsdorf (1989) which was modified as requested by Barrett *et al.* (1994). For the test, the substance under investigation is diluted in tap water and applied to glass plates (40 cm × 18 cm, and 0.6 cm high) in a spray volume equivalent to 200 l/ha using a calibrated spray cabinet. After spray deposits have dried, ten 4–5-day-old second instar larvae are individually confined on each treated plate. At least five replicate plates are prepared for each variable (test and reference treatment) and for the control. An excess of living aphids is provided as a food source. Ladybird beetle larvae and the aphids are confined on the glass plates using safety-glass cylinders with the inner sides coated with 'Fluon' (Zeneca) or talcum. The condition of the larvae is checked daily (except weekends) until they enter the pupal stage. The pupae remain on the treated glass plates until the adult beetles emerge.

After emergence, beetles are sexed and their reproductive performance is assessed in breeding cages (40 × 40 cm, and 54 cm high) under controlled climatic conditions (22–25°C, 16 h photoperiod > 1000 lux, 60–80% air humidity) over a period of 5 weeks. During this phase, beetles are provided with prey by placing aphid-infested broad bean plants in the breeding containers. Black paper sheets rolled into cylinders are provided as substrate for egg deposition: these are checked daily (except weekends) for egg clutches. Areas with egg masses are clipped out; after the eggs have been counted, clutches are stored separately in petri dishes until larval hatch. Adult mortality in the breeding cages is also assessed daily. This allows determination of the number of eggs laid per viable female per day.

When single pairs of beetles are examined for their reproductive performance, they are placed in a petri dish (10 cm diameter) with an excess of aphids. As substrate for egg deposition, black paper rolls are placed in each dish.

Second step: bean plant test

The method used here is similar to that proposed by Bigler and Waldburger (1988) for testing *Chrysoperla carnea* under outdoor exposure conditions. The detailed test methodology is outlined by Schmuck *et al.* (1997). Testing endpoints were identical with those of the glass plate test. In the period from 1993 to 1995, three ring tests were performed using the following compounds: ME 605 (40% w/w methylparathion), 0.3–1 kg/ha; Metasystox R (25% w/v oxydemeton-methyl), 0.3–2 l/ha;

and Afugan EC 30 (30% w/v pyrazophos), 0.01–0.5 l/ha. At their typical use rates, all three products are classified as harmful to plant-dwelling insects (e.g. Theiling and Croft, 1988; Zoebelein, 1988; Grande *et al.*, 1989). In addition, Pirimor Granulat (50% w/w pirimicarb) at 0.3 kg/ha was tested. Pirimor is a selective aphicide which is supposed to be harmless or only moderately harmful to ladybird beetles (Franz *et al.*, 1980; Gräpel, 1982; Boller *et al.*, 1989). This treatment was included to determine whether or not indirect effects via food depletion confound results obtained with this test. All products were dissolved in tap water and applied to the test cages in a water volume equivalent to 300 l/ha.

RESULTS

Glass plate test: number of successful metamorphosed beetles

To evaluate the performance of the method for routine ecotoxicological testing, the data from five research facilities were reviewed. Under the ambient conditions suggested for the laboratory test (i.e. 20 ± 2°C, 50–90% relative humidity and a 16 h photoperiod), control beetles typically emerged within 2 weeks after treatment. In the tests reviewed here, 70% or more of the control larvae successfully completed metamorphosis (Figure 13.1a). According to these data, a failure of metamorphosis of up to 30% of the larvae was used as a base rate in the glass plate test.

In the search for an appropriate reference treatment for this test, several insecticides were tested. For all insecticides examined the slope of the dose–response curve was very steep. Accordingly, these substances were of limited value as reference substances. More promising was the dose–response curve obtained for the fungicide Afugan EC 30 (active ingredient: pyrazophos) (Figure 13.2a). With this compound, a rate of about 8 ml/ha appeared suitable to reach the desired pre-imaginal mortality rate of 40–90%. To adjust for a control mortality of 20–30%, a rate equivalent to 10 ml/ha was chosen for the first inter-laboratory test run. The pre-imaginal mortality rates measured in this experiment are shown in Figure 13.2b. From these results, it seems that the rate of 10 ml/ha reliably caused a corrected pre-imaginal mortality rate of between 40 and 90%. This, then, could be an appropriate indicator for the susceptibility of the strain of species used for this test.

Bean plant test: number of successful metamorphosed beetles

The data from four research facilities were reviewed to obtain information on the value of the bean plant test for routine ecotoxicological testing. Under the outdoor conditions prevailing during the tests, larvae required between 9 and 26 days to enter the pupal stage. Relative to the

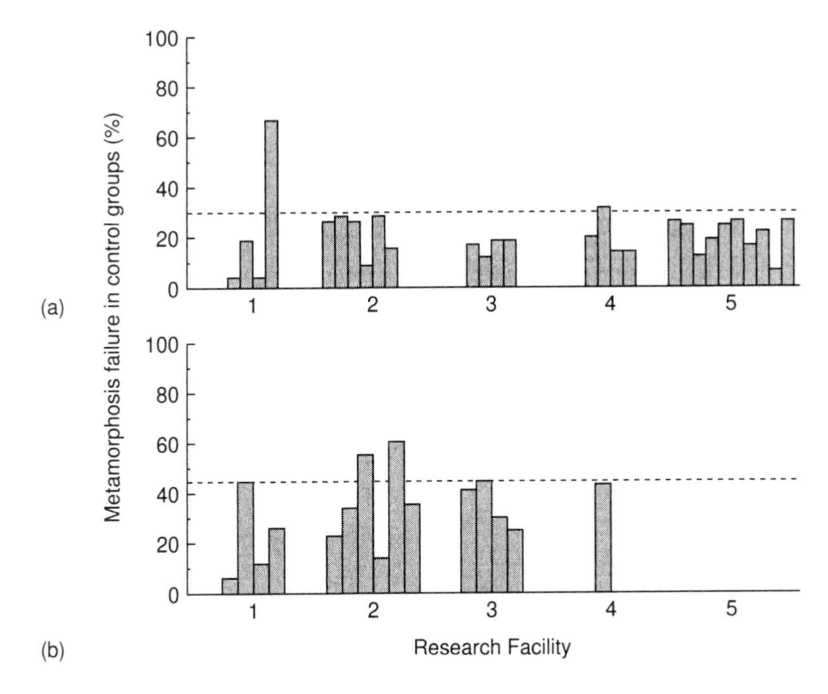

Figure 13.1 Percentage of ladybird beetle larvae which failed to complete metamorphosis in **(a)** the glass plate test and **(b)** the bean plant test.

number of larvae initially added, the number of pupae recovered in the control cages of all four research facilities averaged between 44 and 94% (Figure 13.1b). Accordingly a failure of metamorphosis of up to 45% of the larvae was used as a base rate in the bean plant test.

In one of the four bean plant tests conducted, treatment with Pirimor caused a significant increase in pre-imaginal mortality (Figure 13.3). Of the other compounds examined, Afugan EC 30 gave the most valuable results. At a rate equivalent to 0.1 l/ha, this product reproducibly caused a pre-imaginal mortality rate in the desired range, i.e. between 40 and 90% (Figure 13.3).

Assessing the reproductive performance of emerging beetles

In routine testing where the beetles from individual treatments were all pooled in one breeding container, reproductive performance was frequently much higher among the beetles that had been exposed to the pesticide than in the pertinent controls. This indicated that this testing endpoint may be subject to some variability. For a refined assessment of variability, the lifetime reproductive output of individually maintained

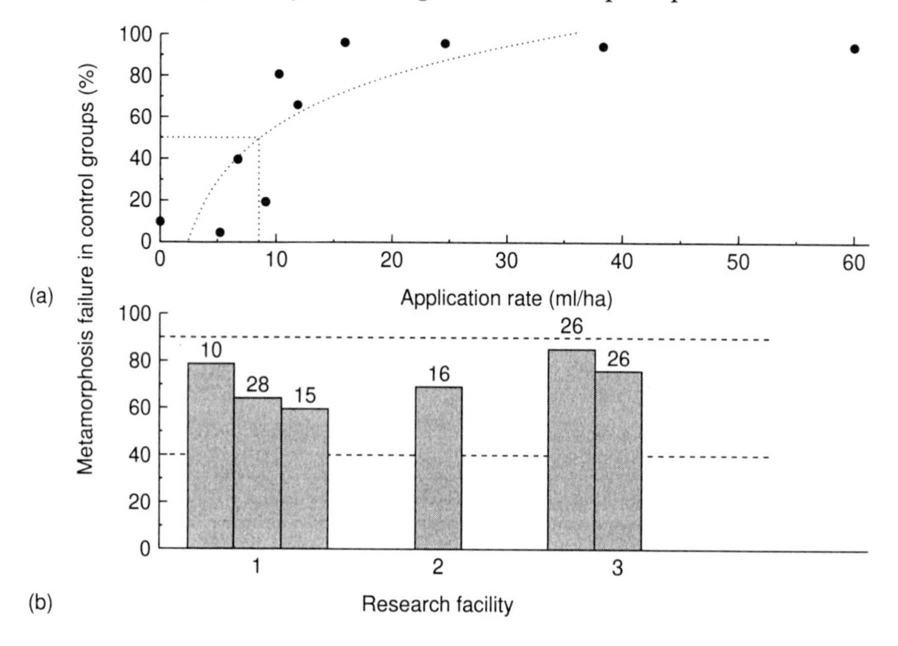

Figure 13.2 **(a)** Dose–response curve for Afugan EC30 in the glass plate test with *Coccinella septempunctata.* **(b)** Pre-imaginal mortality rate observed by three research facilities for a dose of 10 ml Afugan/ha, corrected for control mortality (numbers above columns give mortality rates in pertinent controls).

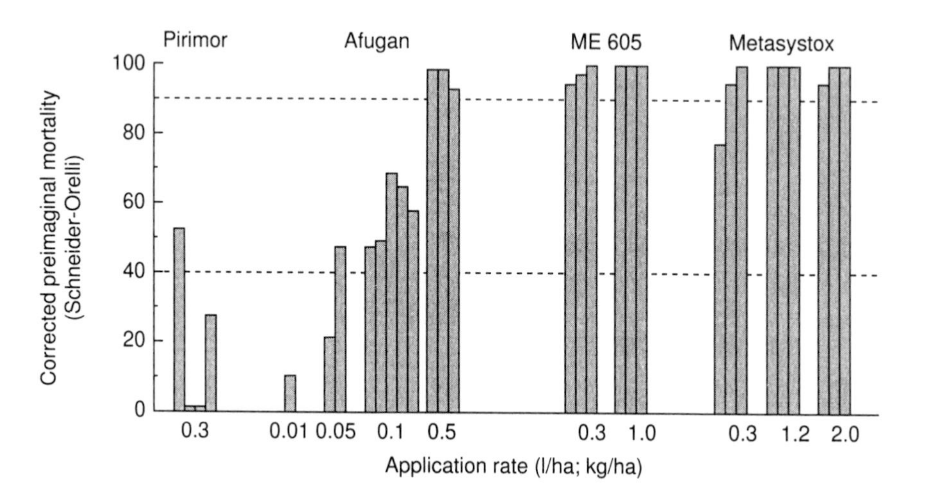

Figure 13.3 Pre-imaginal mortality observed for different products in the bean plant test. The values are corrected for the respective control mortality. Each bar represents an independent test run. Four research facilities contributed to these results.

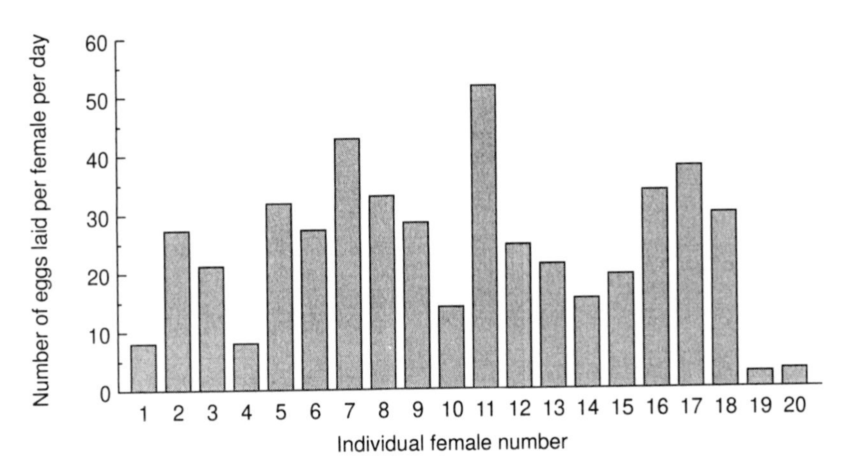

Figure 13.4 Egg-laying capacity of 20 individually maintained pairs of *Coccinella septempunctata* over the entire lifetime of the female. In tests following their metamorphosis, female longevity was between 18 and 92 days.

pairs of beetles was examined. The result of these tests (Figure 13.4) clearly demonstrate that egg production rates between *Coccinella* females are extremely variable.

DISCUSSION

In this chapter, a two-step system is proposed for testing effects of pesticides on the plant-dwelling non-target insect *C. septempunctata*. Data from the tests showed a high rate of metamorphosis failure for control larvae. With the glass plate test, failure was up to 30% (one of 28 tests discarded); and with the bean plant test, failure was up to 45% (two of 15 tests discarded). Metamorphosis failure can perhaps be reduced by technical improvements of test protocols. For detecting a 30% reduction in survival relative to controls with a probability of 0.05, and a power of 0.8, control mortalities of 18 and 31% appear acceptable for the current number of replicates in the glass plate and the bean plant test, respectively (Bailer and Oris, 1996: pseudoreplication not considered). In spite of this problem, the two-step test gave reproducible results for all pesticides tested. All of the research facilities that contributed to the ring tests found comparable mortality rates when recommended reference compounds were used.

The glass plate test proved to be very sensitive. Larvae were 10 times more sensitive to dried spray deposits on glass plates than to a direct spray treatment on bean plants. Based on the results with Afugan EC 30, the glass plate test appears to provide a very high margin of safety for detecting harmful effects of pesticides on plant-dwelling insects such as those represented by ladybird beetles.

The effects observed with Pirimor Granulat in the bean plant test indicate that there is some risk of producing false positive results using this test method. The effects are most probably attributed to food depletion. This emphasizes that special care must be taken with this test to ensure that the requested re-infestation with aphids is successful.

The data from the various laboratories showed that species-inherent variability in the reproductive performance of ladybird beetles is very high. In fact, differences between individuals or subgroups in the control treatments were greater than 30%: this is of particular importance since EU Directive 91/414/EEC gives 30% as the threshold value for treatment-related effects. Accepting the high species-inherent variability in reproductive performance, and in preference to defining a precise threshold value which then gives a treatment-related effect, it would probably be best to set an acceptable range for egg production and viability.

Data from six research facilities on the number of fertile eggs laid per *Coccinella* female are summarized in Figure 13.5. The data show that

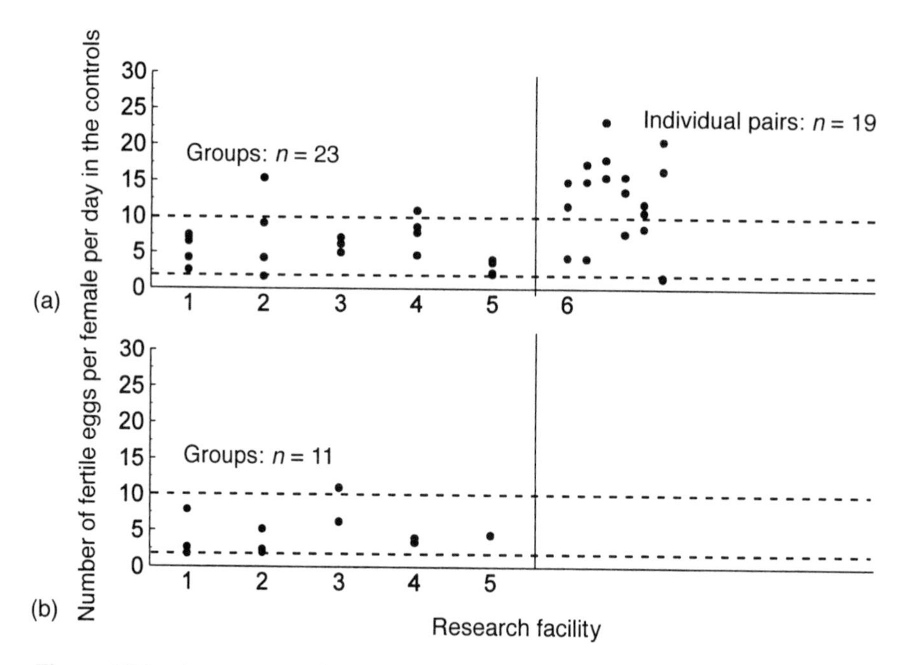

Figure 13.5 Average number of fertile eggs laid per *Coccinella* female per day during a 5-week period following metamorphosis of the beetles: **(a)** glass plate test; **(b)** bean plant test. In the glass plate test, the reproduction assay was performed either by pooling female beetles in groups or by examination of single pairs of beetles.

groups of female *Coccinella* produce between two and 10 fertile eggs per female per day over a 5-week period following metamorphosis. A higher mean egg production rate was observed for females kept individually as pairs in smaller breeding units. This might indicate that parent density has an influence on fecundity. On the other hand, this could simply be due to a higher egg predation rate when beetles are held in groups than when they are held in pairs. Figure 13.5 also shows the rate of egg production of ladybird females in the bean plant tests. The mean number of fertile eggs produced per female beetle per day over a 5-week period was again between two and 10. Thus, preliminary ranges for egg-laying capacity and egg viability that are typical for these tests can be defined. For regulatory purposes, if the reproductive performance falls below the lower limit of these ranges, the effect could be considered as treatment related. This would reduce the need for excessive repetition of tests due to random fluctuations in the reproductive performance of the test organisms.

REFERENCES

Bailer, A.J. and Oris, T. (1996) Implications of defining test acceptability in terms of control-group survival in two-group survival studies. *Environmental Toxicology and Chemistry*, **15**, 1242–1244.

Barrett, K.L., Grandy, N., Harrison, E.G. *et al.* (1994) *Guidance Document on Regulatory Testing Procedures for Pesticides with Non Target Arthropods. From the ESCORT workshop, Wageningen, Netherlands, March 1994*, SETAC-Europe, Brussels.

Bigler, F. and Waldburger, M. (1988) A semi-field method for testing the initial toxicity of pesticides on larvae of the green lacewing, *Chrysoperla carnea* Steph. (Neuroptera: Chrysopidae). *Bulletin IOBC/WPRS*, **XI/4**, 127–134.

Boller, E., Bigler, F., Bieri, M. *et al.* (1989) Nebenwirkungen von Pestiziden auf die Nützlingsfauna landwirtschaftlicher Kulturen. *Recherche Agronomique en Suisse*, **28**, 3–40.

Franz, J.M., Bogenschütz, H., Hassan, S.A. *et al.* (1980) Results of a joint pesticide test programme by the Working Group 'Pesticides and Beneficial Arthropods'. *Entomophaga*, **25**, 231–236.

Grande, C., Ingrassia, S. and Grande, M. (1989) Effeti dei ditiocarbammati sullén-tomofauna utile del vigneto da tavola. *L'Informatiore Agrario*, **42**, 121–123.

Gräpel, H. (1982) The influence of some insecticides on natural enemies of aphids. *Zeitschrift für Pflanzenkrankheiten und Pflanzenschutz*, **89**, 241–252.

Pinsdorf, W. (1989) *Auswirkung von Pflanzenschutzmitteln auf* Coccinella septempunctata *L. Richtlinie für Prüfung von Pflanzenschutzmitteln im Zulassungsverfahren*, Biologische Bundestalt für Land- und Forstwirtschaft, Braunschweig.

Schmuck, R., Tornier, I., Bock, K.-D. *et al.* (1997) A semi-field testing procedure using the ladybird beetle, *Coccinella septempunctata* L. (Coleoptera: Coccinellidae) for assessing the effects of pesticides to non-target leaf-dwelling insects under field exposure conditions. *Journal of Applied Entomology*, **121**, 111–120.

Theiling, K.M. and Croft, B.A. (1988) Pesticide side-effects on arthropod natural enemies: a database summary. *Agriculture, Ecosystems and Environment*, **21**, 191–218.

Zoebelein, G. (1988) Long-term field studies about pesticide effects on ladybird beetles (Coleoptera: Coccinellidae). *Entomologia Generalis*, **13**, 175–187.

14

Side-effects of pesticides on larvae of *Chrysoperla carnea* (Neuroptera, Chrysopidae): actual state of the laboratory method

Heidrun Vogt, P. Degrande, J. Just, S. Klepka, C. Kühner, A. Nickless, A. Ufer, M. Waldburger, Anna Waltersdorfer and F. Bigler

INTRODUCTION

This chapter describes the laboratory method for testing the side-effects of pesticides on *Chrysoperla carnea* (Neuroptera, Chrysopidae), which is the first tier within the sequential testing scheme. This would be followed by the semi-field method (Bigler and Waldburger, 1988, 1994) and then the field method (Vogt *et al.*, 1992; Vogt, 1994) using released organisms.

The laboratory glass plate test presented here is based on Suter (1978) and Bigler (1988), on the test method characteristics defined by the IOBC Working Group 'Pesticides and Beneficial Organisms' (Hassan, 1994) and on the many years of experience of most of the laboratories involved. The test is designed to evaluate the acute residual toxicity of a pesticide as well as sublethal effects on the reproductive performance of the emerging adults.

As data on side-effects of pesticides on beneficial organisms have become obligatory for registration purposes in many European countries, the number of laboratories involved in such tests has increased during

Ecotoxicology: Pesticides and beneficial organisms.
Edited by P.T. Haskell and P. McEwen. Published in 1998 by Chapman & Hall, London. ISBN 0 412 81290 8.

recent years. The need for validation of the *C. carnea* laboratory method by ring-testing has become urgent. Based on the above-mentioned references, each laboratory has worked out its own specific handling and methodology, and consequently some differences exist in the methods used. The methodical differences and points of discussion when the *Chrysoperla* ring-test group started its work to standardize the laboratory glass plate test were:

- application equipment and calibration methods;
- climatic conditions during the test;
- size of test arenas;
- age of *Chrysoperla carnea* larvae when starting the test;
- food supply for the larvae – type of food and feeding intervals;
- clearing old eggs from the test arena;
- food supply for the adults – type of food and feeding intervals;
- type and size of rearing boxes during the reproduction test;
- number of individuals and sex ratio during the reproduction test;
- inclusion or exclusion of adults that hatch later than the majority;
- use of carbon dioxide as anaesthetic when handling adults during the reproduction test;
- number of egg samples and hatching control;
- variability in reproduction;
- sensitivity of different *C. carnea* strains.

Within the joint initiative of IOBC (International Organization for Biological and Integrated Control of Noxious Animals and Plants), BART (Beneficial Arthropod Regulatory Testing group), COMET (Commercial Ecotoxicology Testing Group) and EPPO (European and Mediterranean Plant Protection Organization) for the validation of beneficials testing, the ring-test group responsible for the validation of *C. carnea* started its work in 1995. Differences in handling were determined with the help of a questionnaire. Data from laboratory tests were analysed to check the influence of several parameters, and a common protocol was worked out before starting the first tests. In the meantime two test series have been accomplished and a third one is running. The current test protocol for the laboratory method is described in detail in this chapter, taking into consideration the parameters listed above. Examples of test results are given.

THE TEST SPECIES

Chrysoperla carnea, the common green lacewing, is one of the general entomophagous predators. It is of cosmopolitan distribution (Greve, 1984) and occurs in most crops and natural habitats. The predaceous larvae attack a wide range of phytophagous insects, such as aphids, young larvae of lepidopterans, spider mites and other soft-bodied arthropods, and therefore play an important role in limiting pest arthropods. The adults of *C. carnea* are not predaceous; they feed mainly on honeydew, flower nectar and pollen (Principi and Canard, 1984). *C. carnea* is easily reared in large numbers (Hassan, 1975; Morrison, 1985). As a foliage-dwelling predator it is one of the relevant beneficial test species selected for regulatory requirements as published in SETAC's *Guidance Document on Regulatory Testing Procedures for Pesticides with Non Target Arthropods* (Barrett *et al.*, 1994).

Rearing of *C. carnea* for pesticide testing

Twenty to 40 adults are kept in a plastic or glass container (diameter 11–13 cm, height 13–17 cm) or in a polystyrol box (13 × 18 × 6 cm), which is covered with a cotton or nylon gauze. They are fed at least twice a week with an artificial diet (S.A. Hassan, BBA Darmstadt, personal communication) consisting of 15 ml condensed milk, one egg, one egg yolk, 30 g honey, 20 g fructose, 30 g brewers yeast, 50 g wheatgerm and approximately 45 ml distilled water. The diet should be a viscous pulp, which is easy to spread with a brush. With this diet the green lacewing has a high reproductive capacity (usually between 20 and 30 eggs per female per day). The diet should be stored in small containers in the freezer. The container that is in use can be stored in the refrigerator for up to 14 days at the most.

Water is offered continuously by a cotton plug, which is put into a small water container. The water reservoir should be changed twice a week.

The females deposit their eggs on the gauze, which has to be changed at least twice (preferably three times) a week to prevent hatching of the larvae in the container.

Climatic conditions

Adults and larvae are reared at 23 ± 2°C, 70 ± 15% relative humidity, > 2000 lux and 16 h daylight.

Rearing of the test larvae

The eggs are incubated in boxes, the walls of which are treated with fluon (polytetrafluorethylene). This non-poisonous substance provides a barrier to prevent the larvae from escaping. It is sufficient to apply a small strip of fluon in a thin layer to the walls, preferably with the help of a cellulose tissue. Alternatively, boxes closed with a gauze lid can be used.

Larvae that are 2–3 days old are used for the test. One-day-old larvae are not used, because they are very sensitive and can easily be damaged. We recommend incubating eggs that have been laid within a 24-hour period. Alternatively, eggs laid within 2–3 days of each other are incubated and only the larvae that hatch within 24 hours are used. Carefully shake off the hatched larvae and incubate them for about 2 days. Under the environmental conditions recommended above, approximately 7 days elapse from the incubation of the eggs until the larvae can be used for the test.

The larvae are fed preferably with fresh eggs or UV-sterilized eggs of *Sitotroga cerealella* (Oliv.) or *Ephestia kuehniella*, which can be stored for about four weeks at 2–4°C. There are some indications that frozen eggs are not suitable.

THE GLASS PLATE TEST

Test units

Most laboratories use test units that are in accordance with those described by Bigler (1988). A test unit consists of a glass plate with the pesticide sprayed on one surface. After the pesticide film has dried, the glass plate is covered with a plexiglass plate fastened down with clips. The plexiglass plate has holes of the same diameter (see below). Rings made of glass or polystyrol (height 1.5 cm), treated with fluon, are placed in the holes (Suter, 1978). Thus small arenas for the individual placing of the test organisms are created. The rings fit tightly in the holes, so that they cannot move and injure the larvae when handled. It is also possible to use individual glass plates for each test arena.

The diameter of the test arenas used up to now in the different laboratories varies from 6.5 to 7. cm. A question to be examined by the ring-test group is whether differences in the size of the treated surface area influence the results.

Application of the test substances

The glass plates are treated with the amount of product required for the purpose of the test; for example, for single-application products the maximum recommended rate (i.e. amount of product/ha) for the intended use is applied. To analyse dose-dependent effects, further rates can be tested. The amount of liquid applied should be not less than 1 mg and not more than 2 mg/cm^2, with a variation of 10% at most. In order to produce an even film and a uniform distribution of the residue, laboratory spray equipment has to be used – preferably automatic sprayers. The spraying equipment has to be calibrated before starting each test. Calibration is usually done by weighing the glass plates before and immediately after treatment with water. If irregularities are observed when spraying the test substance, it might be necessary for calibration to be done with the test substance. This point is still under discussion within the ring-test group.

Calculation of the pesticide amount to be applied when a concentration is given

For pesticides used in orchards and vineyards, usually a concentration (%) is given and the amount of product varies according to the volume (l/ha) applied appropriate to the state of the crop. With regard to a worst-case situation, when testing single-application products, the amount of pesticide has to be calculated on the basis of the highest water amount, e.g. 1000 or 1500 l/ha for orchards. Because of the three-dimensional character of orchards and vineyards, the amount per hectare is then corrected by using the following formula.

$$\text{PIEC} = \text{dose rate} \times \text{fd}/100$$

where PIEC is the predicted initial environmental concentration (a.i. or formulated product in $\mu g/cm^2$), dose rate is a.i. or formulated product in g/ha, and fd is the correction factor representing deposits under field conditions (arable crops = 1.0; orchards and vineyards = 0.4 for foliage-dwelling predators or 0.5 for ground-dwelling predators).

For example, for a recommended dosage of 0.05% (= 50 g/100 l), spraying of 1000 l/ha results in 500 g product/ha. PIEC = $(500 \times 0.4)/100 = 2\ \mu g/cm^2 = 200$ g/ha.

The PIEC formula is based on field investigations on residues on leaves, resulting from spraying according to good agricultural practice. In orchards, the residue on the leaf surface ranges between 10 and 40% of the amount applied per hectare (Schmidt, 1993; Ganzelmeier and Osteroth, 1994). Thus, by using this formula, the worst-case field situation (i.e. the 40% amount) is simulated in the laboratory. The formula was proposed by the IOBC Working Group 'Pesticides and Beneficial Organisms' (meeting in Vienna in 1993); it has been agreed by the experts at the workshop on 'European Standard Characteristics of Beneficials Regulatory Testing' (meeting in Wageningen in 1994) and is published in the SETAC *Guidance Document on Regulatory Testing for Pesticides with Non Target Arthropods* (Barrett *et al.*, 1994).

Number of replicates

At least 30 individually kept larvae are used per variant. Each single larva is a replicate. Each experiment, consisting of either one or more pesticides to be tested, includes a control, sprayed with water, and a toxic reference product. Each of these reference treatments also includes at least 30 larvae.

Starting the test

As soon as the pesticide film has dried (normally not later than one hour after spraying) the 2–3-day-old larvae are carefully put on the glass plate with the help of a fine brush (one larva per test arena). Larvae of approximately the same size have to be selected. The larvae are fed with fresh or UV-sterilized eggs of *S. cerealella* or *E. kuehniella*, three times a week (Monday, Wednesday, Friday). Only small quantities of eggs are sprinkled over the test arena, so that the residue is not covered too much. Old eggs have to be removed from the arena, perhaps with the help of a suction apparatus. Removing the eggs has to be done very carefully, without touching the residue and without hurting the larvae. When the lacewing larvae have reached the third larval instar, it is recommended that sterilized instead of fresh eggs are used for feeding so that no *S. cerealella* or *E. kuehniella* caterpillars can hatch, which might penetrate the *C. carnea* cocoons and kill the larva or pupa inside.

Evaluation of the glass plate test

At the beginning, the condition of the larvae is observed daily. Depending on the mode of action of the pesticide, the interval for the observation of effects can be prolonged; for example, three checks per week are recommended. Cocoons are collected 4–6 days after formation, carefully cleaned of *Sitotroga* or *Ephestia* eggs and put into hatching boxes. In the water-treated control it takes 8–10 days for the development of the larvae to the beginning of pupation and 10–15 days from the pupal stage until the emergence of the adults at 23°C and 70% relative humidity. The emergence period of the adults has to be recorded.

Reproduction test

The reproduction test is carried out in tubes or boxes as described above for rearing *C. carnea* for pesticide testing. All adults from a treatment hatching within a period of up to 7 days are put together in one box. Adults with deformations are not included. If only a few adults hatch after the one-week period, they are not used for a further test. If a significant number of adults emerge after this period, the reproduction test is divided into two series. A test unit should contain at least five females and two males.

The adults are fed *ad libitum* (diet according to Hassan's recipe, described above) at least twice (preferably three times) a week. Water is provided permanently from a reservoir.

The first eggs are usually laid 4–5 days after emergence of the adults. The reproduction test is started 5–7 days after the emergence of the last adult included in the reproduction unit. At the beginning of the test, the number of females and males has to be determined. This can easily be done by examining the ventral abdominal end of the insects (Figure 14.1). If test organisms escape during the test, the number of females and males has to be checked again. The egg samples are taken twice a week over a 4-week period, as during this period the fecundity of the lacewings is very constant (Zheng *et al.*, 1993). Each sample covers an egg-laying period of 24 hours, i.e. the rearing boxes are covered with a new gauze for 24 hours. Carbon dioxide can be used to anaesthetize the insects each time when changing the gauze. Usually all or the majority (> 95%) of the eggs are deposited on the gauze. If more than

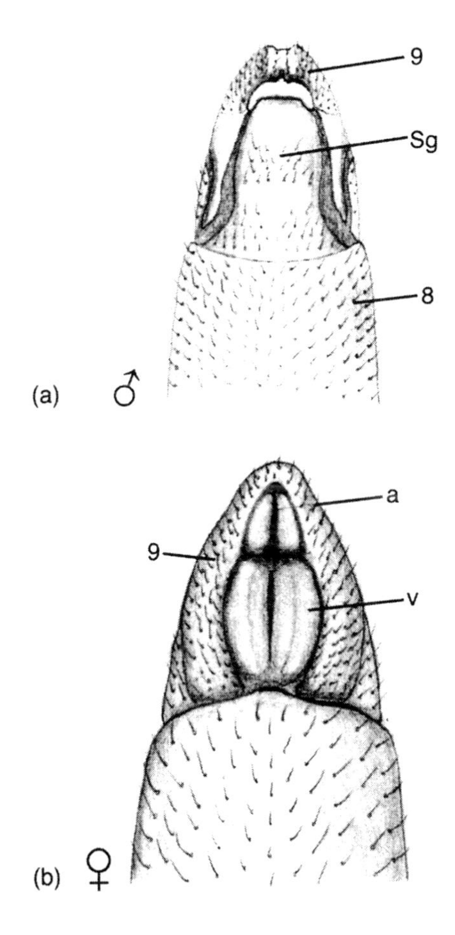

Figure 14.1 *Chrysoperla carnea*. **(a)** Male, ventral view: **8**, genital segment; **9**, anal segment; **Sg**, subgenital plate. **(b)** Female, ventral view: **a**, anus; **v**, vulvae around the genital opening; **9**, anal segment. (Drawings by J. Just.)

10% of the eggs are laid on the wall of the reproduction box, they have to be counted. Adult mortality during the reproduction test has to be recorded.

When incubating the egg samples, food (*S. cerealella* or *E. kuehniella* eggs) must be added before the larvae hatch, usually after 2–3 days, so that young larvae do not predate eggs that have not yet hatched. In order to reduce further the risk that eggs are predated, it is recommended that hatched larvae are removed from the breeding box by shaking out the gauze and the box as soon as significant numbers of larvae have hatched. When no further hatching is observed (normally after 6 days), the gauze can be evaluated at once or, if this is not possible, stored in a freezer until the analysis is carried out. For analysis all eggs on the gauze are counted and classified into the categories 'hatched' and 'not hatched'. Unusual observations must be noted – for example, that a high proportion of larvae died shortly after hatching and these are still sticking to the egg shell.

Final evaluation of the pesticide effect

Mortality

To calculate the mortality, the total number of dead larvae, dead pupae and adults that died within 24 hours after hatching is summed up. The corrected mortality (M_{corr}) is obtained by correcting the values observed with those obtained in the water-treated control (Abbott, 1925).

Morphological abnormalities

If adults with morphological abnormalities (e.g. malformed wings) are present, this has to be noted.

Reproductive performance

The mean number of eggs per female per day (fecundity), the mean hatching rate (fertility) and the mean number of fertile eggs per female per day are calculated for the treated and untreated test organisms. The average production of fertile eggs per female per day of the treated group is compared with that of the untreated group. The quotient of treated and untreated is a measure of how fertility has been affected.

The ring-test group recommends considering the two results of the test – the mortality and the effects on reproduction – separately, and not calculating a total effect (cf. Overmeer and Van Zon, 1985). A total effect conceals the scale of the two results and renders the interpretation much more difficult.

Statistics

The application of recognized statistical procedures will be elaborated within the ring-test group with the support of a statistician.

RESULTS FROM LABORATORY RING TESTS

Mortality in the water-treated group

Within the ring test, the mortality in the water-treated group in most cases did not exceed 10%, but in some cases it was 20% and higher. Based on these results as well as on many tests carried out previously by the test laboratories, it is obvious that the mortality in the untreated group is lowest in laboratories with a long experience. There it rarely exceeds 15%. The acceptable level of the mortality is still being discussed in the ring-test group. It will probably be fixed between 10 and 20%.

Reproductive performance

Influence of food quality during larval development on reproductive performance

Within the ring test a good correspondence of fecundity was registered among the different laboratories where the larvae were fed with either fresh or UV-sterilized eggs and where the adults were fed the synthetic diet described above. The fecundity of the females amounted to 20–30 eggs per female per day and the hatching rate of the eggs was over 75%. If frozen eggs were fed, the fecundity was much lower with only 3–9 eggs per female per day and, in one case, paired with a poor hatching rate (Table 14.1). In order to exclude potential factors other than the food quality influencing the results, comparative studies will be carried out by several laboratories.

At any rate, to screen out sublethal effects of pesticides from nutritional deficiencies, optimal larval nutrition is essential. The influence of larval nutrition on the fecundity of the adult lacewing has also been described by Zheng *et al.* (1993). Their data revealed that unrestricted feeding by adult lacewings on an artificial diet did not compensate for poor feeding regimes previously.

Influence of sex ratio, number of individuals per rearing box, and size and type of rearing box on fecundity

The analysis of 19 experiments, carried out in different laboratories in the years 1995 and 1996 with the same *C. carnea* strain, i.e. BI (with the

Table 14.1 Fecundity of *Chrysoperla carnea* (water-treated group) in reproduction tests following glass plate tests depending on the food regime during the larval development (in all tests the adults obtained the same synthetic diet as described in the main text)

Laboratory	*Type of larval food*	*Fecundity: no. of eggs per female per day* (n = *no. of egg samples*	*Hatching rate of the eggs (%)*	*Date of test*
2	Frozen *Ephestia* eggs	8.9 ± 4.3 (*n* = 8)	74.6 ± 11.2	23.04/14.06.95
8	Frozen *Ephestia* eggs and a few aphids (*Metopolophium dirhodum*)	2.5 ± 1.1 (*n* = 10)	48.4 ± 51.0 (no hatching at all in the first 5 samples)	05.09/08.11.96
5	UV-sterilized *Ephestia* eggs	21.8 ± 3.6 (*n* = 15)	79.7 ± 6.5	26.09/08.12.95
6	Fresh *Sitotroga* eggs	19.6 ± 5.0 (*n* = 8)	88.7 ± 4.2	05.10/05.12.95
3	Fresh *Sitotroga* eggs	30.1 ± 6.2 (*n* = 8)	85.8 ± 3.8	15.11.95/14.01.96
4	Fresh *Sitotroga* eggs	23.0 ± 4.6 (*n* = 7)	77.8 ± 11.8	10.01/15.03.96
1	Fresh *Sitotroga* eggs	26.4 ± 5.0 (*n* = 6)	88.9 ± 4.4	29.04/25.06.96
4	Fresh *Sitotroga* eggs	26.7 ± 8.4 (*n* = 8)	83.8 ± 11.9	03.07/30.08.96
7	Fresh *Sitotroga* eggs	24.3 ± 2.6 (*n* = 8)	81.5 ± 3.1	27.08/11.11.96

exception of one test) according to the present guideline has revealed that sex ratios (males per female) have no significant influence on fecundity (Figure 14.2). The data included in this analysis comprise the results of the ring tests and of tests carried out at the BBA Institute for Plant Protection in Fruit Crops, Dossenheim (cf. Händel, 1996). This result is supported by tests carried out at the same institute in the years 1993 and 1994 (r = 0.099, n = 13).

In contrast to these findings, Schmuck *et al.* (1996) report on data indicating that increasing the number of males per female results in a decrease of the fecundity of the females. With values of 0.8 to 12.8 eggs per female per day, the fecundity in their tests was very low. This is an indication that the food supply was not sufficient, or that the quality of the food was not optimal (see above, and cf. Zheng *et al.*, 1993). Therefore the differences in fecundity are more likely to be due to differences in the nutrition level than in the sex ratios.

Fecundity is not greatly influenced by the number of individuals per rearing box, ranging from 17 to 30. Only where there are higher numbers is there a slight trend that the females lay fewer eggs (Figure 14.3). The results of the ring tests confirm previous findings of Vogt (unpublished), who describes a slightly negative correlation (r = 0.34, n = 13). Further

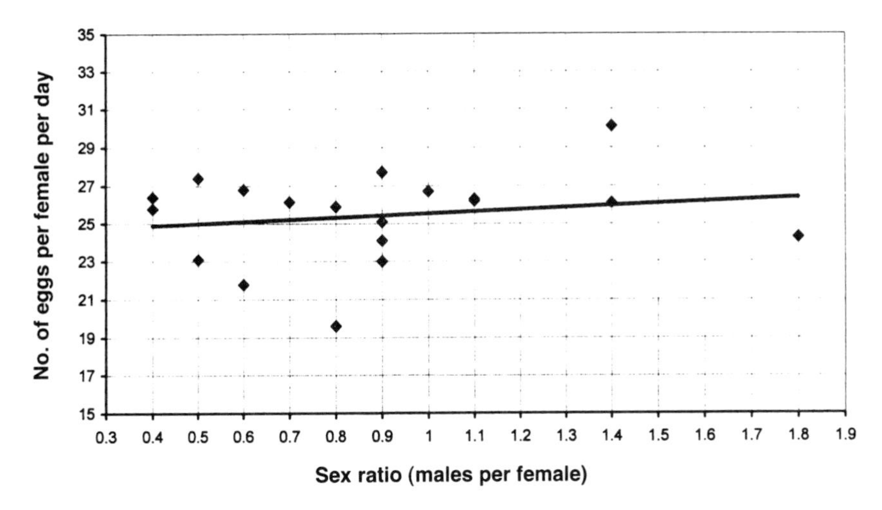

Figure 14.2 Fecundity of *Chrysoperla carnea*: regression analysis of the number of eggs per female per day in relation to the sex ratio (males per female). Data derived from water-treated controls of laboratory tests carried out in 1995 and 1996 in several laboratories according to the same method and, with one exception, the same *Chrysoperla carnea* strain (BI) (n = 19, r = 0.17).

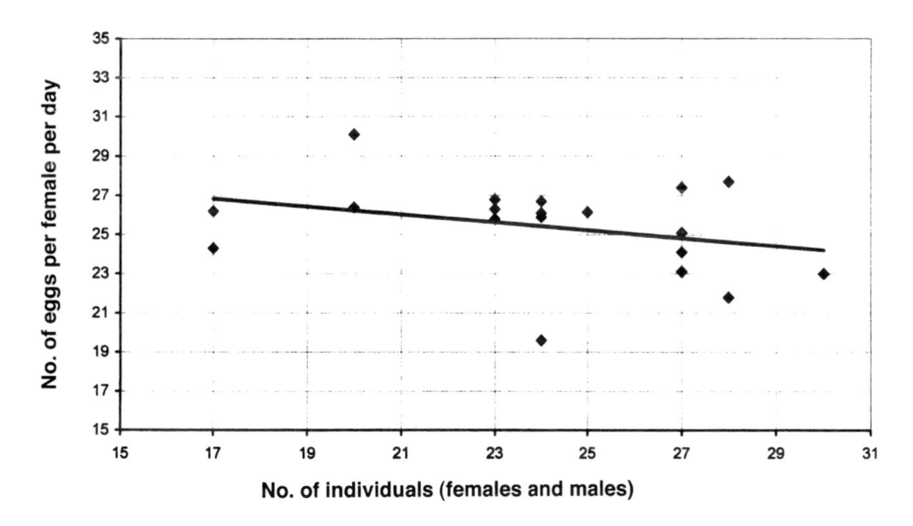

Figure 14.3 Fecundity of *Chrysoperla carnea*: regression analysis of the number of eggs per female per day in relation to the number of individuals per reproduction box. Data derived from water-treated controls of laboratory tests carried out in 1995 and 1996 in several laboratories according to the same method and, with one exception, the same *Chrysoperla carnea* strain (BI) (n = 19, r = –0.31).

Table 14.2 Number of eggs per female per day in relation to the gauze surface available for egg laying

Laboratory number	*Gauze surface (cm^2)*	*Mean number of eggs per female per day*	*Number of females per test unit*
4	204	23.0 ± 4.6	16
6	133	19.6 ± 5.0	13
5	95	21.8 ± 3.6	17
1	78.5	26.4 ± 5.0	14

tests with lower numbers of individuals are necessary to examine this situation.

Nor was there an influence of the different rearing boxes used within the ring test (cf. above), which have different gauze surfaces ranging from 78.5 to 204 cm^2. If there is less space, the eggs are laid more closely together (Table 14.2).

Pesticide effects: comparison of results obtained in different laboratories

Within the ring test the pyrethroid Sumicidin Alpha EC (a.i. 50 g esfenvalerate/l) was chosen as the test substance at two rates: 5.0 g a.i./ha (= 100 ml formulated product/ha) and 3.0 g a.i./ha (= 60 ml formulated producted/ha). The toxic standard Perfekthion (a.i. 400 g dimethoate/l) was fixed at a rate of 45 ml Perfekthion/ha. Six of eight laboratories used the same *C. carnea* strain.

In this summary, only those tests are considered where mortality in the water-treated group did not exceed 10% ($n = 6$). Perfekthion resulted in $M_{corr} > 74\%$, with no significant differences between the strains. With regard to the low rate of Sumicidin, there was some variability in results between the laboratories using the same *C. carnea* strain, with M_{corr} ranging between 36.7 and 85.8%. With the higher rate of Sumicidin, however, a very good correspondence of the results was obvious, with M_{corr} amounting to 90.0 to 92.8%. The higher variability with the lower rate of Sumicidin is probably due to the fact that this rate is close to the LD_{50} rate and that the dose–response curve in this range is rather steep. Thus slight differences in the fitness or age of the test organisms can result in different mortality rates.

One *C. carnea* strain revealed a much lower sensitivity to the pyrethroid. With the low rate of Sumicidin, M_{corr} in two tests was 21.5 and 13.8%, respectively; with the high rate of Sumicidin it amounted to 41.4 and 17.3%. Similar results with this strain were observed in previous experiments with another pyrethroid (Baythroid 50, a.i. 50g cyfluthrin/l, cf. Vogt, 1994; Bigler and Waldburger, 1994). This emphasizes the importance of good characterization of the test strains.

FURTHER ACTIVITIES OF THE RING-TEST GROUP

The ring-test group will carry out further test rounds to verify the test guideline and to enlarge the database. The tests will include dose-response trials with the toxic reference product as well as other test substances. Further studies will be carried out on parameters that might affect the fecundity of the adults (e.g. larval nutrition). The group will discuss the trigger values for categorizing harmful effects, the importance of the reproduction test and in which way the results from the reproduction test should be considered. Until now, the outcome of the laboratory test is the total effect, calculated according to Overmeer and Van Zon (1985). However, as mentioned above, this total effect conceals the scale of the two results and renders the interpretation difficult.

The group also intends to test and validate the semi-field method.

REFERENCES

Abbott, W.S. (1925) A method of computing the effectiveness of an insecticide. *Journal of Economic Entomology*, **18**, 265–267.

Barrett, K.L., Grandy, N., Harrison, E.G. *et al.* (1994) *Guidance Document on Regulatory Testing Procedures for Pesticides with Non Target Arthropods. From the ESCORT workshop, Wageningen, Netherlands, March 1994*, SETAC-Europe, Brussels.

Bigler, F. (1988) A laboratory method for testing side-effects of pesticides on larvae of the green lacewing, *Chrysoperla carnea* Steph. (Neuroptera, Chrysopidae). *IOBC/WPRS Bulletin*, **XI**(4), 71–77.

Bigler, F. and Waldburger, M. (1988) A semi-field method for testing the initial toxicity of pesticides on larvae of the green lacewing, *Chrysoperla carnea* Steph. (Neuroptera: Chrysopidae). *Bulletin IOBC/WPRS*, **XI**(4), 127–134.

Bigler, F. and Waldburger, M. (1994) Effects of pesticides on *Chrysoperla carnea* Steph. (Neuroptera, Chrysopidae) in the laboratory and semi-field. *IOBC/WPRS Bulletin*, **17**(10), 55–69.

Ganzelmeir, H. and Osteroth, H.J. (1994) Sprühgeräte für Raumkulturen – Verlustmindernde Geräte. *Gesunde Pflanzen*, **46**, 2255–233.

Greve, L. (1984) Chrysopid distribution in northern latitudes, in *Biology of Chrysopidae*, (eds M. Carnard, Y. Séméria and T.R. New), Junk Publishers, The Hague, pp. 180–186.

Händel, U. (1996) Evaluierung und Erweiterung von Methoden zur Prüfung der Auswirkungen von Pflanzenschutzmitteln auf die Florfliege, *Chrysoperla carnea* (Stephens.). Diplomarbeit thesis, 121 pp.

Hassan, S.A. (1975) Über die Massenzucht von *Chrysoperla carnea* Steph. (Neuroptera, Chrysopidae). *Zeitschrift für angewandte Entomologie*, **79**, 310–315.

Hassan, S.A. (1994) Activities of the IOBC/WPRS Working Group 'Pesticides and Beneficial Organisms'. *IOBC/WPRS Bulletin*, **17**, 1–5.

Morrison, R.K. (1985) *Chrysoperla carnea*, in *Handbook of Insect Rearing*, Vol. 1 (eds P. Singh and R.F. Moore), Elsevier Science Publisher, Amsterdam, pp. 419–426.

Overmeer, W.P.J. and Van Zon, A.Q. (1985) Standard methods to test side-effects of pesticides on natural enemies of insects and mites: *Amblyseius potentillae* (Garman) (Phytosidae, Acari). *EPPO Bulletin*, **15**, 234–235.

Principi, M.M. and Canard, M. (1984) Feeding habits, in *Biology of Chrysopidae*, (eds M. Canard, Y. Sémérian and T.R. New), Junk Publishers, The Hague.

Schmidt, K. (1993) Einfluss der Geräteeinstellung mit Prüfstand auf die Spritzbelagsverteilung im Baum. *Mitteilungen aus der Biologischen Bundestandt für Land- und Forstwirschaft Berlin-Dahlem*, **292**, 113–120.

Schmuck, R., Mager, H., Künast, C. *et al.* (1996) Variability in the reproductive performance of beneficial insects in standard laboratory toxicity assays – implications for hazard classification of pesticides. *Annals of Applied Biology*, **128** 437–451.

Suter, H. (1978) Prüfung der Einwirkung von Pflanzenschutzmitteln auf die Nutzarthropodenart *Chrysoperla carnea* Steph. (Neuroptera: Chrysopidae). Methodik und Ergenbnisse. *Schweiz. landw. Forschung*, **17**, 37–44.

Vogt, H. (1994) Effects of pesticides on *Chrysoperla carnea* Stepyh. (Neuroptera, Chrysopidae) in the field and comparison with laboratory and semi-field results. *IOBC/WPRS Bulletin*, **17**(10), 71–82.

Vogt, H., Rumpf, C., Wetzl, C. and Hassan, S.A. (1992) A field method for testing effects of pesticides on the green lacewing *Chrysoperla carnea* Steph. (Neuroptera, Chrysopidae). *IOBC/WPRS Bulletin*, **XV**(3), 176–182.

Zheng, Y., Daane, K.M., Hagen, K.S. and Mittler, T.E. (1993) Influence of larval food consumption on the fecundity of the lacewing *Chrysoperla carnea*. *Entomologia Experimentalis et Applicata*, **67**, 9–14.

Part Two B:

Semi-field and Field Testing

15

Field studies with pesticides and non-target arthropods in orchards

Kevin Brown

INTRODUCTION

The majority of new insecticides and acaricides, together with a surprising number of fungicides and herbicides, are found to be harmful to one or more of the four species tested in the initial glass plate laboratory tests (Barrett *et al.*, 1994). Since the glass plate represents extreme exposure, with maximum bioavailability of the substance and with no opportunity for the organism to avoid the residue, these results are considered to be 'worst case'. Whilst harmlessness on the glass plate is taken to mean harmlessness in the real world, a harmful result can only indicate a potential for harmfulness under field conditions. To evaluate the potential for harmfulness requires more realistic testing, at either semi-field or field level.

Whilst the logical step is the semi-field study, there are relatively few recognized semi-field methods. The carabid beetle method (Chapter 11) has been ring-tested and works well, but to date no methods specific to orchard-dwelling species have been ring-tested. This is regrettable since semi-field methods can generate conclusive results for individual species and allow sufficient replication to make meaningful and statistically based conclusions. Without generating some semi-field data the step to the large-scale field study can seem like a leap into the unknown.

The large-scale field study, also referred to as a 'total fauna study' in pome fruit orchards, generates data for a wide range of non-target species and allows evaluation at the community level, as well as an

Ecotoxicology: Pesticides and beneficial organisms.
Edited by P.T. Haskell and P. McEwen. Published in 1998 by Chapman & Hall, London. ISBN 0 412 81290 8.

investigation of indirect effects. The diversity and abundance of species at the start of the study will determine the quality of the results obtained. In high input commercial orchards predators and parasites may be important but present at levels too low to allow the effects of treatments to be determined. Selection of a site that has received low agrochemical input, and that has a population of mites, aphids or psyllids, usually results in the presence of predatory and parasitic species as well as other non-target arthropods.

STUDY DESIGN

Unlike arable ecosystems, where fully replicated designs with large plots are possible (MAFF, 1990), in orchards there is real dilemma of replication versus plot size. How large a field plot needs to be will depend on how long the study is expected to run. Reboulet (1994) advocates single or double tree plots for short-term studies and three 7 m rows for longer-term studies. Brown (1989) found 30 trees per plot too small for studying anything more than short-term acute toxicity. To quantify toxicity from a single application of treatments, determine the duration of that toxicity and then observe subsequent recovery of a population, a study would need to last for 6–8 weeks. A study of this duration will need plots of at least 120 trees each to produce meaningful data. Ideally the study would be designed with four replicates of four treatments, needing almost 2000 trees. The largest orchards suitable for this type of work typically contain about 1000 trees, which results in a study design with only eight plots.

Whilst it is tempting to run four replicates of a water-treated control and the test substance, with only eight plots it is most cost effective to include three replicates of the control and test substance and two of the toxic reference. It is necessary to include a reference treatment to verify that the study is capable of detecting effects.

The species composition of naturally occurring arthropods available for study in orchards will vary between sites, depending on prevailing weather conditions, the presence of certain phytophagous pest species and the history of previous use of the site. The most abundant beneficial taxa in apple and pear orchards are normally Heteroptera, with different species being dominant in different years and at different periods during the season. In southern France, *Heterotoma planicornis* are usually common, as are *Deraeocoris ruber* and *Pilophorus perplexus*. The more widely known genera of *Anthocoris* and *Orius* are usually well represented but may be in lower numbers except when the pest is a psyllid, when they generally dominate. Coccinellids can be numerous in apple and pear orchards, with the species composition again depending on the pest species. When aphids are present then *Coccinella septem-*

punctata is normally the most abundant, whereas with scales present *Chilocorus bipustulatus* may be the most numerous.

SAMPLING

Methods that sample the fauna of whole trees, such as inventory sampling, generate data with a higher sample size and a smaller variance between samples than those that sample branches or parts of trees, such as beating or suction sampling (Brown, 1989). In fruit orchards the tree appears to function as the unit of biology and trees of similar size and condition generally contain similar numbers of any given taxa.

The inventory sampling involves erection of collecting sheets beneath selected trees within each plot. Initial sampling is carried out 2–4 days before the anticipated application of treatments. Perimeter trees are not sampled and three trees are selected at random from within each plot area. A toxic and volatile substance, such as dichlorvos, is applied to the trees to be sampled using a motorized backpack mist blower on a very low setting. Extreme care is taken to direct the dichlorvos only on to the selected trees. All the specimens that fall on the sheets are collected approximately 8 hours after spraying dichlorvos. The trees used for this sampling are marked to prevent them from being used again in the study.

The original collecting sheets were made of cotton, which often resulted in specimens being collected with broken tarsi or missing legs. A more effective material for the collecting sheets is mylar, from which the specimens can be brushed directly into collecting pots. If the sheets are sticky from honeydew (usually from *Psylla*), water can be added to the sheets from a wash bottle, and the water and arthropods brushed through a funnel and sieve.

The prey population is also sampled before treatment, usually by visual observations of 50 leaves from each plot under a binocular microscope or with a hand lens.

Before application of treatments the collecting sheets are moved and erected beneath three fresh trees within each plot. Treatments are then applied at the maximum rate for the test substance in a realistic volume of water.

There is some confusion concerning application rates in orchard field studies. Some manufacturers specify a given weight (g) of active ingredient per hectolitre. This is a concentration, not a rate. Yet a true rate (g/ha) is misleading since a hectare of orchard can contain a few larger trees or many smaller trees. The quantity of test substance to which the organisms are exposed will be very different in small open trees when compared with large dense trees. Since the main purpose of the field study is the simulation of realistic or commercial conditions, it is important that the application results in residues equal to those that will occur when the product is used by growers. This is a thorny issue since it opens the question of how realistic the field study is.

In practice, on the question of treatments, it is worthwhile for the ecotoxicologist to calculate the volume applied per tree and to record the size of the trees in order to make the results comparable. The application itself can have a major effect on arthropods, particularly if high volumes are used with a powerful mist blower which will agitate the trees and may disturb and dislodge many of the arthropods. The control column of Table 15.1 shows those specimens collected per tree eight hours after treatment with water.

Treatments are applied to their respective plots in the order of water first, followed by test substances, and finally the toxic reference. Water-sensitive paper can be used to verify the extent of spray cover obtained with each treatment in different parts of the tree canopy.

The collecting sheets in each plot are then examined at intervals after treatment (typically 8, 24 and 28 h after application) and all dead specimens found on the sheets are collected. These samples contain the specimens that have been killed by the products and are valid for all groups, including flying insects such as parasitoids and predatory Diptera. To give an impression of the number and diversity of taxa collected in sheets after spraying, Table 15.1 shows the summarized results obtained for a new insecticide (test substance) with azinphos-methyl being used as a positive reference. In terms of the acute toxicity to non-target arthropods this product was found to be at least as toxic as azinphos-methyl. A mean of 161.3 predatory hymenopteran individuals were collected per tree from the test substance treated plots, compared with 100.5 in azinphos-methyl and 5.0 in water-treated plots. For other taxa, such as Coccinellidae, the effects of the test substance were similar in magnitude to those of azinphos-methyl.

To verify the validity of the results obtained by collecting arthropods from sheets shortly after spraying, the same trees as those from which specimens were collected after spraying are sampled with dichlorvos 48 h after the application of treatments. Specimens are again collected after a further 8 hours. This sample is considered to contain those specimens that were **not** killed by the test substance.

Some field studies stop short with a comparison of the arthropod population per tree before treatment, the numbers of taxon killed and the numbers surviving each treatment. Indeed, this may be sufficient to evaluate the acute toxicity of a substance to non-target arthropods under field conditions.

It is not surprising to discover in a field study that a new insecticide is acutely toxic to a wide range of insects. Perhaps of greater interest is an investigation into the duration of the toxic effect and the potential

Table 15.1 Mean number of arthropods per tree collected after application of treatments

	Control	*Test substance*	*Azinphos-methyl*
Hemiptera			
Aphis pomi alate	8.7	31	26.5
Aphis pomi apterae	28.8	129	167.2
Dysaphis spp. *alate*	1.8	10.3	2
Dysaphis spp. *apterae*	0.3	2	0.3
Other *Aphidoidea*	0.7	0.7	0.5
Total *Aphidoidae*	40.3	173	196.5
Psylla mali ad	0.2	3.8	0.7
Psylla madi ny.			0.7
Cicadellidae adult	2.5	45.2	31.5
Cicadellidae nymph	1.7	57.8	20.7
Pentatomidae nymph	0.2	1.5	
Predatory Heteroptera			
Pilophorus perplexus adult	0.5	19.8	20.5
Pilophorus perplexus nymph	0.2	6.5	5.5
Atractotomus mali adult	0.8	18	14.5
Orius vicinus adult		0.5	4
Orius vicinus nymph		1.8	3.5
Deraeocoris ruber adult		1	0.2
Deraeocoris ruber nymph		1.71	5
Heterotoma panicornis adult	0.8	6.8	3.8
Heterotoma panicornis nymph	2.7	45.3	43.5
Total predatory Heteroptera	5	101.5	100.5
Coleoptera			
Chilocorus bipustulatus adult		3.5	4.8
Chilocorus bipustulatus larvae		26.8	20
Coccinella 7-punctata adult		0.3	0.2
Coccinella 7-punctata larvae		4.3	2.7
Scymnus spp. adult		0.7	1.2
Stethorus punctillum adult		0.5	1.5
Total Coccinellidae		36.2	30.3
Carabidae	0.2	3.3	1.3
Staphylinidae		0.3	1
Curculionidae	0.5	10.5	7.7
Chrysomelidae		0.5	
Lathridiidae			4
Other Coleoptera		0.5	1.5
Neuroptera			
Chrysoperla carnea adult	0.3	0.3	
Chrysoperla carnea larvae		0.5	0.7
Hymenoptera			
Proctotrupoidae		1	0.8

Table 15.1 *continued*

	Control	Test substance	Azinphos-methyl
Proctotrupoidae ceraphronidae		11.8	3.8
Braconidae		2	0.5
Ichneurmonidae		1	1.2
Chalcidoidae		15	6.7
Total parasitic Hymneoptera		30.8	13
Formicoidae		45.7	51.5
Diptera			
Syrphidae episyrphus adult		0.7	0.3
Syrphidae syrphus adult		0.5	0.5
Syrphidae scaeva adult			
Dolichopdidae medeterus		0.2	
Emphididae		7	3.3
Total predatory Diptera		8.3	4.2
Non-predatory Diptera		150	54.3
Dermaptera			
Forficula auricularia adult		1.5	0.7
Forficula auricularia juvenile	1	314.5	80.3
Other Demaptera			
Lepidoptera			
Lepidoptera adult	0.3	2.2	0.5
Lepidoptera larvae	0.3	2.3	1.3
Orthoptera	1.8	18.8	5.2
Thysanoptera	1	21.7	106.7
Araneae			
Araneae lycosidae	0.5	3.7	2.5
Other Araneae	1.3	24	19.5
Total Araneae	1.8	27.7	22

for the population to recover or re-colonize the treated plots. In most studies, particularly with slow-acting products, or products such as insect growth regulators (IGRs), inventory sampling is continued using fresh trees at increasing intervals for 4–6 weeks after the application of treatments, together with parallel visual observations. This continued post-treatment sampling allows within-season recovery to be observed, permits the development of immature stages through to adults to be examined, and allows a community-based analysis of the results.

Figure 15.1 presents the results for *H. planicornis* nymphs obtained by inventory sampling from the same study that generated the data in Table 15.1.

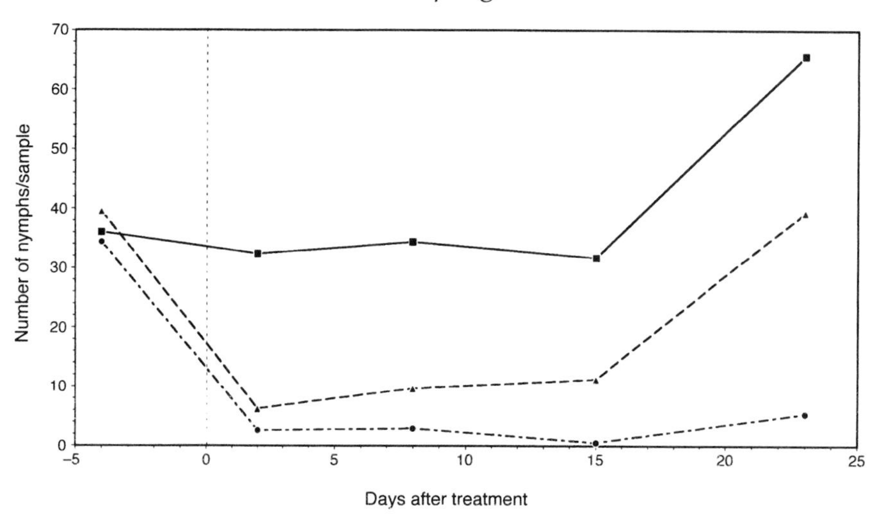

Figure 15.1 Mean number of *Heterotoma planicornis* nymphs per inventory sample. ▪ Control; ▴ test substance; • Azinphos-methyl.

The major decline in numbers between pre-treatment and post-treatment sampling in the test substance and azinphos-methyl treated plots confirms that the test substance was of similar acute toxicity to *H. planicornis* nymphs as the reference substance. By 15 days after treatment there was an indication of some recovery in the test-substance treated plots, with partial recovery occurring by 24 days after treatment. The fact that the recovery at the end of the study coincided with an increase in the numbers of nymphs in control plots indicates that treatment residues were no longer toxic beyond 15 days after treatment. Interpretation of these data is of course more complex than described here since relative prey abundance will also play a part in whether predator numbers increase.

This approach works well for predatory Heteroptera, Coccinellidae, Neuroptera (lacewings) and polyphagous groups such as Dermaptera (earwigs) and Araneae (spiders), but is not appropriate for the highly mobile alate taxa such as Diptera and Hymenoptera.

Note as an aside that repeated applications of azinphos-methyl were widely used in apple orchards in southern France in 1996 in an attempt to control resistant codling moth.

Even with the best preparation and good weather conditions, it is unlikely that a single field study will generate data for the full range of species. However, it is usually possible to include bioassays and semi-field tests in the study plan so that those species which are not naturally occurring at the site can be exposed to appropriately treated plant material within cages. For semi-field investigations arthropods are enclosed in mesh cages over branches within the treated plots. For bioassays branch tips can be cut from the treated plots and arthropods exposed to the treated surfaces in the laboratory. Cages have been used successfully for the parasitoid *Trichogramma*, coccinellids and *Chrysoperla carnea*. Such add-on components to a field study can be extremely valuable and can demonstrate surprising differences in susceptibility between different taxonomic groups.

SINGLE VERSUS MULTI-APPLICATION STUDIES

Ten years ago the demand for large-scale field work in orchards came with the development of IGRs. Since several IGRs demonstrated selectivity in favour of beneficial arthropods, the motivation behind these early studies was to demonstrate or confirm a perceived harmlessness to beneficial arthropods. In other words the early field studies in orchards were conducted to demonstrate suitability of the products for use in integrated pest management. The products in question had one or two applications per season and the studies could run for a period of 6 weeks after treatment, looking at effects on both immature and adult life stages.

Many of the molecules now triggering further work are not insecticides or acaricides, with one or two applications per season, but are fungicides with up to eight or 10 treatments a season. Whilst it is possible to study predatory mites for a whole season's duration, it is not practical to assess effects of more than two or three treatments in a total fauna study. The main reason for this is the question of timing. When the fungicide programme starts, early in the season, there are typically very low numbers of arthropods present in orchards. A site would have to be chosen because of a potential population of invertebrates. Once such a study had begun the hoped-for species might never arrive. A second difficulty with studies of long duration is the question of plot size. To generate valid results for a whole season's treatments would need plots larger than could be found in a European orchard.

CONCLUSIONS

Large-scale 'total fauna' studies can give a valuable insight into the likely effects on the arthropod fauna resulting from commercial use of pesticide. Such studies are only successful when carried out in large and homogeneous orchards which support a rich and diverse arthropod fauna. Low input orchards with only a few sprays a year make ideal study sites, particularly in warmer climatic regions. The subsidy paid by the European Union to remove apple orchards has resulted in the destruction of the majority of low input orchards, making site selection extremely difficult.

Whilst large-scale field studies are still possible, there is now an urgent need for well designed and validated semi-field methods for testing the effects of pesticides on non-target arthropods in orchards.

REFERENCES

Barrett, K.L., Grandy, N., Harrison, E.G. *et al.* (1994) *Guidance Document on Regulatory Testing Procedures for Pesticides with Non Target Arthropods. From the ESCORT workshop, Wageningen, Netherlands, March 1994*, SETAC-Europe, Brussels.

Brown, K.C. (1989) The design of experiments to assess the effects of pesticides on beneficial arthropods in orchards: replication versus plot size, in *Pesticides and Non-target Invertebrates* (ed P.C. Jepson), Intercept, Dorset, pp. 71–80.

MAFF (1990) *Guideline to Study the Within Season Effects of Insecticides on Beneficial Arthropods in Cereals in Summer.* Working Document 7/7, Ministry of Agriculture, Fisheries and Food, London.

Reboulet, J.N. (1994) Impact des produits phytosanitaires sur la fauna auxiliaire. *ACTA Point*, **1**.

16

Accuracy and efficiency of sequential pesticide testing protocols for phytoseiid mites

Frank M. Bakker

INTRODUCTION

Under the regime of European Commission (EC) Council Directive 91/414, data concerning side-effects of pesticides on arthropod natural enemies are now a mandatory dossier requirement for the placement of plant protection products on the European market. By reference to European and Mediterranean Plant Protection Organization (EPPO) decision-making schemes and the Society for Environmental Toxicology and Chemistry (SETAC) guidance document on regulatory testing procedures for pesticides with non-target arthropods (Barrett *et al.*, 1994), sequential testing protocols are explicitly required. Effects on a particular test species are considered to be significant when they exceed the threshold (trigger) values as defined in the EPPO schemes for the environmental risk assessment, unless species-specific threshold values are defined in the respective test guidelines. Whenever significant effects are observed at lower testing tiers (laboratory, semi-field), a field test is required to provide sufficient information to evaluate the risk of the plant protection product for arthropods under field conditions. At this stage appropriate risk mitigation measures can be proposed.

The worst-case character inherent to the lower testing tiers is assumed to provide sufficient safeguard against false negative results, with 'negative' meaning no significant effects. Hence, at lower testing tiers, effects

Ecotoxicology: Pesticides and beneficial organisms.
Edited by P.T. Haskell and P. McEwen. Published in 1998 by Chapman & Hall, London. ISBN 0 412 81290 8.

below the pre-set trigger value are considered conclusive and further testing is not required. Thus, the rationale behind the sequential testing schemes is the reduction in testing effort resulting from those products conclusively classified as harmless at the lower testing tier. The exact setting of the trigger value therefore determines both overall testing effort and the accuracy and efficiency of the sequential testing scheme (Figure 16.1). Accuracy and efficiency are here defined as the proportion of products correctly classified and the proportion conclusively classified at the lowest testing tier, respectively.

Bakker and Jacas (1995) investigated the accuracy of sequential testing schemes under a range of trigger values for several species of phytoseiid mites. This was done by determining the degree of correspondence between conclusions drawn from laboratory trials with conclusions concerning the same compound drawn at higher testing tiers (field/greenhouse). The species combinations analysed are shown in Figure 16.2 and the findings of their analyses are summarized as follows.

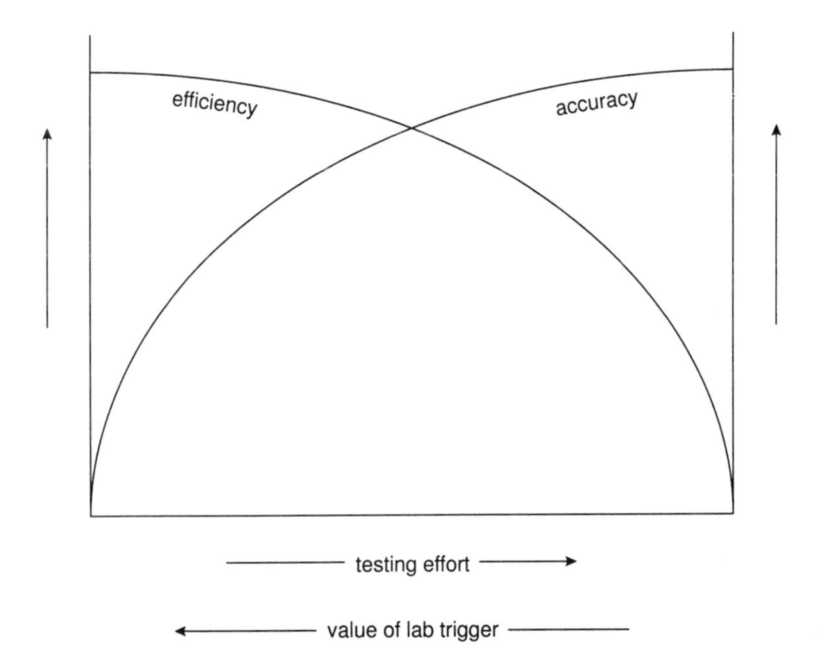

Figure 16.1 Trade-off relationship between efficiency and accuracy of sequential testing schemes in terms of testing effort and their relationship with the laboratory trigger value used to distinguish harmless effects from inconclusive results.

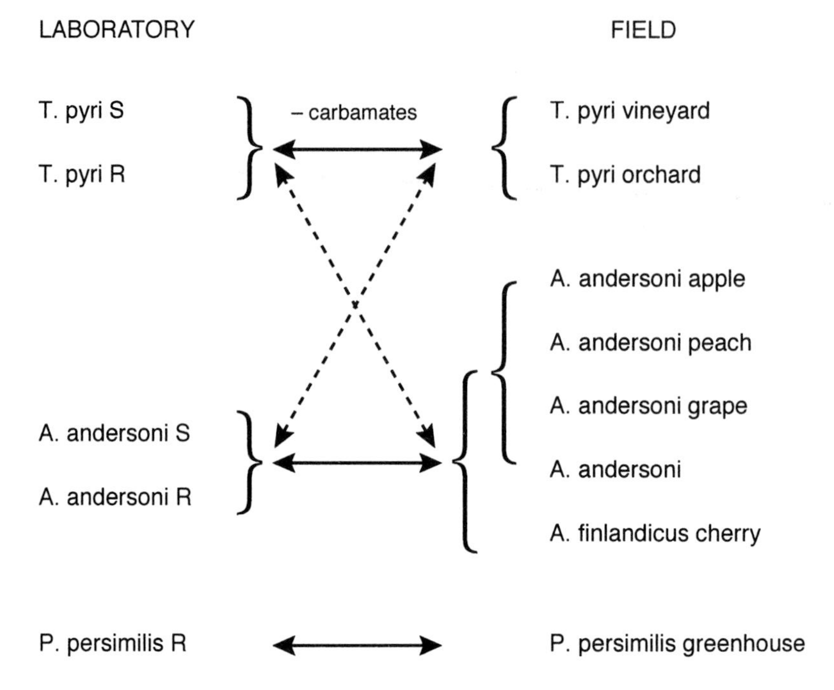

Figure 16.2 Phytoseiid species combinations analysed by Bakker and Jacas (1995). **S** and **R** denote susceptible and resistant strains, respectively. Dashed arrows indicate intraspecific comparisons made to assess the feasibility of using indicator species. For the analyses, data published in the frame of the Joint Pesticide Testing Programme (JPTP) of the IOBC Working Group 'Pesticides and Beneficial Organisms' (PBO) or data obtained with the same methodology were used.

Amblyseius sp.

Laboratory trials can only be used to conclude harmlessness accurately when population reductions in the field of 60% and over are considered to be harmless side-effects. Because this seems to be unrealistically high, the sequential testing approach is considered inappropriate for *Amblyseius*.

Typhlodromus pyri

Results obtained with this species are robust. Irrespective of the strain tested in the laboratory and for a realistic range of population reductions that can be considered to be harmless in the field, laboratory effects between 25% and 55% accurately forecast negligible hazard resulting from field use. However, although the worst-case sequential testing scheme appears accurate for this species, it is not very efficient: 71–75% of all products were triggered to a higher testing tier.

Phytoseiulus persimilis
Of all products listed as harmless at Tier I, 20% received a harmful classification in the semi-field test. The testing scheme was therefore considered inappropriate for this species but when analysing fungicides and insecticides separately, it was found that laboratory results can accurately predict harmless side-effects for fungicides. For insecticides it was demonstrated that concluding a product is harmless, based on laboratory effects ranging from 9 to 59%, implies that 40% of the products will be erroneously classified.

Cross comparisons
For *Amblyseius* and *Typhlodromus* it was found that the accuracy with which field effects are predicted is independent of the species tested in the laboratory. Thus either species can in principle be used as an indicator. The finding remains that field thresholds for *Amblyseius* may be unrealistic, and so in fact only side-effects on *T. pyri* can be accurately forecast.

Hence, the worst-case sequential testing scenario appeared to be valid for one species only, but for this species the efficiency was too low to warrant the approach. To find a way out of this stalemate, the data were re-evaluated in an attempt to identify and mitigate possible causes for the low accuracy and efficiency levels observed.

FACTORS UNDERLYING ACCURACY LEVELS OBSERVED

Methodological flaws and ecological factors that possibly underlie the accuracy levels observed in the testing protocol under study are extensively discussed in Bakker and Jacas (1995). It suffices to say here that although the methodological flaws discussed (test method, test dose, calculation of *E* and reproducibility of the field trials) provided a partial explanation of the accuracy levels observed, they could not explain the interspecific differences found, nor could they explain, as in the case of *P. persimilis*, why results depend on biocidal activity. To identify possible causes for inaccurate predictions of harmlessness, those data where field trials showed a product to be more harmful than the laboratory trial were compiled (Table 16.1).

To elucidate the causes for the discrepancies listed in Table 16.1, the data that scored a 3 or a 4 in the field were distributed for peer review – in most cases to the originators of the data. For *Amblyseius andersoni* it was confirmed that chlorpyriphos is indeed a harmful compound under field circumstances, while in the case of procymidone it might be better to repeat the field trial for a confirmation of the result (C. Duso, personal communication). No reply was received for *Amblyseius finlandicus*. Clearly most conflicting cases were obtained with *P. persimilis*. Common to all

Table 16.1 Products that were classified harmless in the laboratory test but scored higher in the field[a] (harmful–harmless conflicts not included)

Source	*Active ingredient*	*Concentration (ppm a.i.)*	*Laboratory result (%)*	*Semi-field result*
Phytoseiulus persimilis				
IOBC7	Penconazole	250	26.6	'3'
Oomen 1991	Penconazole	160	16	
IOBC5	Dichlofluanid	750	10	'3'
Oomen 1991	Dichlofluanid	1000	13	
IOBC4/Oomen 1991	Bromophos	1000	10	'4'
Oomen 1991	Bromophos	3974	100	
IOBC4/Oomen 1991	Chloryriphos	625	16	'4'
Oomen 1991	Chloryriphos	2500	98	
IOBC4/Oomen 1991	Diazinon	378	7	'4'
Oomen 1991	Diazinon	1512	29	
IOBC4/Oomen 1991	Dimethoate	840	47	'4'
Oomen 1991	Dimethoate	210	22	
Oomen 1991	Dimethoate	3360	98	
IOBC4/Oomen 1991	Mevinphos	580	2	'4'
Oomen 1991	Mevinphos	2320	0	
IOBC4/Oomen 1991	Triazophos	1018	49	'4'
Oomen 1991	Triazophos	4070	100	
IOBC3/Oomen 1991	Azinphosmethyl	500	9	'3'
IOBC3/Oomen 1991	Fenitrothion	550	0	
IOBC3/Oomen 1991	Heptenophos	565 + 2260	0	'4'
IOBC3/Oomen 1991	Etrimphos	1000	25	'3'
IOBC3/Oomen 1991	Vamidothion	500	11	'4'
IOBC3/Oomen 1991	Prochloraz	748	18	'3'
A. andersoni S-strain laboratory vs A. andersoni Duso field[b]				
IOBC6	Procymidone	285	19	'3'
Duso/apple	Chlorpyriphos	500	26	'3'
Duso/vine	Parathion	400	24	'2'
IOBC5	Bitertanol	925	14	'2'
IOBC5	Copperoxychloride	4500	7	'2'
IOBC5	Dithianon	446	22	'2'
IOBC5	Clofentezine	200	42	'2'
A. andersoni R-strain vs A. andersoni Duso field				
IOBC6	Procymidone	285	10	'3'
A. andersoni S-strain vs A. finlandicus field				
IOBC6	Triforine	285	3	'2'
IOBC5	Copperoxychloride	4500	7	'2'
IOBC5	Thiocyclam	150	55	'4'
IOBC4	Fenarimol	140	10	'3'
IOBC4	Folpet	1650	10	'2'

Table 16.1 *continued*

Source	*Active ingredient*	*Concentration (ppm a.i.)*	*Laboratory result (%)*	*Semi-field result*
IOBC4	Penconazole	40	19	'3'
IOBC4	Atrazin	3350	10	'3'
IOBC3	Prochloraz	748	'3'	'4'
IOBC3	Chlormequat	4900	'1'	'2'
IOBC3	Naphtylaceticacid	15	'1'	'4'
Typhlodromus pyri S-strain vs T. pyri orchard				
IOBC7	*Verticillium levanii*	0.4% product	0	'2'

[a] Figures in quotation marks denote IOBC classification.
[b] Note: Duso found on peach and vine = 93% and 95% in laboratory, compared with '2' and '3' in the field.

comments pertaining to this species was the mention of possible strain differences (S. Blümel, P. Ooomen, G. Sterk, personal communications). As mentioned in most IOBC (International Organization for Biological Control of Noxious Animals and Plants) publications, the laboratory data were obtained with an organophosphate (OP) resistant (commercial) strain, whereas most semi-field data came from a susceptible strain. Because all insecticides in Table 16.1 are indeed OPs, the strain differences mentioned may explain most errors found. What remains unexplained are the results obtained with the remaining three fungicides and the fact that not all OPs were classified as harmless in the laboratory (metamidophos, chlorfenvinphos, acephate and heptenophos – the latter in the 7th Joint Pesticide Testing Programme, JTPT – were all scored as harmful). In this respect it is important to report another possible cause of error that was mentioned by some respondents, viz. that strict compliance with established guidelines was not always observed in the past. This issue should not be taken too lightly, since it implies that the validity of the data gathered in the frame of IOBC's JPTPs is questionable.

Finally, it should also be stressed that the higher tier data for *P. persimilis* were all obtained in a semi-field set-up and not in full-scale greenhouse trials. Therefore the question of the real correspondence between semi-field trials and practice remains. As pointed out by Blümel *et al.* (1993) and Stolz (1994), interpretation of the semi-field test is not always clear, due to methodological problems such as the interpretation of escaping test organisms. Enquiries by G. Sterk (personal communication) showed that the OPs listed in Table 16.1, when still available, are perceived by biocontrol practitioners as incompatible with *P. persimilis*. The problem with escapees also occurs in the laboratory assay on detached leaves, which led Blümel *et al.* (1993) to conclude that a test system consisting of ventilated glass cages might be more appropriate. At present, experiments are conducted to test the validity of some of the *P. persimilis* data presented in Table 16.1 (G. Sterk, personal communication).

The discussion can be summarized as follows. There was broad agreement that the approach, as such, was valid; the discrepancies that were found can be explained by strain differences (*P. persimilis*) or differences in the resistance status of field populations, and by limitations in the comparability of data, due to a lack of standardization or, early in the programme, lack of experience with the testing methodology. It remains to be established to what extent the biological/ecological causes for discrepancies as mentioned by Bakker and Jacas (1995) have also played a role.

VALIDATION OF SPECIES- AND PRODUCT-SPECIFIC PROTOCOLS AND TRIGGER VALUES

The data used in the analysis by Bakker and Jacas (1995) are summarized in Table 16.2. Although contingency table analysis led these authors to pool data up to the species level, Table 16.2 indicates that some comparisons may show a better match than others. For example, in the case of *Amblyseius* sp. it seems that the laboratory tests with *A. andersoni* lead to the same effect profile as the field tests with *A. finlandicus* but, strikingly, not in the case of field tests with *A. andersoni*. The former two both originate from northwest Europe (The Netherlands and Belgium, respectively), whereas the latter was tested in Italy. It is conceivable that a common agricultural background has led to toxicological convergence of the two species from northwest Europe, whereas the two *A. andersoni* strains diverted in a toxological sense. Finally, inspection of the *T. pyri* data in Table 16.2 indicates that the toxicological profile of the susceptible laboratory strain is more congruent with the profile obtained in vineyards than with the one obtained in orchards, whereas for the resistant laboratory strain the reverse applies. It has been confirmed that the *T. pyri* in Swiss vineyards is indeed more susceptible than the strain tested in orchards, although susceptibility is lost rapidly at present (E. Boller, personal communication). Thus the low accuracy levels observed previously could have been partly caused by the pooling of strains for the analysis.

As outlined earlier, the rationale underlying sequential testing schemes is the reduction in testing effort resulting from products that can be conclusively classified after the laboratory assay, here defined as the efficiency. In most currently used sequential testing protocols, such as those proposed in the IOBC, the SETAC guidance document and hence Directive 91/414/EEC, harmless classifications obtained after the laboratory assay are assumed to be conclusive because the laboratory assay represents a worst-case situation. By the same token, harmful effects are considered to be inconclusive. Thus efficiency is limited by the proportion of products that are in fact harmless. As Table 16.2 illustrates, the unbiased

Table 16.2 Proportion of products classified as (1) harmless, (2) slightly harmful, (3) moderately harmful and (4) harmful, in trials performed in the frame of IOBC Joint Pesticide Testing Programmes 2–7

(a) *Phytoseiulus persimilis*

Classn	*Fungicides*		*Insecticides*		*Herbicides*	
	Semi-field	*Laboratory*	*Semi-field*	*Laboratory*	*Semi-field*	*Laboratory*
1	67%	76%	21%	54%	100%	63%
2	17%	11%	5%	6%	0%	16%
3	13%	5%	8%	7%	0%	0%
4	4%	8%	66%	33%	0%	21%
N	24	37	38	54	4	19

(b) *Amblyseius* sp.

Classn	*Fungicides*				*Insecticides*				*Herbicides*			
	Field		*Laboratory*		*Field*		*Laboratory*		*Field*		*Laboratory*	
	finl	*ander*	*R*	*S*	*finl*	*ander*	*R*	*S*	*finl*	*ander*	*R*	*S*
1	52%	33%	63%	55%	23%	22%	17%	14%	0%		67%	50%
2	16%	33%	25%	8%	0%	27%	17%	12%	25%		0%	7%
3	12%	33%	0%	15%	6%	22%	17%	13%	25%		0%	7%
4	20%	0%	13%	23%	71%	30%	50%	61%	50%		33%	39%
N	25	9	8	40	31	37	6	77	4	0	6	28

(c) *Typhlodromus pyri*

Classn	*Fungicides*				*Insecticides*				*Herbicides*			
	Field		*Laboratory*		*Field*		*Laboratory*		*Field*		*Laboratory*	
	Orch	*Vine*	*R*	*S*	*Orch*	*Vine*	*R**	*S**	*Orch*	*Vine*	*R*	*S*
1	57%	65%	52%	44%	34%	31%	16%	15%	0%	0%	69%	48%
2	14%	6%	10%	8%	10%	5%	16%	6%	0%	0%	0%	22%
3	7%	12%	17%	23%	3%	18%	13%	6%	100%	100%	19%	4%
4	21%	18%	21%	26%	52%	46%	56%	74%	0%	0%	13%	26%
N	14	17	29	39	29	39	45	53	1	1	16	27

Data shown obtained in laboratory trials or at higher testing tiers.
Two *Amblyseius* species were tested in the field: *A. andersoni* (*ander*) in Italian orchards, and *A. finlandicus* (*finl*) in Belgian cherry orchards.
T. Pyri field trials were performed both in apple orchards and in vineyards.
N = total number of trials used for comparison
R = resistant strain
S = susceptible strain
R* = TpR lab
S* = TpS lab

emphasis on harmlessness is not necessarily the most efficient approach. This is because the number of products that can maximally be sorted with Tier I tests is limited by the number of products that are actually harmless. Consequently, although most fungicides will be conclusively identified in the laboratory, this applies only to a small proportion of insecticides. For these reasons, testing effort may be reduced to the maximum extent when biocidal activity is also taken into account when outlining the testing strategy.

To test the viability of this approach and to compensate for biases that might have resulted from pooling strains/species, the IOBC data set used by Bakker and Jacas (1995) was re-analysed for *Amblyseius* and *T. pyri*, but this time separately for insecticides and fungicides and without pooling the *Amblyseius* species. This more detailed approach resulted in a lower number of replicates for the analyses. Consequently, single deviant observations had a large influence on the outcome. The analytical approach should still be suitable to reveal common trends, but the results should be interpreted more in a qualitative than in a quantitative sense. The *T. pyri* data were analysed separately for the two field strains (orchard and vineyard) and also pooled. Because the results obtained with the two strains were nearly identical, only pooled results are presented here. This has the advantage that replication is enhanced. In all cases field data were compared with the standard test strain, i.e. the susceptible laboratory strain. The approach taken was similar to the one taken by Bakker and Jacas (1995), but now only a subset of the parameter space was explored and no error isoclines were derived. The specific objective of the analysis was to validate species-specific and product class-specific trigger values and to investigate their effect on the following.

1. The accuracy with respect to conclusions drawn on harmlessness and harmfulness; in particular, on the occurrence of false negative results (products incorrectly identified as harmless). In the following, false negatives will be referred to as type A errors. Products classified as harmful in the laboratory that appear harmless in the field will be referred to as type B errors.
2. The efficiency of the test protocol (defined as total proportion of products conclusively identified at Tier I).
3. The proportion of false positive results (defined as the proportion of products re-tested in the field that were unnecessarily triggered to a higher tier).

Although the analysis was performed for all field classifications, only the data for categories 1 and 2 (25–50% reduction of field populations) will be discussed. This is because the author feels, albeit intuitively, that at present it is not justified to consider reductions exceeding 50% as a harmless side-effect. Finally, to assess how maximum efficiency could be obtained, three test scenarios were explored: one in which the laboratory is used to sort out harmless products, one to sort out harmful ones and the third to sort out both.

AMBLYSEIUS FINLANDICUS

Harmless insecticides can be accurately identified in the laboratory, using a trigger value of up to 50%. Changing the trigger from its currently proposed value of 30% to 50% reduces the number of false positives from 14.3% to 6.7% without altering the accuracy. Consequently efficiency is doubled. However, the highest proportion of insecticides that can be identified in the laboratory (with a field trigger of 50%) is still only 13.3%. As Table 16.2 shows, the maximum attainable would be 23% (proportion of insecticides harmless in the field). Efficiency can be greatly enhanced by considering harmful outcomes as conclusive. With the laboratory trigger for harmfulness at 99%, 80% of the insecticides are conclusively identified in the laboratory, albeit that the classification would be erroneous in 4.2% of the cases. The proportion of false positives (now meaning products re-tested that are harmful after all) under this scenario would be 16.7%. Because acceptable type A and type B error rates can both be achieved with the same field trigger value, the combined harmless/harmful scenario can be applied, in which case 93.3% of all insecticides will be sorted out in the laboratory.

For fungicides the conditions are more restrictive. To avoid false negatives the laboratory trigger value has to be brought down to 10%. With the current 30% T-value an unacceptably large proportion of the fungicides would be incorrectly classified as harmless (31.3% and 12.5% for field triggers of < 25% and < 50%, respectively). This is not very efficient, because only 27.3% of the fungicides are identified, while 56.3% are false positives (unnecessarily re-tested). As Table 16.2 shows, the maximum potential efficiency for identifying harmless fungicides is as high at 71%. Because harmful fungicides are easily identified without error, the maximum efficiency with current methods could be 45.5% of all fungicides identified in the laboratory, provided a 50% reduction of field populations is accepted as the demarcation between harmless and harmful results. The results are summarized in Table 16.3.

AMBLYSEIUS ANDERSONI

Because not enough trials were available to analyse fungicide data, the discussion will be restricted to insecticides. With the currently proposed trigger value of 30%, the proportion of false negatives was unacceptable high. Reducing it to 24% eliminated the errors (parathion and chlorpyriphos; Table 16.1). With this trigger the efficiency was 9.7%, while

Table 16.3 Accuracy and efficiency of sequential testing scheme for *Amblyseius* species – three scenarios indicating protocols where laboratory results are considered conclusive when: (I) smaller than the Laboratory 1 trigger (harmless); (II) larger than the Laboratory 3 trigger (harmful); and (III) both

(a) Species combination: *Amblyseius andersoni* susceptible (laboratory) and *A. finlandicus* (field)

Product class	*Trigger values*			*Scenario I (harmless)*			*Scenario II (harmful)*			*Scenario III (both)*
	Field	*L1*	*L2*	*err*	*eff*	*fls*	*err*	*eff*	*fls*	*eff*
Fungicides	< 25%	30	99	31%	–	–	0%	18%	33%	–
(N = 22)	< 25%	0	94				0%	18%	33%	–
	< 50%	30	99	13%	–	–	0%	18%	17%	–
	< 50%	10	99	0%	27%	56%	0%	18%	17%	45%
Insecticides	< 25%	30	99	0%	7%	14%	4%	80%	17%	87%
(N = 30)	< 25%	50	99	0%	13%	8%	4%	80%	17%	93%
	< 50%	30	99	0%	7%	14%	4%	80%	17%	87%
	< 50%	30	99	0%	13%	8%	4%	80%	17%	93%

(b) Species combination: *Amblyseius andersoni* susceptible (laboratory) and *A. andersoni* (field)

Product class	*Trigger values*			*Scenario I (harmless)*			*Scenario II (harmful)*			*Scenario III (both)*
	Field	*L1*	*L2*	*err*	*eff*	*fls*	*err*	*eff*	*fls*	*eff*
Insecticides	< 25%	30	99	40%	–	–	0%	39%	68%	–
(N = 31)	< 25%	24	99	0%	6%	14%	0%	39%	68%	45%
	< 50%	30	99	20%	–	–	17%	–	–	–
	<50%	24	99	0%	10%	46%	17%			

Laboratory triggers 30 and 99 are the currently proposed values.
err = error
eff (efficiency) = total proportion of products conclusively classified in the laboratory.
fls (falses) = proportion of unnecessarily retested products (false positives).
– indicates that error levels exceeded acceptable limits (5%).

five times more trials (46.4%) yielded false positive results at a field threshold of 50% (for the 25% field level 6.5% and 13.8%, respectively). With the field threshold of 50% the accuracy level for harmful classifications could not be brought within acceptable levels (16.7% erroneous classifications). Since this implies that only the harmless scenario can be applied, the maximum efficiency of the sequential scheme is 9.7%. A summary of the results is given in Table 16.3.

TYPHLODROMUS PYRI

The fungicide data analysed indicate that the sequential testing scheme is very robust against false negative results for this product class. This robustness is clearly illustrated by the finding that the laboratory trigger can be raised to 80% without loss of accuracy. However, the vast safety margin of the scheme comes at the expense of efficiency. With currently proposed trigger values, 38.5% of the fungicides re-tested in the field appeared to be false positives. This situation can be only slightly improved by raising the laboratory trigger (Figure 16.3 shows the relationship between false positives and trigger values). Because the accuracy level for harmful classifications obtained in the laboratory is unacceptably low (14.3% errors) at a field threshold of 50% reduction in mite numbers, the maximum efficiency attainable for fungicide classification is 50%.

The findings for insecticides were qualitatively similar (Table 16.4). Because harmful classifications obtained in the laboratory cannot be considered conclusive, only the harmless scenario can be applied, yielding a maximum efficiency of 23%. The proportion of false positives is again rather high (around 20% for current trigger values; Figure 16.3), which in the light of the robustness of the scheme means an unnecessary loss of efficiency. The false negative found at the field threshold of 25% relates to a field trial with *Verticillium lecanii* that yielded a score of 2 (in the 7th JPTP), whereas laboratory trials convincingly demonstrated non-toxicity of this plant protection product. In this light it is proposed to forestall conclusions pertaining to the 25% field trigger until replicate experiments with *Verticillium* in the field have been performed.

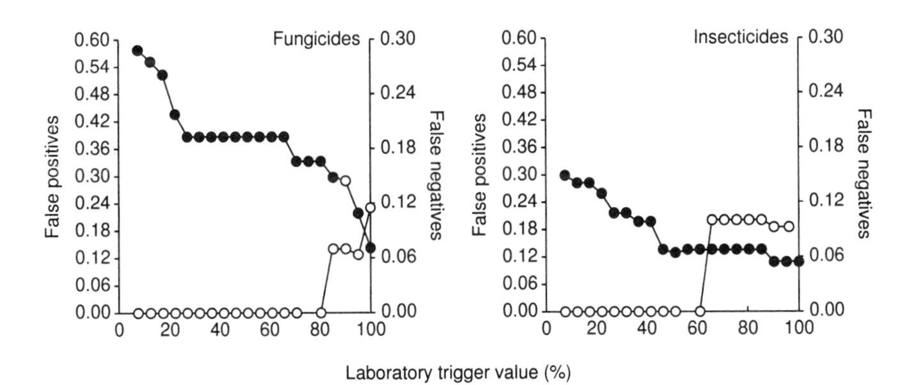

Figure 16.3 The effect of varying the laboratory trigger separating harmless from others on (●) false positives (proportion of products classified harmless after being retested in the field) and (○) false negatives (proportion of products classified harmless in the laboratory that appear to be harmful in the field).

Table 16.4 Accuracy and efficiency of sequential testing scheme for *Typhlodromus pyri* – three scenarios indicating protocols where laboratory results are considered conclusive when: (I) smaller than the Laboratory 1 trigger (harmless; (II) larger than the Laboratory 3 trigger (harmful); and (III) both

Product class	*Trigger values*			*Scenario I (harmless)*			*Scenario II (harmful)*			*Scenario III (both)*
	Field	*L1*	*L2*	*err*	*eff*	*fls*	*err*	*eff*	*fls*	*eff*
Fungicides (N = 22)	< 25%	30	99	0%	46%	23%	0%	29%	18%	75%
	< 25%	80	94	0%	50%	17%	0%	29%	18%	79%
	< 50%	30	99	0%	46%	39%	14%	29%	12%	–
	< 50%	80	99	0%	50%	33%	14%	29%	12%	–
Insecticides (N = 30)	< 25%	30	99	17%	15%	18%	7%	72%	–	–
	< 25%	60	99	11%	23%	10%	7%	72%	–	–
	< 50%	30	99	0%	15%	21%	11%	–	–	–
	< 50%	60	99	0%	23%	13%	11%	–	–	–

Laboratory triggers 30 and 99 are the currently proposed values.
err = error
eff (efficiency) = total proportion of products conclusively classified in the laboratory.
fls (falses) = proportion of unnecessarily retested products (false positives).
– indicates that error levels exceeded acceptable limits (5%).

SPECIES COMPARISONS

The toxicological profiles of the sensitive laboratory strains of *A. andersoni* and *T. pyri* are very similar. Therefore the finding that for *A. andersoni* the test system is not sensitive enough (trigger values had to be lowered), whereas for *T. pyri* it is oversensitive, either indicates differences in the resistance status of field populations of these species or points to ecological differences that render the former species more sensitive. Although in this case the former explanation might be adopted for reasons of parsimony, the *A. finlandicus/T. pyri* comparison is more troublesome. The insecticide trials with these species yielded qualitatively similar results in that the test was too sensitive for both (trigger values had to be raised to reduce false positives without compromising accuracy) but the fungicide trials had completely opposite outcomes. For *T. pyri*, laboratory effects of up to 80% still mean harmlessness in the field; for *A. finlandicus*, it was demonstrated that it is virtually impossible to forecast harmlessness for fungicides.

It is unclear how these biocidal activity-related differences can be explained by differences in the susceptibility of field populations. In this case it is more parsimonious to assume that ecological differences underlie the phenomena observed. An explanation could be that fungi are an important dietary factor for *A. finlandicus* but not for *T. pyri*. Consumption of mildew spores has been demonstrated for *A. finlandicus* (Kropczynska-Linkiewicz, 1971) and some other species of phytoseiids (e.g. Bakker and Klein, 1992; Zemek and Prenerova, 1997) but not all species can utilize fungal food (Bakker, unpublished data). It is interesting to note that Kropczynska-Linkiewicz (1971) found on apple that *A. finlandicus* could consume mildew, whereas *T. pyri* could not.

What this exercise demonstrated is that the sequential testing approach critically depends on knowledge concerning the susceptibility of the field population, relative to the laboratory test strain. It is this ratio alone that determines the setting of the trigger values and the accuracy of the testing scheme. Not only the *P. persimilis* data but also the trials with R-strains of *T. pyri* and *A. andersoni* yielded unacceptable levels of false negatives (Bakker, unpublished data). A higher tolerance of *A. finlandicus* field populations, relative to *A. andersoni*, might well explain the different outcomes obtained with these two species. Thus there is no principle reason why direct toxic effects on one phytoseiid species could not be forecast using another species in the laboratory assay – for example, Bakker and Jacas (1995) demonstrated that effects on *T. pyri* could be accurately forecast from data obtained with *A. andersoni* – but this analysis has pointed out some important pitfalls.

Firstly, the tolerance status of the laboratory strain must be well defined and documented. Secondly, the use of indicator species is only possible when the susceptibility of the field population can reasonably be assumed to be similar to that of the species for which the testing scheme was validated – for example, because they have a similar background of exposure to pesticides. If tolerance levels differ, trigger values must be adapted. Thirdly, as illustrated by the *A. finlandicus*/*T. pyri* comparison, ecological characteristics (such as dietary range) may complicate comparisons as a result of indirect pesticide effects that differentially affect species. In such cases it may be necessary to consider biocidal activity of the test product in the evaluation. Finally, as with the setting of trigger values, it may well be that different geographical regions or different habitats will require indicator species with different tolerance levels. Using the most susceptible species for all cropping systems will in many cases lead to an exaggerated degree of safety at the expense of an economical testing effort. This is demonstrated by the *P. persimilis* data set, where, according to Oomen *et al.* (1991), results obtained with the (commercial) R-strain in the laboratory could accurately be used to screen products for use in IPM systems. In this case results obtained with the S-strain would clearly lead to an unacceptable number of false positives. However, the biocontrol practitioners interviewed by Sterk (personal communication) would clearly be misled by the laboratory results obtained with the R-strain (cf. Table 16.1).

DISCUSSION OF TESTING METHODOLOGY

The validation exercise demonstrated that trigger values must be tuned to the species combination under study and that it should take into account the biocidal activity of the test product. Rigid implementation of currently used triggers would lead either to unacceptably high levels of false negatives (*Amblyseius*) or to unacceptably high levels of false positives (*T. pyri*). Bearing in mind that laboratory test results with the latter species were obtained with the method described by Overmeer and van Zon (1982), which at present is not considered to be a worst-case method, and with a test dose that nowadays is considered far too low, even for a 'realistic' worst-case scenario (0.1–0.2 × recommended field rate), the question arises whether the margin of safety obtained by the worst-case approach does not come at too high a cost.

A standard characteristic of worst-case laboratory tests is that products are typically applied to inert substrates such as glass. Glass substrates are advantageous in terms of standardization, e.g. reproducibility of results and high bioavailability of the test product and are therefore generally required in Tier I laboratory test (e.g. Barrett *et al.*, 1994). This explains why laboratory testing methods for phytoseiids currently used or being ring-tested all employ glass substrate (e.g. Overmeer, 1988; Bakker *et al.*, 1992; Louis and Ufer, 1992; Chapter 10 of this volume). However, the use of glass also implies that exposure may involve components of the formulation that would otherwise penetrate into the leaf or be made otherwise unavailable. Residual toxicity of such components might well explain why the level of false positives remains unacceptably high (e.g. in th case of *T. pyri*) even when the laboratory trigger is set at 80% effect.

Given the broad safety margin and the level of false positives observed in the tests using glass substrate, it is recommended that the use of natural substrates for Tier I laboratory testing should be considered. The momentum attained by the ongoing ring-testing effort could well be exploited to evaluate such substrates in terms of reproducibility of results, accuracy and their effect on the efficiency of the testing protocol, expressed as the level of false positive results. That an accurate and efficient laboratory test on leaf substrate is not ideal is illustrated by Oomen *et al.* (1991). These authors evaluated the EPPO guideline for a laboratory test on bean leaves with *P. persimilis* by comparing laboratory results with practical experience and found an excellent match (but note Sterk's observations mentioned earlier). Another reason to propose a natural substrate rather than glass is that the latter can only be used for the evaluation of products with a low application frequency. Multiple application products, such as many fungicides, require a different approach – for example, bioassays on leaves collected from intact plants treated according to the maximum proposed rate and frequency. By choosing natural substrates for all Tier I tests, at least all products will be evaluated in the same standard way.

In their evaluation, Oomen *et al.* (1991) reported an 83% efficiency of the testing protocol. This large proportion of products conclusively classified at the first testing tier is due to their strategy of considering both harmless and harmful outcomes as conclusive results. The match with practical experience justifies their approach. Inspection of the data in Table 16.2 shows that, particularly for insecticides, the effects of reducing false positives on the efficiency of the testing scheme will only be marginal when compared with the tremendous gain that would result from developing a testing strategy that enables conclusive identification of both harmless and harmful products in the laboratory (e.g. Tables 16.3 and 16.4). This definitely applies to those who have to screen products for use in IPM. Although pesticide manufacturers would also clearly benefit from recognizing at an early stage when further testing of a harmful product is futile, the risk of a type B error (incorrect assignment of harmfulness) may have economic repercussions that overrule the extra testing effort that results from considering harmful outcomes as inconclusive. Therefore the identification of harmful products in the laboratory assay should be optional.

Strategies to identify harmful and harmless products may have contradictory requirements. Whereas the worst-case approach accommodates the need to avoid false negative results (type A errors) for harmless classifications by choosing a test dose equivalent to the predicted initial environmental concentration (PIEC), harmfulness is more conclusively demonstrated in a 'soft case' approach with a discriminating dose lower than the PIEC. Such a test should be on a natural substrate. Choosing too low a dose results in an unacceptable proportion of products unnecessarily triggered to a higher testing tier (cf. Bakker and Jacas, 1995). Given that natural substrates may be required to reduce false positives in the harmless scenario and that they are required for the harmful scenario, the only difference between the two scenarios seems to pertain to the choice of test dose. Consequently, a dose-response study on natural substrate, limited by the PIEC on the one hand and the lowest expected environmental concentration (LEEC) on the other, seems to be a logical compromise between the need to reduced false positives and the need to create the option to identify harmful products in a conclusive way.

SUMMARY OF RECOMMENDATIONS

To render future testing protocols accurate and efficient, deviations from the worst-case sequential testing protocol may be required. Natural substrates, such as leaves, may reduce the tremendous amount of false positives seen at present and will enable straightforward standard evaluation of all fungicides and insecticides, irrespective of whether or not they are multiple application products. Multiple rate testing on natural substrate will pave the way for efficient schemes that can identify harmless and (optionally) harmful products in a single assay. In fact only two rates have to be considered: the PIEC at the target site and the discriminating dose for harmful products. Trigger values (or test doses) must be tuned to the tolerance levels of the system under study and take into account the biocidal activity of the test product. Validation of testing schemes, particularly for groups other than phytoseiids, is therefore desperately needed.

Finally, what 'harmful' and 'harmless' really mean is as yet unclear. Fixed levels as presented in IOBC schemes presumably lack realism. Expert advice on realistic settings of species-specific field triggers is urgently required.

ACKNOWLEDGEMENTS

I am indebted to the participants of the Ecotox meeting for a lively discussion. Particularly helpful were the comments I received from Pieter Oomen, Sylvia Blümel, Guido Sterk and Carlo Duso on the data presented in Table 16.1 while preparing this chapter. Many of the arguments presented here were discussed during stimulating meetings of the BART group. I thank all my colleagues in BART for their frank and open discussions, which really altered my view on this topic. However, all the views expressed in this chapter should be considered entirely my own.

REFERENCES

Bakker, F.M. and Jacas, J.A. (1995) Pesticides and phytoseiid mites: strategies for risk assessment. *Ecotoxicology and Environment Safety*, **32**, 58–67.

Bakker, F.M. and Klein, M.E. (1992) Transtrophic interactions in cassava. *Experimental and Applied Acarology*, **14**, 293–311.

Bakker, F.M., Grove, J.A., Blümel, S. *et al.* (1992) Side-effect tests for phytoseiids and their rearing methods. *IOBC/WPRS Bulletin*, **15**(3), 61–81.

Barrett, K.L., Grandy, N., Harrison, E.G. *et al.* (1994) *Guidance Document on Regulatory Testing Procedures for Pesticides with Non Target Arthropods. From the ESCORT workshop, Wageningen, Netherlands, March 1994*, SETAC-Europe, Brussels.

Blümel, S., Bakker, F. and Grove, A. (1993) Evaluation of different methods to assess the side-effects of pesticides on *Phytoseiulus persimilis* A.-H. *Experimental and Applied Acarology*, **17**, 161–169.

Duso, C. (1994) Comparison between field and laboratory testing methods to evaluate the pesticide side-effects on the predatory mites *Amblyseius andersoni* and *Typhlodromus pyri*. *IOBC/WPRS Bulletin*, **17**(10), 7–19.

Hassan, S.A., Albert, R., Bigler, F. *et al.* (1987) Results of the third joint pesticide testing programme by the IOBC/WPRS Working Group 'Pesticides and Beneficial Organisms'. *Z. angew. Entomol.*, **103**, 92–107. (IOBC3)

Hassan, S.A., Bigler, F., Bogenshütz, H. *et al.* (1988) Results of the fourth joint pesticide testing programme carried out by the IOBC/WPRS Working Group 'Pesticides and Beneficial Organisms'. *Z. angew. Entomol.*, **105**, 321–329. (IOBC4)

Hassan, S.A., Bigler, F., Bogenschütz *et al.* (1991) Results of the fifth joint pesticide testing programme carried out by the IOBC/WPRS Working Group 'Pesticides and Beneficial Organisms'. *Entomophaga*, **36**, 55–67. (IOBC5)

Hassan, S.A., Bigler, F., Bogenschütz *et al.* (1994) Results of the sixth joint pesticide testing programme of the IOBC/WPRS – Working Group 'Pesticides and Beneficial Organisms'. *Entomophaga*, **39**, 107–119. (IOBC6)

Kropszynska-Linkiewicz, D. (1971) Studies on the feeding of four species of phytoseiid mites (Acarina: Phytoseiidae), in *Proceedings 3rd International Congress Acarology, Prague, 31 August–6 September 1971* (eds M. Daniel and B. Rosicky), Academia, Czechoslowak Academy of Sciences, Prague, pp. 225–227.

Louis, F. and Ufer, A. (1992) Raubmilben fressen nicht nur Spinnmilben. *Der Deutsche Weinbau*, **1**, 23–27.

Oomen, P.A., Romeijn, G. and Weigers, G.L. (1991) Side-effects of 100 pesticides on the predatory mite *Phytoseiulus persimilis*, collected and evaluated according to the EPPO Guideline. *OEPP/EPPO Bulletin*, **21**, 701–712.

Overmeer, W.P.J. (1988) Laboratory method for testing side-effects of pesticides on the predacious mites *Typhlodromus pyri* and *Amblyseius potentillae* (Acari: Phytoseiidae). *IOBC/WPRS Bulletin*, **XI**(4).

Overmeer, W.P.J. and van Zon, A.Q. (1982) A standardized method for testing the side-effects of pesticides on the predacious mite *Amblyseius andersoni* (Acarina: Phytoseiidae). *Entomophaga*, **27**, 357–364.

Sterk, G., Hassan, S.A., Baller, F. *et al.* (submitted) Results of the 7th joint pesticide testing programme carried out by the IOBC/WPRS Working Group 'Pesticides and Beneficial Organisms'. *Entomophaga*.

Stolz, M. (1994) Efficacy of different concentrations of seven pesticides on *Phytoseiulus persimilis* A.-H. (Acarina: Phytoseiidae) and on *Tetranychus urticae* K. (Acarina: Tetranychidae) in laboratory and semi-field test. *IOBC WPRS Bulletin*, **17** (10), 49–54.

Zemek, R. and Prenerova, E. (1997) Powdery mildew (Ascomycetes: Erysiphales) – an alternative food for the predatory mite *Typhlodromus pyri* Scheuten (Acari: Phytoseiidae). *Expermental and Applied Acarology*, **21**, 405–414.

17

Using a sequential testing scheme under laboratory and field conditions with the bumble bee *Bombus terrestris* to evaluate the safety of different groups of insecticides

Burkhard Sechser and B. Reber

INTRODUCTION

Bumble bees are now used on a large scale for pollinating various glasshouse crops. At present an estimated 200 000 to 250 000 colonies are produced annually and used in 30 different countries world-wide on some 25 different crops, not just in glasshouses but in plastic tunnels, structures covered with screens and in open fields as well (Griffiths and Robberts, 1996). The total pollination activity takes place over an average of 12 weeks, during which time the bumble bees may be exposed to repeated applications of pesticides for the control of arthropod pests and fungal diseases. It is of utmost importance for a glasshouse grower to know the impact of the pesticides used on the bees.

Standard methods to test the side-effects of pesticides on predators and parasitoids were developed by the International Organization for Biological Control of Noxious Animals and Plants (IOBC) some time ago (Hassan, 1985). The system consists of a combination of tests that includes laboratory,

Ecotoxicology: Pesticides and beneficial organisms.
Edited by P.T. Haskell and P. McEwen. Published in 1998 by Chapman & Hall, London. ISBN 0 412 81290 8.

semi-field and field methods to be carried out in a tier sequence according to the degree of toxicity to the natural enemies at the three levels. Harmlessness of a pesticide at the laboratory level would render further testing futile at the next stage. All these methods refer to solitary species.

Bumble bees are considered to be beneficial organisms but require a different approach, due to their social behaviour. The situation is further complicated by the appearance of new types of pesticides (metamorphosis inhibitors, chitin synthesis inhibitors, moult-accelerating compounds, etc.) with different modes of action to the still dominant conventional products like organophosphates, carbamates and pyrethroids. A testing scheme for bumble bees has to cover all these aspects in a system as flexible as possible in order to produce reliable recommendations to the grower for the combined use of chemicals and bees. The development of such a scheme by using insecticides with different modes of action was the purpose of this study.

MATERIALS AND METHODS

The products and rates tested were:

- pymetrozine (Chess®, Plenum®) (0.02% a.i.), a pyridine azomethine;
- lufenuron (Match®) (0.005% a.i.), a chitin synthesis inhibitor.

For comparison, the following standards were included:

- organophosphates – heptenophos (Hostaquick®) (0.05% a.i) and dimethoate (Perfekthion®) (0.04% a.i.);
- inhibitors of metamorphosis – fenoxycarb (Insegar® (0.01% a.i.) and diofenolan (Aware®, Context®) (0.02% a.i.);
- chitin synthesis inhibitor – diflubenzuron (Dimilin®) (0.01% a.i.).

Five groups of methods, with some variants, have been developed and evaluated. A flow diagram of the sequence of these methods is shown in Figure 17.1.

Laboratory

Test A: Stomach activity by feeding contaminated food (fructose solution)

The test units are four plexiglass cylinders (20 × 11 cm), placed on a glass plate (45 × 30 cm) and closed on top with a permeable lid. Each cylinder is provided with a plastic cup (3 × 1.5 × 1 cm), which is filled up daily with 2 ml of fructose solution, mixed with the pesticide at the recommended rate (product per fructose solution, equivalent to 20 g a.i./hl). In each cylinder five adult worker bees are exposed for 5 days. One cylinder represents one replicate.

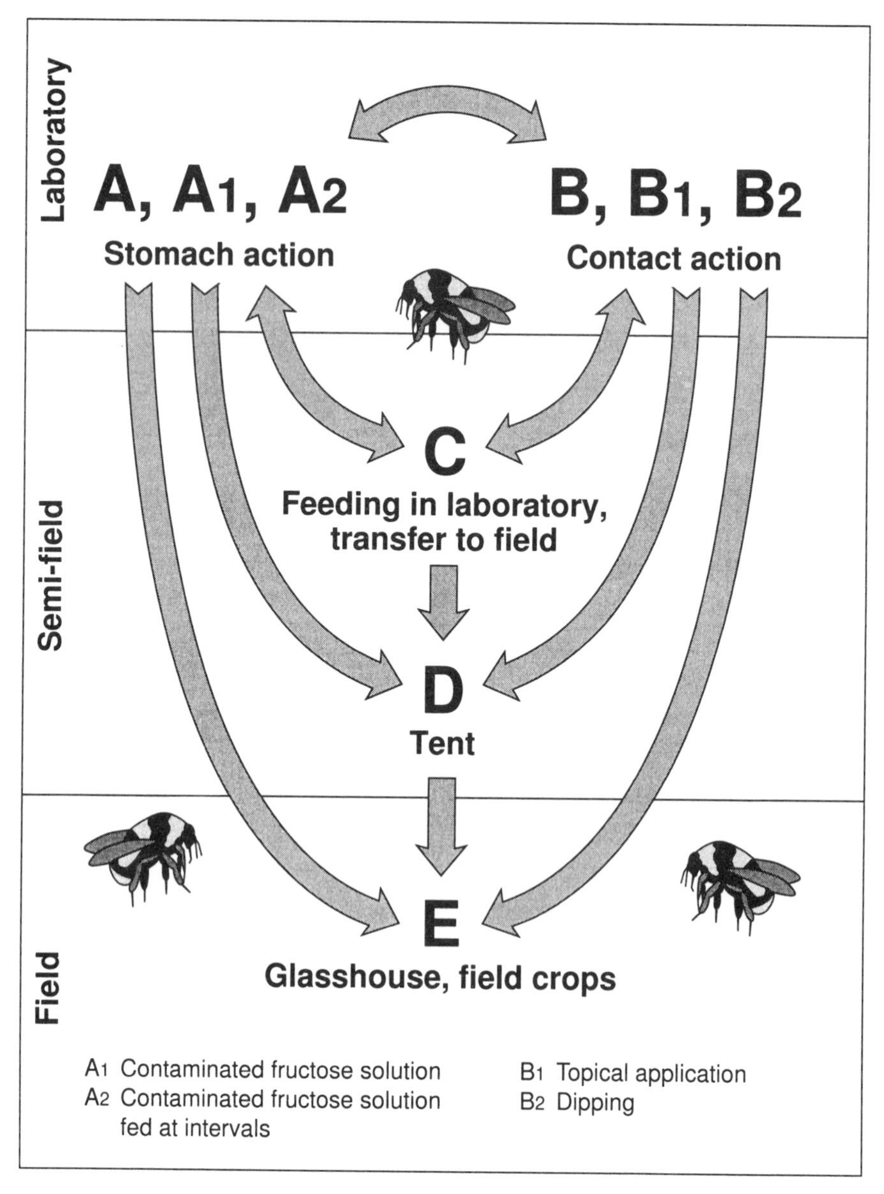

Figure 17.1 Flow diagram for testing the selectivity of pesticides against the bumble bee *Bombus terrestris*.

The evaluation criterion is the mortality, supplemented by observations on the behaviour of the bees. Two variations of this test are possible:

- Test A1: the same trial outline as in Test A but with aged contaminated fructose solution (2 days). Therefore the whole testing period is extended to 7 days.
- Test A2: the same trial outline as in Test A but with contaminated fructose solution just once or every 2nd, 3rd, 4th or 5th day. The testing period is extended to 12 days.

Test B: Contact activity by exposing bee adults to treated surfaces

The trial arrangement is identical to Test A, except that the fructose solution is not treated. Instead, the glass plate surface ($0.2\ m^2$) is sprayed with 4 ml of the pesticide solution (= 200 l/ha). The adult workers are exposed once the spray deposit has dried. The evaluation criterion is the mortality, recorded until day 5, and again supplemented by observations on the behaviour of the bees.

The contact test can be modified in two ways:

- Test B1: topical application – 1 ml of the product solution is applied to the thorax of each bumble bee. The number of bees and the further testing in the plexiglass cylinders follows that described for Test B.
- Test B2: dipping – bumble bees are dipped into the spray solution for 5 seconds and transferred thereafter into the plexiglass cylinders and kept there as in Test B.

Semi-field

Test C: Exposure of bee hive in the field with continuing feeding of contaminated food (fructose solution) inside the hive; adults allowed to be on the wing in untreated field

In this testing procedure, hives are opened and all adult bees are removed after anaesthetization with carbon dioxide. The recommended rate of the pesticide solution is injected into the bag with the fructose solution inside the hive and the bees are reintroduced into the hive. The hives are transferred to the field with foraging plants and protected against rainfall and sunshine. Adults are allowed to circulate freely. The treatment is replicated three times. The duration of the trial is 5 to 6 weeks. All stages in the hives are evaluated.

Test D: Contact activity by exposing bee adults to treated foraging plants within a tent

Potted foraging plants (*Phacelia* sp.) are sprayed at the recommended rate, allowed to dry and placed inside a tent. The tent size varies from 50 to 9 m^2. Each treatment is replicated three times. The duration of the trial is 5 to 6 weeks. As in Test C, all hives are transferred to the laboratory at the end and all stages of the bumble bees in the hives are evaluated.

Field

Test E: Exposing bee adults to contaminated foraging plants or contaminated field tree crops

For glasshouse pesticides, exposure is to contaminated foraging plants in a glasshouse; for non-glasshouse pesticides, exposure is to contaminated field tree crops. The number of hives in a treatment is replicated at least three times. The duration of the trial is at least 5 weeks. As in Tests C and D, all stages in the hives are evaluated in the laboratory.

RESULTS

Pymetrozine

It is a general recommendation by IOBC to start selectivity tests in the laboratory (Tier I). In case of unfavourable findings, the tier scheme suggests further testing under semi-field (Tier II) and/or field conditions (Tier III), which are normally less harsh for the beneficial arthropods. Given the very good selectivity rating against all tested predators and parasitoids (Sechser *et al.*, 1994; Sechser and Reber, 1996), testing at the laboratory level was considered superfluous in this experiment and was started directly at the semi-field level (Test C) in order to get data close to the realistic conditions. The findings were surprising: when hives were opened after 6 weeks of exposure to contaminated fructose inside the hive, in an untreated meadow with flowering foraging plants, only a few surviving bumble bees were found in one replicate. The highest number of dead bees was found in the contaminated fructose solution. Heptenophos was used as a reference and caused total mortality.

To find out the cause of this unexpected mortality, the testing was restarted on Tier I, i.e. in Tests A (stomach action) and B (contact action).

When the bumble bees were fed contaminated fructose solution in plexiglass cylinders (Test A), pymetrozine did not cause any mortality – only a temporary disorientation in the behaviour. After 5 days all bumble bees were back to normal behaviour again. This result explains the mortality caused in the semi-field test: because of the deranged flight behaviour, the bumble bees were unable to find the exit hole inside the complicated hive system and drowned in the fructose solution.

Heptenophos caused 100% mortality. This is an extremely severe test, demonstrating potential risks, but is not very relevant to the use of bumble bees under practical conditions in a glasshouse.

More relevant is Test B, with exposure to sprayed glass surfaces. Neither pymetrozine nor heptenophos caused any mortality. This test conditions corresponds more to the situation in a glasshouse with treated plant surfaces.

Test A was repeated in a modified form (A1), where bumble bees were fed with 2-day-old treated fructose solution. The result is virtually the same as with the freshly contaminated food. In addition, the consumption of fructose solution was measured in this test. Consumption in the pymetrozine treatment was about half of the one in the untreated control and negligible in the heptenophos treatment.

Finally, both pymetrozine and heptenophos were applied in two glasshouses of several hundred square metres in a commercial tomato production. Two to eight hives were exposed in each treatment. The analysis of the hives at the end of the growing season demonstrated a normal development of the bumble bee population. The glasshouse growers were highly satisfied with the performance of the bumble bees concerning pollination activity, fruit setting and yield of tomatoes in both treatment schemes.

Lufenuron

The compound was tested following claims by farmers growing tomatoes in tunnels, in the Valais region of Switzerland, about bad pollination by bumble bees after supposed treatments of surrounding fruit orchards with lufenuron. Lufenuron was tested first according to Test A (feeding test) and compared with a blank formulation (without active ingredient), the unformulated active ingredient in acetone solution, and several pesticides (fenoxycarb, diofenolan, dimethoate) in two series. Only dimethoate proved to be harmful.

As a next step, the same materials were tested for contact activity in two series according to Test B. Again, only dimethoate proved to be detrimental to bumble bees under these laboratory conditions.

Although it had become clear from these two types of trials that lufenuron could hardly be blamed for the failures of pollination by the bumble bees in the indicted region, further tests were carried out in the laboratory to exclude the slightest possibility of a risk of lufenuron to the bumble bees.

In a topical test (Test B1), lufenuron was applied as a formulated product (EC 050) and blank formulation (without active ingredient) as well as the active material, dissolved in acetone, at the normal field rate (0.005%) and 10, 100 and 1000 times this rate, respectively. Mortalities occurred only with the blank and the normal formulation at 1000 times the normal rate, but not with the pure active ingredient. From these findings it became clear again that lufenuron is harmless to bumble bees.

This result could be verified in a further very severe testing of the formulated product by dipping bumble bee workers (Test B2) into the product solution. The result was the same as in Test B1, with additionally some mortality of the acetone formulation at 100 times the normal rate.

Since lufenuron is a chitin synthesis inhibitor, the argument could be raised that, under the long-term exposure of bumble bees in the field, hitherto unknown modes of action of the product could show up there. Therefore lufenuron was tested further under semi-field conditions according to Test D. Flowering *Phacelia* sp. plants were sprayed and nine of them placed in a tent. Diflubenzuron was used as a standard, which belongs to the same chemical class of the benzoylphenylureas. Water-treated plants served as an untreated control. Each treatment was replicated three times. The bumble bee population developed normally in both the lufenuron and diflubenzuron treatment.

Finally, in response to the claims of the tomato growers, lufenuron was also tested under field conditions (Test E for non-glasshouse products). Three hives were placed in an apple orchard of 10 ha, 24 hours after a commercial spray application of lufenuron at 0.005% a.i. pre-blossom. Another set of three hives was placed in a non-insecticidal apple orchard of 0.9 ha, about 3 km away from the first one. The hives were left in the orchards for 5 weeks and were then analysed in the laboratory. Probably due to the excessive offer of blossoms and flowers, all hives had very high numbers of larvae and adults, compared with what would be found in a glasshouse. There was not the slightest sign of any damage in the lufenuron plot, which verifies the results obtained in the preceding tests.

DISCUSSION

The suggested sequence of testing pesticides for their effects on bumble bees proved to cover all possible risks of contamination by and exposure to this chemical group. Only adult workers were used in the experiments. A bee brood cannot be treated directly under practical conditions, and possible impact of pesticides can be simulated by the suggested tests through the feeding of contaminated fructose solution and/or pollen by the workers to the larvae. The demonstration of potential effects is guaranteed if the semi-field and field tests continue over several weeks, giving the pesticides time to cause deviations from normal development.

Glasshouse growers are mainly interested in having a properly functioning bumble bee worker population over the average pollination period of 6 to 12 weeks between installation and removal of a hive. The contact tests are the most relevant ones as they most closely simulate the situation of contact between the bumble bees and treated plant surfaces in a glasshouse. The feeding of contaminated fructose solution is an extreme case of pesticide uptake and highly improbable in real commercial crop production and protection in glasshouses. This sort of testing demonstrates a potential rather than an actual risk for the most extreme cases and gives information about the possible mode of action of a compound.

In the test series described in this chapter, five of the six tested compounds (pymetrozine, lufenuron, heptenophos, fenoxycarb and diofenolan) can be used safely either in the glasshouse (pymetrozine, lufenuron, heptenophos) or in surrounding field crops (fenoxycarb, diofenolan). This second group is not intended for glass house use. All these five substances proved to be harmless in the contact tests in the laboratory. Only dimethoate was harmful and is not recommended for use in glasshouses. It would be wrong to recommend only the contact test for rating the safety of a compound to bumble bees, because we had one case of 70% mortality in a contact laboratory test with imidacloprid (unpublished data), but no further testing has been done so far with this substance. A continued sequential testing is required to clarify whether the product can be used in glasshouses.

Pymetrozine has been on the market since 1994 (Fuog *et al.*, 1995) and has proved its good compatibility with the use of bumble bees in glasshouses over a period of three years of commercial sales without any complaint. Lufenuron being a benzolyphenylurea compound, some doubts may be justified because of its chemistry since some compounds of this chemical class have detrimental effects on bumble bees. In a tent test at the higher rates of 0.025 and 0.05% a.i., diflubenzuron caused the death of nearly all the brood after 2 days and no new brood could build up over a period of 2 weeks (Gretenkord and Drescher, 1995).

When teflubenzuron, another chitin synthesis inhibitor, was administered in 1 : 1 sucrose solution to young bumble bee colonies kept in the dark in a controlled climate chamber, larvae died and egg development was arrested (de Wael *et al.*, 1995). After 5 weeks there was no developing brood in this colony.

Another question is whether bumble bees require an additional separate testing if results on honey bees are already available. This necessity can be demonstrated with the metamorphosis inhibitor fenoxycarb, which proved to be harmless to bumble bees in our laboratory contact and feeding tests. When it was administered at 100 ppm in a 1 : 1 sucrose solution to young bumble bee colonies, they developed normally for the next 5 weeks (de Wael *et al.*, 1995). Conversely, the product has detrimental effects on honey bees, and so in orchards several countermeasures are necessary, such as mowing and mulching of flowering plants in the plantation, no treatment of blossoms and avoidance of drift (Gerig, 1991).

The suggested sequence of experiments has proved its value for the tested spectrum of products from four chemical groups and should be verified with more compounds from other chemical groups.

REFERENCES

de Wael, L., de Greef, M. and van Laere, O. (1995) Toxicity of pyriproxyfen and fenoxycarb to bumble bee brood using a new method of testing insect growth regulators. *Journal of Apicultural Research*, **34**(1), 3–8.

Fuog, D., Fergusson, S. and Flückiger, C. (1995) Reduction of virus transmission after application of pymetrozine. *European Journal of Plant Pathology*, XIII International Plant Protection Congress, The Hague, The Netherlands, 2–7 July 1995, Abstracts: no. 819.

Gerig, L. (1991) Die Rolle von Insegar für Bienenzucht und Obstbau. *Allgemeine Deutsche Inker Zeitung*, **4**, 12–18.

Gretenkord, C. and Drescher, W. (1995) Laboratory and cage test methods for the evaluation of the effects of insect growth regulators (e.g. Insegar and Dimilin) on the brood of *Bombus terrestris*. *Apidologie*, **26**(4), 331–332.

Griffiths, D. and Robberts, E.J. (1996) Bumble bees as pollinators of glasshouse crops, in *Bumble Bees for Pleasure and Profit*, (ed. A. Matheson), International Bee Research Association, Cardiff, pp. 33–39.

Hassan, S.A. (1985) Standard methods to test the side-effects of pesticides on natural enemies of insects and mites developed by the IOBC/WPRS Working Group 'Pesticides and Beneficial Organsisms'. *Bulletin OEPP/EPPO*, **15**, 214–255.

Sechser, B. and Reber, B. (1996) Pymetrozine: a case of perfect selectivity. *Proceedings XX International Conference of Entomology, Firenze, Italy, August 23–31*, p. 592.

Sechser, B., Bouregeois, F., Reber, B. and Wesiak, H. (1994) The integrated control of whiteflies and aphids on tomatoes in glasshouses with pymetrozine. *Med. Fac. Landbouww. Univ. Gent*, **59/2b**, 579–583.

18

Testing effects of pesticides on adult carabid beetles in semi-field and field experiments

Udo Heimbach

INTRODUCTION

Carabid beetles are abundant in agricultural crops and are important predators of invertebrate pests, and their activity-density can easily be recorded using pitfall traps. Consequently, these beetles are often chosen to monitor side-effects of pesticides in field experiments and were selected in a SETAC guidance paper as one of the test organisms for registration purposes (Barrett *et al.*, 1994). Laboratory and semi-field methods for *Poecilus (Pterostichus) cupreus* (Heimbach, 1992; Heimbach *et al.*, 1992a), for example, have been in use for several years. International validation of these methods will be concluded in the near future.

Data derived from semi-field and field studies are likely to be highly variable due to the number of factors influencing the physicochemical properties of agrochemical products as well as the activity and behaviour of carabid beetles (Heimbach and Abel, 1994). Effects of pesticides on carabid beetles depend on the level of uptake of the active ingredient into their body and on the sensitivity of the tested species and life stage. Exposure (body burden) is related to the application rate, but also to factors such as type, duration and intensity of activity of the beetles, rate of detoxification in the beetles, and persistence, volatilization, metabolization and bioavailability of the active ingredient. All these may vary depending on the species tested, the soil type, structure and moisture,

Ecotoxicology: Pesticides and beneficial organisms.
Edited by P.T. Haskell and P. McEwen. Published in 1998 by Chapman & Hall, London. ISBN 0 412 81290 8.

crop type and condition, and climatic conditions. The amount of the pesticide reaching the area in which the beetles are active is influenced by application technique, crop density and stage, and weather conditions. In field tests, re-invasion from field boundaries or untreated plots and changes in activity pattern due to the application of the pesticide or to changes of climatic conditions may mask effects on carabid beetles (Sprick, 1992) in addition to the lack of easily interpreted sampling methods for this insect group.

MATERIAL AND METHODS

P. cupreus was used as the test organism in laboratory and semi-field experiments. In laboratory tests, 30 beetles per product were exposed to pesticides on quartz sand and their mortality and behaviour were recorded over the following two weeks (Heimbach, 1992).

In semi-field tests, 40 beetles (*P. cupreus*, 3 to 10 weeks old) per product tested were released shortly before treatment into open frames (0.5 × 0.5 to 1 × 1 m metal enclosures) (Heimbach *et al.*, 1992b) and into metal boxes with a gauze bottom and a wire mesh lid (0.24 × 0.26 m) (Abel and Heimbach, 1992). In one of the experiments, overwintered beetles were introduced into frames and boxes after treatment with chlorpyrifos and lindane (see Table 18.3).The frames were dug into the ground, with little disturbance to the crop other than at the edge of the frames, while the metal boxes were filled with sandy soil and sunk into the ground, with the crop only outside the boxes. In both semi-field tests, beetles were usually sprayed under field conditions according to good agricultural practice. In summer experiments the studies were ended after about 2 weeks; in autumn studies this period was prolonged, in some cases up to 6 months.

Field experiments were carried out in summer in approximately 1 ha plots, and in autumn in plots of at least 600 m^2. Summer assessments were made using five to 10 pitfall traps per plot for at least 1 week before and 3 weeks after the application. The species composition of the trap catch was influenced by the sampling period and crop type. If sufficient beetles were caught before application, the efficiency was calculated using the formula of Henderson and Tilton (1955). If numbers were too low, Abbott's formula (1925) was used. In autumn, four to eight photoeclectors (emergence traps covering 0.25 m^2) per plot were dug into the ground. They were repositioned every 2 weeks after their catch had been collected.

RESULTS AND DISCUSSION

Results from the two semi-field experiments, the open frames and the more standardized boxes, are compared in Table 18.1. Similar or higher percentages of *P. cupreus* were affected or dead in metal boxes than in frames after treatment with the two organophosphorous pesticides pyrazophos and parathion, whereas a higher percentage of healthy beetles was found in the boxes than in frames at the end of the experiments, after treatment with the pyrethroid, λ-cyhalothrin. In the protected situation in the boxes more sublethally affected beetles are likely to recover, in the absence of exposure to predation, than in open frames in which predators such as birds, healthy carabids or mice are active. This is supported by the number of beetles that were missing from the frames at the completion of the experiment and the number of dead and sublethally affected beetles found in metal boxes. On the other hand more effects were detected in boxes than in frames after treatment with lindane (see Table 18.3). This was probably due to the influence of soil type on the bioavailability of this product (Heimbach *et al.*, 1992a) and the use of soil with low organic material in boxes compared with soil rich in clay and organic matter, into which the frames were set.

Comparison of effects of λ-cyhalothrin on *P. cupreus* in summer and autumn semi-field experiments (Table 18.2) revealed more apparent effects in the autumn. On completion of the experiments, effects at the reduced rate of 2.5 g a.i./ha of λ-cyhalothrin in autumn were similar to those at the rate of 10 g a.i./ha in summer. This may have been due to differences in exposure under crops at different stages of development in autumn than in summer, but it is more likely to have been due to differences in ambient temperature. Lower temperatures reduce the speed of recovery drastically after exposure to this pyrethroid (Heimbach and Baloch, 1994). In autumn semi-field experiments, a high percentage of affected beetles were still visible in the frames after about 1 week, while the period for recovery of beetles of the same species at 20°C in the laboratory situation on pure sand and with 100% exposure was less than a week (Römbke and Heimbach, 1996).

Assessment of the effects of pesticides on the abundance of carabid beetles in field tests is problematic (e.g. Sunderland *et al.*, 1995). Many beetles may be buried in the soil or inactive, and dead beetles are often removed by predators. Depending on the sampling method used, during the same sampling period and at the same location different species will be monitored and different effects of pesticides on carabids are likely to be determined. For example, chlorpyrifos reduced the activity-density of carabids, measured by pitfall traps, to a greater extent than lindane (Table 18.3), but a lower number of dead beetles was collected from treated fields after treatment with chlorpyrifos than with lindane. Banded applications, in which only a third of the area of conventional applications was treated, led to a greater reduction in effects on carabid beetles after treatment with lindane than with chlorpyrifos. This might be explained by the short-term and fumigant effect of lindane compared with the longer lasting contact toxicity of chlorpyrifos. Effects recorded

Table 18.1 Comparison of effect of pesticides on *Poecilus cupreus* in semi-field enclosure ($n = 24–40$) or in boxes ($n = 30–36$)

Crop[a]	*GS*[b]	*Pesticide*[c]	*Test*			*Results*				
			Ended after	*Type*[d]	*Rate*[e] *(g/ha)*	*% Aff or visible*[f]	*Hlthy (%)*	*Dead (%)*	*Aff (%)*	*Miss (%)*
WRye	39	λ-cyh	21/24d	FR	Untr	0	95	0	0	5
					10	92.3	25	0	0	75
				BX	Untr	0	91.7	2.8	0	5.6
					10	33.3	88.9	11.1	0	0
SBt	46	λ-cyh	15/17d	FR	Untr	0	91.7	4.2	0	4.2
					10	0	58.3	0	0	41.7
				BX	Untr	0	91.7	0	0	8.3
					10	17.6	86.1	9.7	1.4	2.8
WBly	11	λ-cyh	28d	FR	Untr	0	72.5	0	10	17.5
					10	100	2.5	12.5	62.5	22.5
				BX	Untr	17	90	0	10	0
					10	100	0	16.7	83.3	0
SBt	31	pyraz	17d	FR	Untr	0	91.7	4.2	0	4.2
					588	33	58.3	0	0	41.2
				BX	Untr	0	91.7	0	0	8.3
					588	50	58.3	30.6	0	11.1
WBly	12	parat	28d	FR	Untr	0	72.5	0	10	17.5
					105	91	30	37.5	2.5	30
				BX	Untr	17	90	0	10	0
					105	100	16.7	75	8.3	0

[a] **WRye** = winter rye; **SBt** = sugar beet; **WBly** = winter barley
[b] **GS** = EC crop growth stage
[c] **λ-cyh** = lambdacyhalothrin; **pyraz** - pyrazophos; **parat** = parathion
[d] **FR** = frames; **BX** = boxes
[e] Rate of pesticide used (g a.i./ha); **Untr** = untreated
[f] Percentage affected or visible after 6/7 days

in field tests after treatment with λ-cyhalothrin were quite variable, ranging from more than 50% to less than 10% reductions in carabid activity-density after treatment with 10 g a.i./ha and varied from less than 10% to 50% with 5 g a.i./ha (Heimbach and Abel, 1994). This variability may have been due to differences in crop cover, temperature, soil type, species composition and activity patterns of carabid beetles.

Comparison of the effects determined by semi-field and field tests usually revealed greater effects in semi-field studies, with few exceptions, e.g. lindane (Table 18.3). Semi-field studies were more often representative of worst-case situations, particularly if the number of healthy beetles recaptured at the end of the experiment is chosen as endpoint (Tables 18.3 and 18.4).

Table 18.2 Comparison of effects of λ-cyhalothrin on *Poecilus cupreus* in semi-field enclosures (n = 24–40) in summer and autumn experiments

Crop[a]	*GS*[b]	*Test ended after*	*Rate (g/ha)*	*Results*				
				% Aff or visible[g]	*Healthy (%)*	*Dead (%)*	*Affected (%)*	*Miss (%)*
Summer								
WRye	39	21/24d	Untr	0	95	0	0	5
			10	92.3	25	0	0	75
SBt	46	15/17d	Untr	0	91.7	4.2	0	4.2
			10	0	58.3	0	0	41.7
SBt	41	15/16d	Untr	0	45	0	0	55
			10	0	42.5	2.5	0	55
WBly	11	6 mths	Untr	0	72.5	15	0	12.5
			10	96	0	30	0	70
WBly	12	28d	Untr	0	72.5	0	10	17.5
			10	100	2.5	12.5	62.5	22.5
WBly	13	6 mths	Untr	0	82.5	2.5	0	15
			2.5	100	55	5	0	40
			5	100	25	17.5	0	57.5

[a] **WRye** = winter rye; **SBt** = sugar beet; **WBly** = winter barley
[b] **GS** = EC crop growth stage
[g] Percentage affected or visible after 6/7 days

Table 18.3 Effects of soil insecticides on carabid beetles (all species, mainly *Pterostichus* spp.) sprayed (2 April 1990) before sowing of sugar beet in a field and semi-field (*Poecilus cupreus*) experiment in parallel

Treatment	*Rate (g/ha)*	*Field*			*Semi-field (n = 40) (residual effects only) dead (%)*
		Pitfall efficacy (%)[a]		*No. of dead Pterostichus sampled*	
		All carabids	*Pterostichus only*		
Control[b]		(395)	(105)	0	2.5
Chlorpyrifos	960	89	78	23	95
Banded	320	78	60	6	47.5
Lindane	800	93	90	94	15
Banded	267	59	28	2	2.5

[a] (Abbot, average 5 weeks)
[b] (Number in 5 weeks)

Table 18.4 Effects of different types of aphid control in winter cereals on carabid beetles (< 95% *Trechus quadristriatus*) in a field and semi-field (*Poecilus cupreus*) experiment in parallel (seeding 16 September; application 16 October, growth stage EC 11)

Treatment	*Field: efficacy (%) (traps)*		*Semi-field (n = 40) (1 month after application)*			
	Av 10 w after drilling (Abbott)	*Av 6 w after application (Henderson and Tilton)*	*Hlth (%)*	*Dead (%)*	*Aff (%)*	*Miss (%)*
Control (no./m²)	Av of 10 w = 23.5		72.5	0	10	17.5
Imidacloprid (seed dressing, 70 g a.i./100 kg @= 104 g a.i./ha)	49		55	2.5	15	27.5
1-Cyhalothrin (10 g a.i./ha)		7.5	2.5	12.5	62.5	22.5
Parathion 105 g a.i./ha)		32.5	30	37.5	2.5	30

In conclusion, field studies should always be carried out in parallel with semi-field studies due to the variability and difficulty in interpretation of field tests. Semi-field studies aid interpretation of results and allow a more reliable evaluation of pesticide effects for registration purposes. In addition, field tests should be repeated at different locations or seasons, depending on the use pattern of the pesticide tested, especially if information on what forms a reasonable worst-case situation is missing. Further studies are needed to define appropriate worst-case situations more clearly, depending on the physicochemical properties and the biological efficacy patterns of the pesticide tested.

ACKNOWLEDGEMENTS

The author would like to thank Dr P. Kennedy, Rothamsted, for critical comments and correcting the English.

REFERENCES

Abbott, W.S. (1925) A method of computing the effectiveness of an insecticide. *Journal of Economic Entomology*, **18**, 265–267.

Abel, C. and Heimbach, U. (1992) Testing effects of pesticides on *Poecilus cupreus* (Coleoptera: Carabidae) in a standardised semi-field test. *IOBC/WPRS Bulletin*, **XV**(3), 171–175.

Barrett, K.L., Grandy, N., Harrison, E.G. *et al.* (1994) *Guidance Document on Regulatory Testing Procedures for Pesticides with Non Target Arthropods. From the ESCORT workshop, Wageningen, Netherlands, March 1994*, SETAC-Europe, Brussels.

Heimbach, U. (1992) Laboratory method to test effects of pesticides on *Poecilus cupreus* (Coleoptera: Carabidae). *IOBC/WPRS Bulletin*, **XV**(3), 103–109.

Heimbach, U. and Abel, C. (1994) Comparison of effects of pesticides on adult carabid beetles in laboratory, semi-field and field experiments. *IOBC/WPRS Bulletin*, **17**(10), 99–111.

Heimbach, U. and Baloch, A.A. (1994) Effects of three pesticides on *Poecilus cupreus* (Coleoptera: Carabidae) at different post-treatment temperatures. *Environmental Toxicology and Chemistry*, **13**, 317–324.

Heimbach, U., Abel, C., Siebers, J. and Wehling, A. (1992a) Influence of different soils on the effects of pesticides on carabids and spiders. *Abstracts of Applied Biology*, **31**, 49–59.

Heimbach, U., Büchs, W. and Abel, C. (1992b) A semi-field method close to field conditions to test effects of pesticides on *Poecilus cupreus* L. (Coleoptera: Carabidae). *IOBC/WPRS Bulletin*, **XV**(3), 159–165.

Henderson, C.F. and Tilton, E.W. (1955) Test with acaricides against the brown wheat mite. *Journal of Economic Entomology*, **48**, 157–161.

Römbke, J. and Heimbach, U. (1996) Experiences derived from the carabid beetle laboratory test. *Pesticide Science*, **46**, 157–162.

Sprick, P. (1992) Problematik der Erfassung der Auswirkungen von Pflanzenschutzmitteln auf epigäische Coleopteren am Beispiel von *Loricera pilicornis* (Carabidae). *Mitteilungen der Deutschen Gesellschaft für allgemeine und angewandte Entomologie*, **8**, 161–168.

Sunderland, K.D., De Snoo, G.R., Dinter, A. *et al.* (1995) Density estimation for invertebrate predators in agroecosystems. *Acta Jutlandica*, **70**(2), 133–162.

19

The value of large-scale field trials for determining the effects of pesticides on the non-target arthropod fauna of cereal crops

Michael Mead-Briggs

INTRODUCTION

The short-term (i.e. within-season) risk to non-target arthropods posed by broad-spectrum insecticides is arguably best evaluated in large-scale field trials, as only these can provide quantitative assessments of the impact of products at the population level. However, although they provide realism, field trials have their limitations. From a biological viewpoint, the most important of these is that the arthropod populations being sampled need to be both bountiful and relatively homogeneous across a trial site. If they are not, it is often difficult to demonstrate a statistically significant treatment effect, even with a harmful product. Unfortunately, the density and heterogeneity of populations cannot be predicted accurately, even with extensive pre-treatment sampling. Furthermore, since there may be considerable variation in the numbers of arthropods present between sites and between years, there is a need for caution when interpreting data from single-site studies. From a purely practical viewpoint, the greatest limitation on the design of field trials is often the labour resources required to take, process and analyse a large number of samples. In designing a sampling strategy, it is often necessary to compromise between what is considered ideal (i.e. high sample replication to measure

Ecotoxicology: Pesticides and beneficial organisms.
Edited by P.T. Haskell and P. McEwen. Published in 1998 by Chapman & Hall, London. ISBN 0 412 81290 8.

levels of within-plot variance more accurately), and what is feasible in terms of the time and labour available.

In 1990, at the invitation of Ministry of Agriculture, Fisheries and Food (MAFF), a group of UK experts produced a field trial protocol entitled *Guideline to study the within-season effects of insecticides on non-target terrestrial arthropods in cereals in summer* (Anon., 1991). The intention of this was to help to standardize trial design between researchers and it took account of the experience gained from earlier studies using a variety of experimental designs. The MAFF guideline recommended two possible strategies: the use of either small barriered plots (each no more than a tractor boom in width) or large unbarriered plots (each > 1 ha in area). With both approaches, the treatments were to include the test product, an unsprayed control, a toxic reference product and, if practical, a soft standard such as a selective insecticide. The latter was to demonstrate indirect effects that might result from the removal of prey or host species. Each treatment was to be applied to a minimum of four replicate plots. Within each, arthropods were to be sampled both before and after product application by means of pitfall trapping, suction sampling (using a Dietrick vacuum insect net or 'D-vac') or sweep netting.

To assess the impact of insecticides on the non-target arthropod fauna of cereal crops, four separate field trials were carried out between 1988 and 1992. These studies were on three farm sites in southern England and will be referred to as 1988, 1991, 1992a and 1992b. All followed the test design for unbarriered plots recommended in the MAFF guideline, with each treatment being assigned to four replicate plots of 1 ha (96 m × 100 m). Treatments were arranged in a randomized block design and although the test product differed for the individual studies, an unsprayed control and a toxic reference treatment of dimethoate 40 EC were included in each trial. The dimethoate was applied at a rate of 340 g a.i./ha in 200–224 l water/ha. The intention in the four studies was to treat the crop at around the time of crop flowering. In practice, treatment dates and crop growth stages were 27 May (GS 67), 3 July (GS 70-73), 10 June (GS 65-69) and 15 June (GS 65-69), respectively (Zadoks *et al.*, 1974).

A retrospective analysis of the data from these studies has indicated that, even within a standardized trial design, several factors appear to have influenced the results obtained for the toxic reference treatment. If the effects detected for dimethoate can vary between studies, so might those detected for the test products.

INFLUENTIAL ASPECTS OF TRIAL DESIGN

Sampling strategies

In all the studies, the epigeal arthropod fauna was sampled using pitfall traps and the crop-active fauna was sampled using a D-vac suction sampler. These two sampling methods differ in terms of how they sample populations but, more importantly, with regard to how efficiently individual species are sampled. Pitfall traps gather fauna over an extended period (for 2-day intervals in these trials) and they provide an estimate of the 'activity-density' of the species collected, rather than their actual density. The species-specific nature of the efficiency of trap has been shown for both carabid beetles (Halsall and Wratten, 1988) and linyphiid spiders (Topping and Sunderland, 1992; Dinter, 1995). Indeed, for the spiders, the ratio of trap encounter to capture may vary as much as fourfold between some species (Topping, 1993). Without an understanding of the relationship between trap catches and population density for the individual species, it is difficult to make interspecific comparisons. It is also important to consider whether sublethal treatment effects (e.g. hyperactivity) might influence either trap encounter rates or trapping efficiency.

The D-vac theoretically provides an absolute measure of arthropod population density by collecting fauna from within a fixed area of crop. In reality its efficiency is variable and some faunal groups are sampled more effectively than others. The time of day that D-vac samples are taken is also important as the activity patterns of individual species will vary. Although not used in these studies, night-time collections may be more efficient for sampling certain species (Vickerman and Sunderland, 1975).

When sampling arthropod populations in cereals, the number of sample units required for a given degree of precision is a species-specific phenomenon (Vickerman, 1985). Nevertheless, in choosing the number of samples to be taken, it is often necessary to compromise between the statistical requirements and the resources available in terms of labour. For the four field studies described here, nine pitfall trap samples and six D-vac samples were collected per replicate plot on each sampling occasion. For any given sample date, the taxa chosen for analysis were those that reached a mean of at least 1.0 individual per pitfall trap or D-vac sample. In the four studies, 12 species of carabid beetle reached this threshold for analysis, but only four of these occurred in three or more of the studies (Table 19.1).

In all the studies, the total number of carabid beetles caught in the first week after product application was reduced by over 50% in the dimethoate treatment, relative to the control. Despite this, only five of

Table 19.1 Species of Carabidae, Staphylinidae and Linyphiidae that reached the threshold for analysis (mean > 1.0 individual per trap) in pitfall trap samples taken during the four trials (in all the studies, nine pitfall traps were set in each of four replicate plots per treatment)

Family	*Species*	*No. trials in which catches reached the threshold for analysis on any sample date*	*No. trials in which statistically significant reductions were detected in the dimethoate treatment*
Carabidae	*Pterostichus melanarius*	****	
	Nebria brevicollis	****	*
	Pterostichus madidus	***	
	Loricera pilicornis	***	
	Amara spp.	**	*
	Bembidion lampros	**	*
	Trechus quadristriatus	**	
	Agonum dorsale	*	
	Harpalus aeneus	**	
	Harpalus rufipes	**	*
	Bembidion obtusum	*	
	Demetrias atricapillus	*	*
Staphylinidae	*Tachyporus hypnorum*	****	****
	Tachyporus chrysomelinus	**	**
Linyphiidae	*Erigone atra*	****	****
	Meioneta rurestris	****	***
	Milleriana inerrans	****	*
	Lepthyphantes tenuis	***	***
	Oedothorax spp.	**	*
	Bathyphantes gracilis	*	
	Savignia frontata	*	

the 28 statistical comparisons carried out indicated significant treatment effects for dimethoate ($P < 0.05$). This would suggest that, at these trial sites, the spatial heterogeneity of the populations of carabid beetles was too great to allow a detection of statistically significant treatment effects at the species level. When one considers the size of field being used for such studies (19–41 ha), it is not surprising that these epigeal beetles tend to be unequally distributed. Many species disperse from overwintering sites in the hedgerows and do not spread uniformly across a large field. For staphylinid beetles such as *Tachyporus hypnorum* and for linyphiid spiders such as *Erigone atra*, *Meioneta rurestris* and *Lepthyphantes tenuis*, significant treatment effects were detected with dimethoate in most of the studies. These groups are able to disperse aerially and this may help with more even colonization of large fields. This would make these groups more reliable as indicator species in such studies.

These results indicate that large-scale field trials are not ideal for assessing treatment effects on natural populations of carabid beetles. An alternative approach, such as semi-field bioassays in which beetles are released into barriered areas of crop, is therefore to be recommended.

The timing of studies and its influence on the treatment effects detected

To control grain aphids in cereals, insecticides are typically applied between the onset of flowering and milky ripe grain (growth stages 61–73). In our four trials, spray dates ranged from 27 May to 3 July, a period during which the numbers of non-target arthropods within the crop were at their highest but were in a state of flux. Marked changes in the densities of arthropod populations occur during the summer months. These are often linked to movements to and from overwintering sites in the soil, the field margins or beyond. In addition, the death of post-reproductive insects and the emergence of their progeny, often as a synchronized cohort, can dramatically alter the numbers being found in samples. For instance, numbers of both adult and larval *Tachyporus* spp. show a marked decline in samples collected around the middle of June (Figure 19.1). This makes it difficult to determine the impact of products on this group when treatments are carried out at this time of year. Strong seasonal trends are also seen in the abundance of aerial fauna, such as the aphid-specific parasitoid *Aphidius rhopalosiphi* and the predatory fly *Platypalpus* spp. (Figure 19.2). Where marked changes in population levels occur naturally, just after a product is applied, they can influence whether significant treatment effects are detected. For instance, the only study in which dimethoate did not have a statistically significant effect on numbers of *Platypalpus* spp. was in 1988, where product application (on May 27) coincided with the early phase of immigration of flies into the crop.

Choice of plot size

Where populations of non-target arthropods are depleted by a pesticide, they will normally recover through immigration from surrounding untreated areas. Some species may also re-establish themselves from protected life stages developing within the crop, such as beetle larvae in the soil or aphid parasitoids within their mummified host. Although dispersal rates are very much a species-specific phenomenon, epigeal groups such as carabids, staphylinids and linyphiids tend to re-colonize treated areas from the edge towards the middle (Thomas *et al.*, 1990;

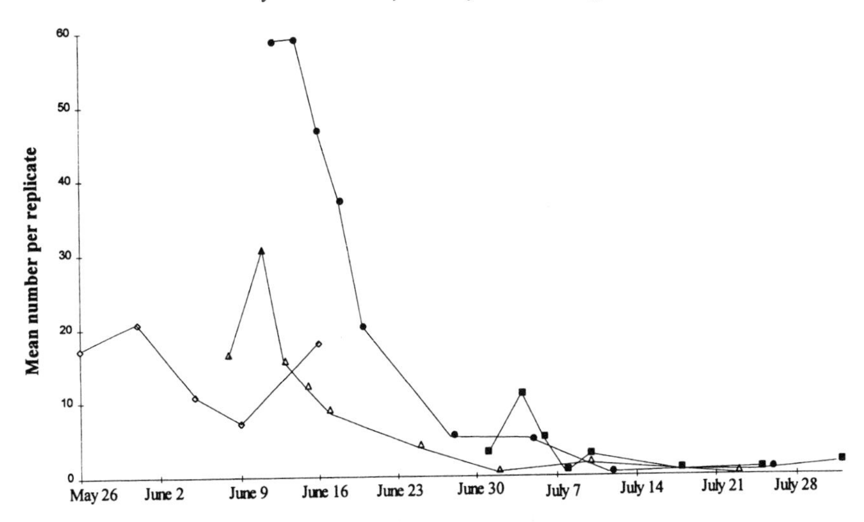

Figure 19.1 Numbers of adult *Tachyporus hypnorum* (Staphylinidae) in pitfall trap samples taken from control treatment in four separate studies, illustrating the sudden decline in the population of post-breeding adults during June, and incidentally reflecting a similar trend in the numbers of *Tachyporus* larvae in samples at this time. ◇, 1988; ■, 1991; △, 1992a; ●, 1992b.

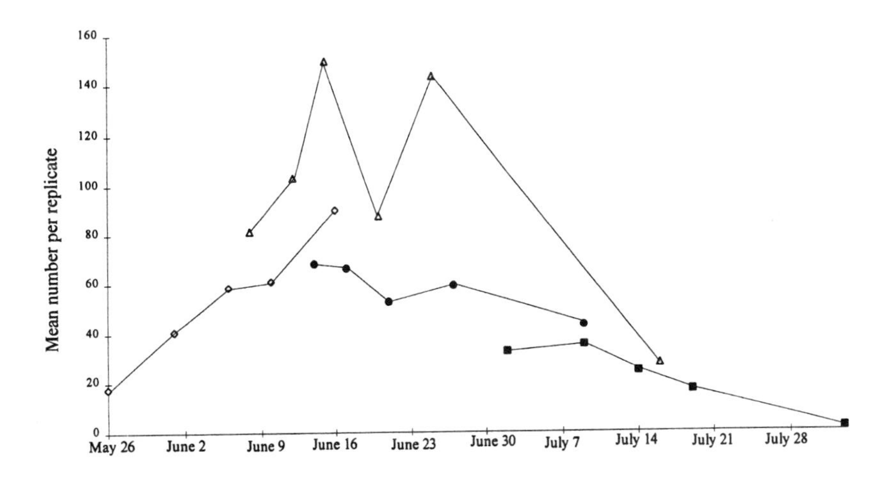

Figure 19.2 Numbers of the predatory fly, *Platypalpus* spp., in D-vac samples taken from control treatment in four separate studies, illustrating the marked seasonal change in numbers, which can influence whether significant treatment effects are detected in some years. Key as for Figure 19.1.

Thacker and Jepson, 1993; Duffield and Aebischer, 1994), whilst aerial fauna such as aphid-specific parasitoids demonstrate a less directional approach (Longley and Jepson, 1997).

Data from field trials of the design described in this chapter are often used to estimate the duration of pesticide treatment effects. Regardless of the plot size being used, this type of extrapolation is likely to be unsafe. Re-colonization processes are influenced by numerous factors, most of which remain unquantified in such studies. These factors include the size of any reservoir population, the distance over which colonization must take place, the rate of decline of harmful or repellent residues within the treated area and the availability of food (Jepson, 1989; Jepson and Sherratt, 1991). As discussed above, the time of year may also have a considerable effect on rates of re-colonization, as may the weather. Without understanding the importance of these individual factors, and how they interact, it is not possible to extrapolate accurately the recovery rates seen in small plots to whole fields or to larger treatment areas.

On this basis, there seems to be little merit in extended periods of post-treatment sampling for the purpose of estimating the time for arthropod populations to recover.

DISCUSSION

Large-scale field trials of this nature are useful for demonstrating under true crop conditions both the spectrum of activity of a product (i.e. the faunal groups at risk) and the extent of any initial effects. They are less reliable for predicting the duration of any effects. Since these studies require considerable resources, it might be advantageous to simplify the sampling strategy and then supplement the results with more directed semi-field experiments. The latter can be designed to answer specific questions, such as how long the crop environment remains harmful to particular species. Such smaller semi-field experiments might also allow the evaluation of more than one treatment rate (e.g. half, one and two times the normal application rates), which would provide useful data on the margin of safety with regard to a test product.

There are several ways in which the large-scale field trial could be simplified to save resources and yet still provide important data. The post-treatment sampling period could be reduced to 6–8 days, which would be sufficient to demonstrate which faunal groups were at greatest risk. The identification of arthropods in the samples could be limited to indicator species from susceptible groups that are abundant in most seasons. For example, for sites in Hampshire, England, these might be *Tachyporus* spp. for the Staphylinidae, *Erigone* spp. for the Linyphiidae, *Platypalpus* spp. for the predatory flies and *Aphidius* spp. for the aphid-specific parasitoids. In the long term, it would be constructive to develop ecological models for these indicator species. These could take account of the natural seasonal fluctuations in the size of populations and also their normal dispersal patterns. Then using field-derived data for the gross initial impact of the test product on the population and by including measurements of the natural decline in product residues over time, it would be possible to make more accurate predictions of the long-term effects of treatments applied at the field or farm scale. There would also be the advantage that, in using such models, between-study comparisons would be less influenced by year-to-year variations in the ambient weather conditions or in the timing of product applications.

REFERENCES

Anon. (1991) *Guideline to study the within-season effects of insecticides on non-target terrestrial arthropods in cereals in summer.* Part Three/A3/Appendix 2 of *The Registration Handbook* (Pesticides Safety Directorate), formerly WD 7/7 in the Data Requirements Book.

Dinter, A. (1995) Estimation of the epigeic spider population densities using an intensive D-vac sampling technique and comparison with pitfall trap catches in winter wheat. *Acta Jutlandica,* 1–12.

Duffield, S.J. and Aebischer, N.J. (1994) The effect of spatial scale of treatment with dimethoate on invertebrate population recovery in winter wheat. *Journal of Applied Ecology,* **31**, 263–281.

Halsall, N.B. and Wratten, S.D. (1988) The efficiency of pitfall trapping for polyphagous predatory Carabidae (Coleoptera). *Ecological Entomology,* **13**, 293–299.

Jepson, P.C. (1989) The temporal and spatial dynamics of pesticide side-effects on non-target invertebrates, in *Pesticides and Non-target Invertebrates,* (ed. P.C. Jepson), Intercept, Wimborne, Dorset, pp. 95–128.

Jepson, P.C. and Sherratt, T.N. (1991) Predicting the long-term impact of pesticides on predatory invertebrates, in *Proceedings Brighton Crop Protection Conference 1991 – Weeds,* British Crop Protection Council, Thornton Heath, Surrey, UK, pp. 911–919.

Longley, M. and Jepson, P.C. (1997) Temporal and spatial changes in aphid and parasitoid populations following applications of deltamethrin in winter wheat. *Entomologia Experimentalis et Applicata,* **83**, 41–52.

Thacker, J.R.M. and Jepson, P.C. (1993) Pesticide risk assessment and non-target invertebrates: integrating population depletion, population recovery and experimental design. *Bulletin Environmental Contamination and Toxicology,* **51**, 523–531.

Thomas, C.F.G., Hol, E.H.A. and Everts, J.W. (1990) Modelling the diffusion component of dispersal during the recovery of a population of linyphiid spiders from exposure to an insecticide. *Functional Ecology,* **4**, 357–368.

Topping, C. (1993) Behavioural responses of three linyphiid spiders to pitfall traps. *Entomologia Experimentalis et Applicata,* **68**, 287–293.

Topping, C.J. and Sunderland, K.D. (1992) Limitations to the use of pitfall traps in ecological studies exemplified by a study of spiders in a field of winter wheat. *Journal of Applied Ecology,* **298**, 485–491.

Vickerman, G.P. (1985) Sampling plans for beneficial arthropods in cereals. *Aspects of Applied Biology,* **10**, 191–198.

Vickerman, G.P. and Sunderland, K.D. (1975) Arthropods in cereal crops: nocturnal activity, vertical distribution and aphid predation. *Journal of Applied Ecology*, **12**, 755–766.

Zadoks, J.C., Chang, T.T. and Konzak, C.F. (1974) A decimal code for the growth stages of cereals. *Weed Research*, **14**, 415–421.

20

Comparing pesticide effects on beneficials in a sequential testing scheme

Gerhard P. Dohmen

INTRODUCTION

The primary interest in beneficial insects originates from organizations like the IOBC working group 'Pesticides and Beneficial Organisms' who attempted to identify plant protection products that could be used together with the introduction of insects for biological control (Hassan *et al.*, 1985; Hassan, 1989). Nowadays, a broader approach is common ('European Standard Characteristics of Beneficial Regulatory Testing (ESCORT) workshop': Barrett *et al.*, 1994). The application of a pesticide should have no unacceptable effect generally on non-target organisms, including beneficials as well as other non-target species.

It is obvious that not all species can be investigated under all conditions. Therefore a simple and cost-effective system was introduced by the International Organization for Biological and Integrated Control of Noxious Animals and Plants (IOBC). Their main aim was to produce reproducible and comparable results, which could only be achieved by standardized laboratory tests (Franz, 1974). These tests were hence developed to sort out those substances that are very unlikely to pose a significant risk to non-target organisms. If this could not be concluded from the standardized laboratory tests, further and more refined investigations had to be conducted to allow an assessment of the potential

Ecotoxicology: Pesticides and beneficial organisms.
Edited by P.T. Haskell and P. McEwen. Published in 1998 by Chapman & Hall, London. ISBN 0 412 81290 8.

effects. A more elaborate sequential testing scheme along these lines is given by Samsoe-Petersen (1990).

A similar tiered approach is used in other ecotoxicological areas and is described, for example, in the European and Mediterranean Plant Protection Organization (EPPO) *Decision-making scheme for the environmental risk assessment of plant protection products* (EPPO, 1993, 1994). It is also used by national regulators. The German Biologische Bundesanstalt für Land- und Forstwirtschaft (BBA) (Brasse, 1990) recommended it for beneficial testing. In other terrestrial and aquatic testing also, the authorities in Europe and the USA first rely on standard laboratory tests before more complex studies are required.

SEQUENTIAL TESTING SCHEME

The first tier in the sequential testing scheme for beneficials (Figure 20.1) is a very standardized laboratory experiment conducted under 'worst-case' conditions, which aims to ensure a maximum exposure of the organisms to the test substance. This is usually achieved by treating an inert substrate at the highest field rate and confining the animals to the treated area. In some test systems, the insects and their feed or prey are also exposed directly to the substance. In the first tier many factors that might influence the experiment are very well defined. This does not necessarily reflect realism, but helps to achieve consistent and compa-

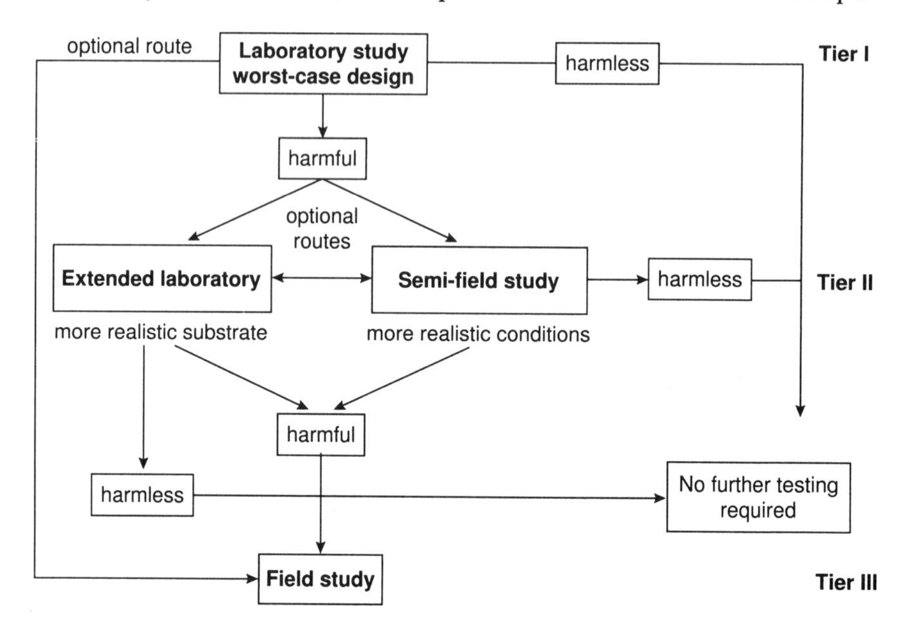

Figure 20.1 The sequential testing scheme for beneficials.

rable results which allow clear cause/effect interpretation. If no effects are observed in such a test, it is very unlikely that a substance would have an unacceptable impact on non-target insects in the field. However, if effects occur in the first-tier test, it is not possible – due to the extreme conditions in the laboratory test – to assess the actual impact in the field. Therefore these products are further investigated in a second step in 'extended laboratory' or 'semi-field' tests. Due to the very different species and test systems used in the first tier, the level of effect that triggers further testing should be adapted to the respective system.

Extended laboratory and particularly semi-field tests are conducted under more realistic conditions, particularly with respect to the exposure to the test substance. In this tier, plant leaves or natural soils are used as substrate rather than glass plates or quartz sand. Considering the different possible use conditions, however, the parameters should still reflect a realistic worst case with respect to plant cover, soil properties and time of application. If significant and inconclusive effects are still observed in this tier, a more complex field study may help to assess the impact of the test substance under realistic field conditions.

PRINCIPAL DIFFERENCES BETWEEN THE MAIN TIERS

The laboratory test is comparatively reliable and reproducible, and can be carried out with normal laboratory skills. It eases the direct comparison between different test substances and allows an assessment of safety in a cost-effective way. On the other hand, it has little realism in terms of its complexity. It greatly overestimates the hazard of a substance, but may not always be able to indicate possible indirect effects.

The extended laboratory and semi-field tests increase the relevance of the findings considerably. In particular, the semi-field test can provide a good indication of what might happen in real field situations. But due to the higher number of uncontrolled parameters (like climate), these tests will also show a greater variability. More experienced personnel are needed, both to conduct the test and to interpret the results. These tests are often more laborious and have a greater chance of failure (for instance, because of extreme weather conditions). In the semi-field test no extreme exposure, as in the laboratory test, is applied; however, a realistic worst case (with the emphasis on 'realistic') with respect to exposure should be simulated by using, for example, rather sandy soils or applying the test substance at a time when interception by crop plants is low. In this way, the semi-field tests allow a safe but much more realistic hazard assessment than the laboratory results and are much more reproducible than the field test (Figure 20.2).

In a final step, extensive field tests may be employed. This usually involves large sites treated with the test substance, equivalent sites

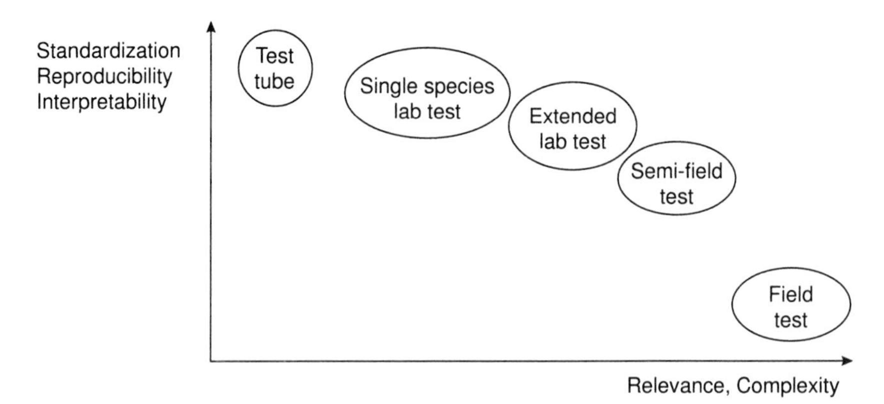

Figure 20.2 The relation of standardization, reproducibility and interpretability versus complexity and relevance for the different tiers of the sequential testing scheme.

that serve as controls and sites that are treated with a toxic reference substance. The species occurring naturally on the plots (with varying population densities) are usually investigated. This obviously leads to greater variability than in the semi-field test which uses a distinct number of organisms introduced into the experimental units. Due to the fluctuations in populations, it may become difficult to distinguish between natural fluctuations, pesticide impact and any other possible causes which may result from climate, field structure, external predators, pathogens, etc. Therefore careful observations of many parameters may be essential, including climate, plant cover, soil parameters and population densities of prey species.

FACTORS AFFECTING THE RESULTS

A series of parameters affect the results in all tiers but to a very different extent (Table 20.1). In the laboratory test the species is usually well defined but, even then, variability may be observed due to the age of the organism or developmental stage (Theiling and Croft, 1988), the physiological stage (Jagers op Akkerhuis, 1993), sex, behaviour and origin (Bigler, 1989). Particularly for predatory mites, there may be significant differences in sensitivity between different resistant or susceptible strains of the same species (Duso *et al.*, 1992), that can be much larger than differences between species. Some of these factors, like behaviour or physiological stage and endpoints like growth and reproduction, are influenced again by other variables such as temperature, humidity (Everts *et al.*, 1991), light cycle, feeding rate and disturbances or handling. An important feature of the laboratory test is that many of these factors

Table 20.1 Factors that influence results of various tiers of beneficial testing; the relative importance of the different factors is mentioned as – (not relevant) + (low impact) +++ (high impact), value in brackets if dependent on type of study design

Factor	*Laboratory*	*Semi-field*	*Field*
Species	–	–	+++
strain	+	+	+++
age	+	+	+++
handling	++	+	–
Climate			
temperature	–	++	+++
humidity	–	+	+++
wind	–	+	+++
light	–	+	+++
Exposure	–	+	+++
surface structure	–	++	++
interception	–	+	++
bioavailability	–	+	++
avoidance	–	+	+++
Complexity			
competition	(+)	+	+++
predation	–	+	+++
feed availability	–	+	+++
recovery	(+)	+	+
recolonization	–	–	++

Relative importance of factors:
–, not relevant
+, low impact
+++, high impact
(+), value depends on study design.

can be manipulated and standardized to a great extent. However, inherent variability in biological systems can still result in considerable variability, particularly when parameters like reproduction or parasitization are assessed as well as mortality (Schmuck *et al.*, 1996). Obviously, most effort should be directed to those parameters that have the highest impact on the outcome of the results, whilst others might be neglected if standardization would take unreasonable resources.

In the semi-field test, too, a single, well defined species is usually investigated, but as this test in normally conducted in the field (in contrast to the extended laboratory test), it is influenced by varying climatic conditions – though some manipulation can be provided: for example, watering, shading, or rain protection. The most important difference compared with the laboratory test is the more realistic pesticide exposure of the test organisms, particularly due to the natural substrate. In experiments with carabid beetles we have shown that this reduces average mortality of 70% on quartz sand to about 30% on natural sandy soils and an even greater reduction if heavy organic-rich soils are used (Chapter 11). The important impact of the substrate on test results has also been shown by Heimbach *et al.* (1992). In addition to the properties of the substrate, the habitat structure modifies the exposure. Plants will intercept a considerable proportion of the spray deposit. The actual surface area, i.e. the area that competes for the test substance, is much larger in a structured natural environment as compared with an even glass surface.

Reproducibility and interpretability are greatly enhanced in the semi-field test by dealing with a single species and a limited, well defined number of individuals. This is the main difference from the full-scale field experiment. Here, a large number of naturally occurring species are investigated whose population densities fluctuate strongly in both time and space. Mitigating factors like refuges, escape chances, or alternative food will be possible in this tier. All this, in addition to the varying climatic conditions, makes a field test a unique experiment (Hussein *et al.*, 1990), which cannot be reproduced in exactly the same way. One field test will not be sufficient to assess a substance when a medium effect is observed, as this effect might be enhanced or lost under different conditions. On the other hand, if a test substance has a considerable impact on various non-target populations, this should be detected regardless of when or where the experiment has been carried out (provided that the test was conducted according to appropriate experimental design using expert knowledge). The same is true if a field test shows no impact on the investigated non-target organisms: it is then very unlikely that an unacceptable effect would occur under different conditions. In these cases, field tests allow the most conclusive risk assessment, as direct and indirect effects as well as short-term and long-term effects and the duration of an effect can be investigated.

THEORETICAL BACKGROUND FOR THE SEQUENTIAL TESTING SCHEME

The first tier in the sequential approach has to detect potentially hazardous substances with a high level of safety. It is therefore conducted under worst-case conditions, which indicate the intrinsic toxicity of a compound but do not allow one to conclude the actual hazard of a test substance. The test design has to be harsh enough to cover the variability of the results as well. The less reliable a laboratory test is, the higher should be the inherent test safety (worst conditions). This means that average effects should be detected in the laboratory at concentrations lower than the concentration that would cause severe effects in the field (Figure 20.3) in order to guarantee that the variability is included.

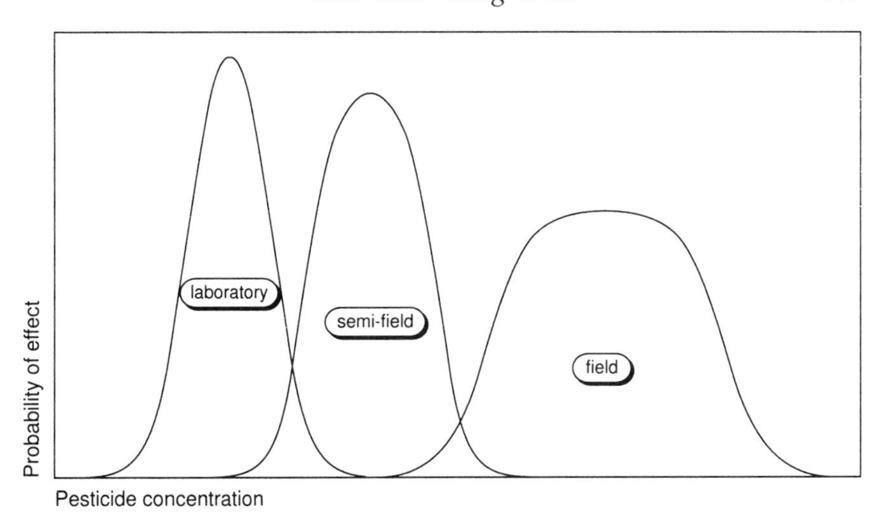

Figure 20.3 Illustration of the likelihood that specific pesticide concentration will cause an effect. (See text for details.)

It may thus be possible to conduct a fairly rough and short-term experiment with relatively high variability that still fulfils the main aim of detecting potentially harmful substances. However, this will considerably increase the number of 'false positive' results, which means that a great number of low-risk pesticides will have to be tested in higher tiers. Therefore, when designing an experiment, the cost/benefit of an easy first tier versus a high number of additional higher tier tests should be considered. In addition, the laboratory results and the higher tiers should be compared to validate the design of the first tier and eventually modify it to give either higher safety, where necessary, or the opposite to reduce the inherent harshness if too many false positive results are obtained.

Figure 20.3 shows, for certain types of tests, a scheme of a probability distribution that a given pesticide concentration will cause an effect. In some experiments some individuals will be rather sensitive and react at a lower concentration, while others will only react to high concentrations; the majority will usually fall in the group that reacts to medium concentration. This kind of normal distribution will not always be correct but usually reflects the real situation quite well. The tiered testing scheme demands that the laboratory curve should be 'left' of the field curve; that is, effects should be observed in the laboratory earlier (at lower concentrations) than in the field. The overlap between the curves should be small and preferably none of the laboratory tests should miss an effect that causes a significant harmful impact in the field. There are different possibilities for laboratory tests. A high precision test would give a steep curve with low standard variation. Such a test may be more elaborate, costly and time consuming, but it would allow us to come close to the field curve, thus requiring a lower safety margin. A low precision test may be easier to perform but, due to its larger variability, it must be far left of the field curve to avoid too much overlap. In some test systems it may be observed that high precision tests are also far left of the field curve and include an unnecessarily large safety margin. This may be avoided by reducing the harshness of the test or it should be taken into account when the trigger values for the test system are set, otherwise an unreasonably high number of false positive results will occur. The further the areas under the curve are to the left of the field curve, the more is unnecessary further testing triggered. If these curves are transformed to concentration versus cumulative effects, typical sigmoid shapes are obtained (Figure 20.4).

COMPARING LABORATORY AND (SEMI-)FIELD DATA: CASE STUDIES

Probably the largest database that allows comparisons between standard laboratory data and results from more realistic situations in the field or semi-field has been elaborated by the IOBC working group 'Pesticides and Beneficial Organisms' (Franz *et al.*, 1980; Hassan *et al.*, 1994). Their experience has shown that when effects in the laboratory were below

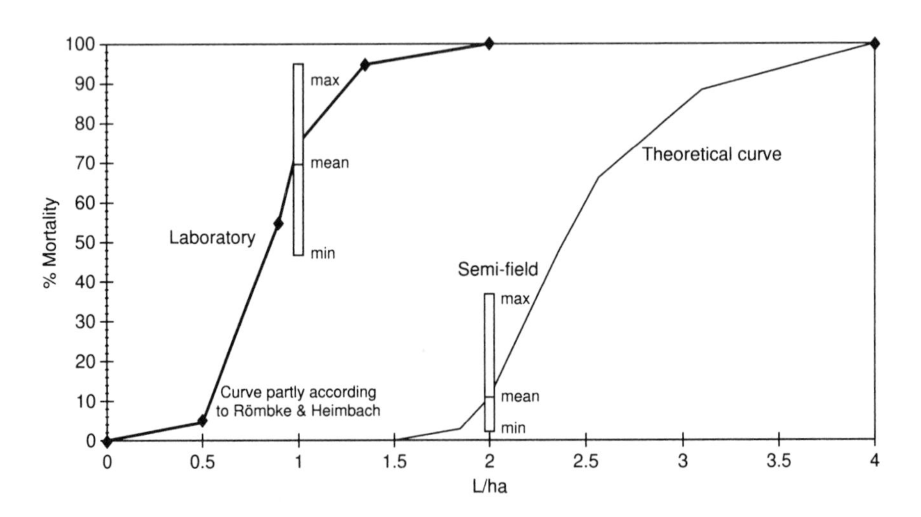

Figure 20.4 Effect of Afugan on the carabid beetle *Poecilus cupreus* in laboratory and in semi-field tests. Curves give effect of different concentrations; bars indicate variability of a series of tests at one concentration.

50%, the product was unlikely to present a significant risk and could be classified as harmless (Samsoe-Petersen, 1990). (This classification has been changed recently, mainly for political or general perception reasons.)

Experiments with carabid beetles in the laboratory and semi-field with the same substances have also demonstrated the usefulness of the scheme and the validity of the laboratory test (Figure 20.4). When, in addition, real field experiments, which include recovery and recolonization, are compared with the laboratory data, it may be difficult to detect consistent and lasting effects on carabids even for substances that have insecticidal activity (Chapter 28) and that would result in considerable mortality (usually 100%) in the laboratory.

Similar comparisons can be made for other species and test systems as well. Investigations of the effects of an insecticide on *Typhlodromus pyri* have shown that there are significant differences between two strains (a very sensitive one and a more resistant one) in the laboratory. However, even the apparently less sensitive strain reacted in the laboratory test at about 1/20 of the concentration that caused effects in the semi-field, which again showed about a twofold higher sensitivity than the field results (Figure 20.5; A. Ufer, personal communication).

Duso (1994) compared several species of predatory mites in laboratory and field and found that in the majority of cases the laboratory test was either more sensitive or produced the same classification as the field test. Bigler and Waldburger (1994) have compared the effects of 55 pesticides on *Chrysoperla carnea* in the laboratory and in the semi-field.

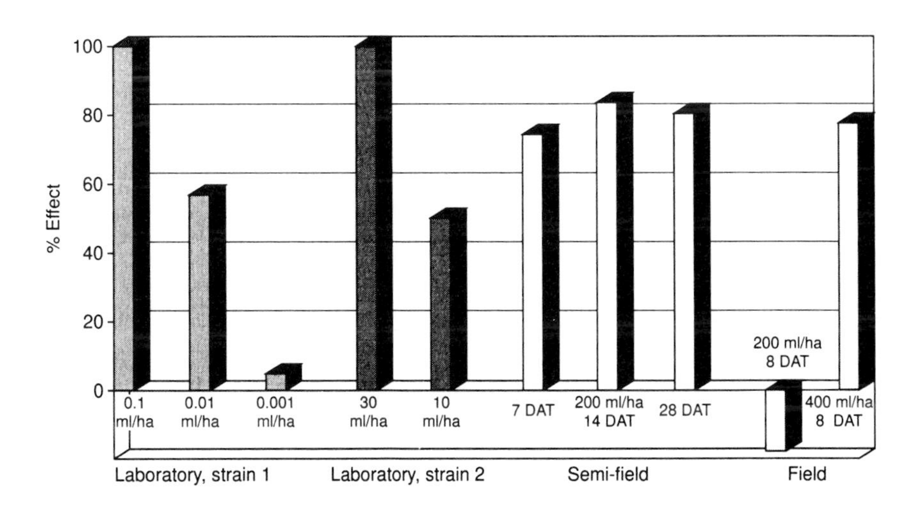

Figure 20.5 Effect of different pesticide rates on *Typhlodromus pyri* in the laboratory, semi-field and field.

They conclude that the principal scheme – which means that no effects will occur in the (semi-)field if the laboratory test has shown no significant effect – is confirmed in general. A few exemptions have to be addressed case by case. Exemptions may be caused by a specific mode of action that may not be detected in the standard laboratory test. This is true, for example, for insect growth regulators: their effects will often not be apparent in tests that have mortality and parasitization rate or feed consumption as endpoints. Such specific cases obviously need specific attention but they do not reduce the general value of a sequential testing scheme for assessing the effects of pesticides on beneficial organisms.

REFERENCES

Barrett, K.L., Grandy, N., Harrison, E.G. *et al.* (1994) *Guidance Document on Regulatory Testing Procedures for Pesticides with Non Target Arthropods. From the ESCORT workshop, Wageningen, Netherlands, March 1994,* SETAC-Europe, Brussels.

Bigler, F. (1989) Quality assessment and control in entomophagous insects used for biological control. *Journal of Applied Entomology,* **108**, 390–400.

Bigler, F. and Waldburger, M. (1994) Effects of pesticides on *Chrysoperla carnea* Steph. (Neuroptera, Chrysopidae) in the laboratory and semi-field. *IOBC/WPRS Bulletin,* **17**(10), 55–69.

Brasse, D. (1990) Einführung der obligatorischen Prüfung der Auswirkung von Pflanzenschutzmitteln auf Nutzorganismen in das Zulassungsverfahren. *Nachrichtenblatt des Deutschen Pflanzenschutzdienstes,* **42**(6), 81–89.

Duso, C. (1994) Comparison between field and laboratory testing methods to evaluate the pesticide side effects on the predatory mites, in *Pesticides and Beneficial Organisms,* (ed. H. Vogt), *IOBC Bulletin* **17**(10).

Duso, C., Camporese, P. and van der Geest, L.P.S. (1992) Toxicity of a number of pesticides to strains of *Typhlodromus pyri* and *Amblysius andersoni* (Acari: Phytoseiidae). *Entomophaga,* **37**, 363–372.

EPPO (1993) Decision-making schemes for the environmental risk assessment of plant protection products. *Bulletin OEPP/EPPO* **23**, 1–165.

EPPO (1994) Decision-making schemes for the environmental risk assessment of plant protection products – arthropod natural enemies. *Bulletin OEPP/EPPO* **24**.

Everts, J.W., Willemsen, I., Stulp, M. *et al.* (1991) The toxic effect of deltamethrin on linyphiid and erigonid spiders in connection with ambient temperature, humidity and predation. *Archives of Environmental Contamination and Toxicology,* **20**, 20–24.

Franz, J.M. (1974) Die Prüfung von Nebenwirkungen von Pflanzenschutzmitteln auf Nutzarthropoden im Laboratorium – ein Sammelbericht. *Zeitschrift für Pflanzenkrankheiten und Pflanzenschutz,* **81**(2/3), 141–174.

Franz, J.M., Bogenshütz, H., Hassan, S.A. *et al.* (1980) Results of a joint pesticide test programme by the Working Group 'Pesticides and Beneficial Arthropods'. *Entomophaga,* **25**, 231–236.

Hassan, S.A. (1989) Vorstellungen der IOBC-Arbeitsgruppe 'Pflanzenschmutzmittel und Nutzorganismen' zur Erfassung der Nebenwirkungen von Pflanzenschutzmitteln auf Nützlinge. *Gesunde Pflanzen,* **41**(8), 295–302.

Hassan, S.A., Bigler, F., Blaisinger, P. *et al.* (1985) Standard methods to test the side-effects of pesticides on natural enemies of insects and mites developed by the IOBC/WPRS Working Group 'Pesticides and Beneficial Organisms'. *Bulletin OEPP/EPPO*, **15**, 214–255.

Hassan, S.A., Bigler, F., Bogenschütz, H. *et al.* (1994) Results of the sixth joint pesticide testing programme of the IOBC/WPRS – Working Group 'Pesticides and Beneficial Organisms'. *Entomophaga*, **39**, 107–119.

Heimbach, U., Abel, C., Siebers, J. and Wehling, A. (1992) Influence of different soils on the effects of pesticides on carabids and spiders. *Abstracts of Applied Biology*, **31**, 49–59.

Hussein, I.A.A., Lübke, M. and Wetzl, T. (1990) Nebenwirkungen von Insektiziden auf Kurzflügalkäfer (Col., Staphylinidae) in Winterweizenfeldern. *Journal of Applied Entomology*, **109**, 226–232.

Jagers op Akkerhuis, G.A.J.M. (1993) Physical conditions affecting pyrethroid toxicity in arthropods. PhD thesis, Wageningen.

Samsoe-Petersen, L. (1990) Sequences of standard methods to test effects of chemicals on terrestrial arthropods. *Ecotoxicology and Environment Safety*, **19**, 310–319.

Schmuck, R., Mager, H., Künast, C. *et al.* (1996) Variability in the reproductive performance of beneficial insects in standard laboratory toxicity assays – implications for hazard classification of pesticides. *Annals of Applied Biology*, **128**, 437–451.

Theiling, K.M. and Croft, B.A. (1988) Pesticide side-effects on arthropod natural enemies: a database summary. *Agriculture, Ecosystems and Environment*, **21**, 191–218.

Part Three

Risk Assessment and Management

21

Introduction

Pieter A. Oomen, Gavin B. Lewis and Richard A. Brown

The need to control the risks arising from the use of pesticides is nearly as old as the need for pesticides in crop protection, and it has been met through legislation and subsequent regulation of pesticides. Effective control of risks through such regulatory measures cannot be achieved without first conducting adequate risk assessment and management, based on appropriate data concerning the environmental fate and effects of a pesticide.

DEFINITIONS

Definitions used are as follows.

- **Risk**: probability that undesired effects ('hazards') of pesticides will occur in practice. In the context of this book, the subjects of these undesired effects and risks are limited to beneficial arthropods and other non-target terrestrial arthropods – in particular, insects, spiders and mites.
- **Risk assessment**: identification and classification of risks caused by the authorized uses of pesticides.
- **Risk management**: regulation of pesticide use in such a way that identified risks are reduced to an acceptable level while maintaining their effective use in crop protection as far as possible.

Ecotoxicology: Pesticides and beneficial organisms.
Edited by P.T. Haskell and P. McEwen. Published in 1998 by Chapman & Hall, London. ISBN 0 412 81290 8.

REGISTRATION PROCEDURE

Where pesticide legislation exists, in most countries the use of any pesticide in a particular country is forbidden unless it is first registered. Registration has developed into a long and very expensive procedure, requiring a large amount of data for each pesticide, including its physical and chemical properties, efficacy, mammalian toxicology, crop residue profile and environmental fate and effects. Briefly, the registration process:

- identifies the uses for which the pesticide is efficacious;
- identifies and assesses the risks arising from the proposed uses;
- evaluates the acceptability of these risks;
- authorizes the uses for which the pesticide is efficacious and for which the risks are acceptable;
- prescribes the way in which the pesticide should be used (these prescriptions, also known as good agricultural practice, or GAP, are generally given as label instructions).

In practice, many different risks are addressed in the registration process. These include agricultural risks, as well as risks to the user and to the consumer of treated crop products, and to the environment. In the context of this book attention is given only to risks to beneficial and other non-target terrestrial arthropods.

NON-TARGET ARTHROPODS IN EUROPEAN PESTICIDE REGISTRATION

Until recently little attention was given in pesticide registration to protection of terrestrial non-target arthropods, with the notable exception of honey bees. This is in contrast to, for example, aquatic organisms. It is only in Germany and the UK that for a few years measures have been taken in pesticide registration to protect certain beneficial arthropods, or at least to warn users about the risks of the pesticide to these organisms. These were the first initiatives to meet the generally felt need to fill this gap in risk management and also protect these organisms from undesired pesticide effects.

Recently the European Union (EU) has decided, in its well known Directive 91/414/EEC (concerning the placing of plant protection products on the market) and in its Directive 94/43/EC (establishing Annex VI to Directive 91/414/EEC, better known as the Uniform Principles), to include this area of risk assessment in the European registration procedure.

THE HISTORICAL PICTURE

Much development work has already been done, in particular by the International Organization for Biological and Integrated Control of Noxious Animals and Plants (IOBC) and the European and Mediterranean Plant Protection Organization (EPPO).

IOBC started work in this area in 1974 with its working group, 'Pesticides and Beneficial Organisms'. Its objective was to facilitate integrated pest management, but not explicitly through registration – rather, by providing growers with relatively simple, comparative information on the effects of different products on predatory and parasitic arthropods. The main achievements of IOBC have been the development of testing methodology, the establishment of evaluation principles and the standardization of methods.

EPPO started in this field in about 1984, with the objective of providing harmonized guidelines for registration, and published several IOBC-based testing guidelines for beneficial organisms. Later (1993) EPPO, together with the Council of Europe (CoE), also prepared a harmonized environmental risk assessment approach, covering all areas of ecotoxicology. This included a detailed decision-making scheme for the risk assessment of natural enemies. This EPPO/CoE environmental risk assessment scheme has been the main scientific base for the European regulations described above.

In addition to these two organizations, the Beneficial Arthropod Regulatory Testing (BART) group was formed in 1989 in response to the changing regulatory climate with respect to beneficial arthropods. This group has brought together experts, principally from industry, with the aim of contributing to the development of appropriate regulatory test methods and the associated risk assessment scheme.

In March 1994 a three-day meeting – the European Standard Characteristics of Beneficial Regulatory Testing (ESCORT) Workshop – was held in Wageningen under the auspices of BART, IOBC and EPPO/CoE, in conjunction with SETAC-Europe and with the support of the European Commission. This workshop brought together nearly 40 experts, principally from European countries, to advise on specific aspects of non-target arthropod testing, risk assessment and management. These contributions ensured that the resulting proposals were practical, feasible and realistic.

The SETAC–ESCORT report has provided an important input into the European regulations, in addition to that of EPPPO/CoE, but this combination does not yet appear to be adequate for the satisfactory risk assessment and management of non-target arthropods in an internationally harmonized manner. A number of problems with the implementation of the guidelines in their current form have been identified. These include extrapolation from the single rates tested to other exposure scenarios (different application rates, off-crop exposure, etc.), extrapolation from the indicator species tested to broad groups of non-target arthropods, and assessment of multiple applications. In particular, there is no clear guidance in risk management and labelling beyond the national level. Thus improvement and a final agreed version of the risk assessment scheme for non-target arthropods are still needed. These issues are discussed in the chapters of Part Three.

THE CURRENT SITUATION

Currently, the risk assessment and management situation in Europe is as follows.

- Test methodology, risk assessment and risk management are well developed and validated for one species, the honey bee (a representative of the pollinating insects) and have been implemented in the European regulations. Honey bees provide an instructive example of how risk assessment and management with beneficial organisms can be effectively and efficiently implemented in pesticide regulation. In this respect there may well be lessons to be learnt from the honey bee example.
- Test methodology for other regulatory species is well developed but certain improvements are needed (e.g. statistics of evaluations, endpoints of laboratory tests, field testing methodology). Validation of the laboratory test methods for most important species is in progress.
- No definitive decision has yet been made on whether other non-target species need to be tested or if sufficient protection is offered to this group through the use of the present test species.
- The risk assessment approach for some specific species has been developed but consensus is still needed on some issues – for example, the use of dose-response testing, laboratory trigger values for further testing, etc.
- Risk assessment for groups of species through extrapolation is in principle possible by using extrapolation factors in accordance with the approach of other ecotoxicological risks, but experience is lacking.
- Consensus about the interpretation of data at the field level and, in particular, threshold values for acceptability of effects is yet to be achieved.
- Risk management options are still largely open. There are the examples provided by the honey bee and other ecotoxicological risk management, and there is limited experience with label instructions for protection of beneficial organisms as used in the UK and Germany.

QUESTIONS ADDRESSED

It appears that most of these aspects need further attention. The main problems are as follows.

- What are the appropriate options for risk management for non-target arthropods and, in particular, for label instructions then to be used? Experience from the UK and Germany are reported in Chapters 23 and 24.
- What effects are acceptable, and what are the important factors in assessing acceptability? This should be seen as related to consequences for the practicality of crop protection. Input from NGOs, long-term research studies, and reports from the few countries with experience in this field provide a welcome contribution to this section (Chapter 27).

These are some of the issues that are addressed in Part Three of this book. Results and suggestions will be taken up primarily by the EPPO/CoE working group 'Natural Enemies', which continues to act as an expert panel for the EU and the implementation of the Uniform Principles. In addition, a follow-up meeting of the ESCORT workshop will take place early in 1998 which, it is hoped, will enable a consensus view to be reached on these unresolved issues.

Part Three: A

Strategy

22

Aims and consequences of regulatory risk management in Europe: a discussion

Pieter A. Oomen

INTRODUCTION

Assessment of pesticide risks to natural enemies and other non-target arthropods is certainly possible, and since 1974 the International Organization for Biological Control of Noxious Animals and Plants (IOBC) has developed basic principles, testing methodology and an approach for evaluating the results. Standardized test methods on many different natural enemies are available, have been in use for many years and are being ring-tested for good reliability. A sequential evaluation scheme for regulatory purposes, based on these principles and testing methods, has been developed by the Joint Panel on Environmental Risk Assessment of the European and Mediterranean Plant Protection Organization (EPPO) and the Council of Europe (CoE). The European Union (EU) in its turn requires that plant protection products will cause no unacceptable impact on beneficial arthropods, referring to the EPPO/CoE scheme for guidance on how to carry out risk assessments.

The scheme is not yet fully operational. In 1995 the Panel's Working Group on Natural Enemies started a new effort to elaborate the scheme. Progress appears to be difficult, as it is becoming apparent that the scheme has two or three separate and conflicting risk management aims.

There are both environmental and agricultural reasons for protecting terrestrial non-target arthropods. This can be illustrated by the designations

Ecotoxicology: Pesticides and beneficial organisms.
Edited by P.T. Haskell and P. McEwen. Published in 1998 by Chapman & Hall, London. ISBN 0 412 81290 8.

used for these organisms. The terms 'beneficial arthropods' and 'natural enemies' refer to the function these animals have in crop protection – contributing to pest control, i.e. to integrated pest management (IPM). The designation 'non-target arthropods' refers to them as a group of organisms in the environment that should be protected for environmental reasons but not for maintaining a specific function. These IPM and environmental aims are in principle different, and even conflicting, but usually they are not separated, either by the CU or by EPPO/CoE. No adequate risk management can be developed for such mixed aims. Only separation of the different aims can make risk management feasible in this field.

This chapter was prepared in order to analyse this problem and to stimulate a discussion on this subject in particular. Clarification is necessary before progress can be made towards making the European legislation work in this field. As the text of this book was prepared after the chapter was given as a lecture paper, results of discussions at the Cardiff Conference are included here.

LEGAL CONTEXT

The EU adopted in 1991 the well known Directive 91/414/EEC (CoE, 1991) in order to bring the registration of plant protection products for use in Europe under one common regulation. This Directive has since been extended with a number of annexes that specify data requirements and principles for evaluation and decision making. One of these (Annex VI) is the *Uniform Principles for Evaluation and Authorization of Plant Protection Products* (CoE, 1994). In its sections on 'Influence on the Environment' and 'Impact on non-target species', requirements for protecting beneficial arthropods are given and these are summarized here as follows.

- No authorization shall be granted if > 30% effect is observed at the maximum application rate in laboratory tests, unless it can be clearly established under field conditions that no unacceptable impact on those organisms will occur.
- Any claims for selectivity and proposals for use in IPM systems shall be substantiated by appropriate data.

The European member states shall bring these regulations into force in their territory not later than one year after the date of publication in the *Official Journal of the European Communities*.

POSSIBLE AIMS OF REGULATORY RISK MANAGEMENT

Three possible aims of regulatory risk management have been inventoried by the SETAC/ESCORT Workshop in Wageningen. This workshop produced a guidance document (Barrett *et al.*, 1994) to which the Uniform

Principles (CoE, 1994) refer for further guidance. The three aims, which are elaborated in Table 22.1, are as follows:

1. IPM
2. Environmental purposes within-crop
3. Environmental purposes off-crop.

This chapter systematically compares the requirements and consequences of the implementation of these three aims or scenarios.

COMPARISON OF IMPLEMENTATION REQUIREMENTS

A view of the requirements for implementing each of the three aims is given in Table 22.2. For IPM purposes, precise information is needed to allow the grower a tailor-made integration of biological and chemical control measures. Relevant species are usually many, and side-effect information is necessary for each of the relevant species, often for different rates and at the level of field or semi-field situations. Extrapolation from one species to another is not usually possible at the level of precision required for IPM. Hence the data requirements as indicated by the Uniform Principles (two standard species and two to four relevant species only) are insufficient for this purpose.

The risk assessment can be done according to the usual IOBC approach, or according to the more formalized risk assessment schemes produced by EPPO (e.g. EPPO, 1989). The work for preparing the required data for the IPM aim will be huge and difficult as a consequence of the many variables involved.

Much less precision is required for the two environmental aims. Data about effects on a few species that are representative of the most important groups of non-target arthropods are sufficient for extrapolation to

Table 22.1 Three possible aims of regulatory risk management in protecting arthropods

Question	*1. For IPM*	*2. Within crop*	*3. Outside crop*
Why	To improve crop protection	To protect environment	To protect environment
What	Natural enemies of pests	Non-target arthropods	Non-target arthropods
Where	Within crop	Within crop	Off-crop
How	Advisory instructions, (e.g. label information)	Advisory or statutory restrictions, e.g. not granting authorization	Advisory or statutory restrictions, e.g. emission reductions through buffer zones

Table 22.2 Practical implementation of the three regulatory risk management aims

	1. For IPM	*2. Within crop*	*3. Outside crop*
Species data requirements	All relevant species	Two standard species. Two to four relevant species	Two standard species. Two to four relevant species
Test data requirements	Laboratory, semi-field, field	Laboratory, unless . . .	Laboratory, unless . . .
Exposure estimate	100% of field rate	100% of field rate	Drift % of field rate
Evaluation thresholds for acceptability	30% unless . . .	30% unless . . . ; EPPO scheme	30% unless . . . ; EPPO scheme
Extrapolation factors	No	Yes: to all species	Yes: to all species
Risk assessment	IOBC approach	EPPO/CoE approach	EPPO/CoE approach
Work load	Extensive and difficult	Limited if . . .	Limited if . . .

all non-target arthropods. Use of extrapolation factors is necessary, as is usual in ecotoxicological risk assessments, in order to cover the uncertainty of differences between species. These data can be collected in the laboratory, unless an applicant is confident that information from more representative situations (semi-field, field) is likely to change the risk assessment into one of the less harmful classifications.

The estimated exposure is the only difference between the two environmental aims. The expected within-crop exposure of arthropods is of course equal to the highest recommended field rate, while expected off-crop exposure can be estimated as a drift percentage of the highest recommended field rate. Drift percentages are also being quantified and formalized for the assessment of risk to aquatic organisms. These same figures can be used for terrestrial arthropods.

The work for preparing the required data for the two environmental scenarios can remain limited, as laboratory tests on some species are sufficient to run the risk assessment.

As a decision threshold for evaluating acceptability, the value of 30% effect in the laboratory is being widely prescribed (IOBC, EPPO and the Uniform Principles) but this value should be considered as preliminary. As indicated by EPPO/Coe (1994), the decision value has yet to be made species specific. The decision value will depend on how acceptability will be defined for either the IPM or the environmental protection aims.

COMPARISON OF IMPLEMENTATION CONSEQUENCES

For each of the three risk management scenarios in Table 22.1, the consequences of the availability of pesticides on the feasibility of chemical crop protection, and of IPM on the environment, are likely to be as described in Table 22.3.

For the IPM scenario, there are two possibilities for risk management. The first is by not granting authorization to products that are harmful to any of the relevant natural enemies. This is the approach indicated by the Uniform Principles. The consequences are very much like the environmental within-crop scenario: very few insecticides and acaricides will have a chance to be registered. This would be counterproductive for IPM, which needs a wide scope of pesticides to enable a tailor-made scheme for controlling pests and diseases while maintaining important natural enemies.

The second possibility for IPM is that no pesticides need to be prohibited. This is because the most effective risk management is through informing the users, and those responsible for developing IPM systems, about the effects on natural enemies. Hence no consequences on feasibility of chemical crop protection are foreseen. Feasibility of IPM will be stimulated because of the availability of side-effect data due to the EU data requirements. However, this stimulation will be limited because the side-effect data from regulatory sources will always be more limited than necessary for implementing IPM – in particular because not all different relevant species can be covered, and dose-effect relationships are not covered either. Further, the regulatory information will be too late to keep pace with the fast developments in IPM. On the other

Table 22.3 Supposed consequences of implementing the three risk management aims

Consequences for	*1. For IPM*	*2. Within crop*	*3. Outside crop*
Pesticides prohibited	None	Many	Few or none
Feasibility of chemical crop protection	No consequences	Chemical control seriously hampered	Chemical control slightly hampered
Feasibility of IPM	Some stimulation, but information too simplified and too late	IPM impossible for lack of selective pesticides; bio-control stimulated	IPM stimulated
Environment	Positive through IPM	Positive through use limitations	Positive through emission reductions

hand, it is likely that data on side-effects will also be available through other, informal channels (for example, from IOBC or from companies producing natural enemies) faster than from the regulatory source. The slight stimulation of IPM will have an equally slight positive effect on the environment.

If the within-crop environmental aim is realized, very few insecticides and acaricides (which are specifically intended to kill insects or mites, respectively) will have a chance to be registered. Nearly all – possibly with the exception of *Bacillus thuringiensis* and the pheromones – will have a strong effect on one or more of the non-target arthropods. Use of the proposed extrapolation factors will worsen this loss of pesticides. Registration of other pesticides than insecticides and acaricides may also be seriously affected. Chemical crop protection will become seriously hampered by lack of effective products. (This applies also to the feasibility of IPM, as mentioned above.) Only biological control measures will thrive, as a consequence of lack of other solutions for controlling pests. Consequences like these will have a positive direct effect on the environment for a strong reduction in pesticide use, but may have negative indirect effects as agriculture as a whole will become economically more marginal.

Realization of the environmental off-crop scenario does not need to lead to such drastic consequences for crop protection. Risk management measures that reduce emission to off-crop areas are sufficient, such as use of low-emission sprayers and of barriers or buffer zones at crop borders. Few pesticides need to be prohibited and chemical control is only slightly hampered. IPM is stimulated for better conservation of the reservoirs of natural enemies. The environmental quality increases as a consequence of the reduced emission to the environment.

These comparisons lead to the conclusion that the first two risk management aims in Table 22.1 – IPM, and within-crop non-targets – are conflicting. Improving IPM cannot be done with the instruments provided by the EU Uniform Principles (CoE, 1994); on the other hand, within-crop environmental protection would take away the instruments necessary to make IPM work, i.e. the necessary wide choice of selective and selectively applicable pesticides. The third aim – off-crop non-targets – conflicts with neither the first nor the second.

COMPARISON OF COSTS AND BENEFITS

Each scenario has its own advantages and disadvantages. These are shown in Table 22.4, together with feasible alternatives and a conclusion on costs and benefits.

Arguments in favour are obvious: all scenarios benefit the non-target arthropods intended to be protected. The arguments against are more

Table 22.4 Evaluation of costs and benefits of the three different regulatory management aims

Evaluation	*1. For IPM*	*2. Within crop*	*3. Outside crop*
Arguments pro (benefits)	Good for IPM and natural enemies	Good for within-crop non-targets	Good for non-targets in environment
Arguments contra (cost)	Large work load and costs, nevertheless too simplified and too late	Work load and costs reasonable. No actual environment protection; crop protection disabled	Work load and costs reasonable; slight impediments to crop protection
Conclusion	Not well feasible and not sensible – go for an alternative!	High costs, few real benefits – go for an alternative!	Reasonable and feasible proposal – go for this aim!
Possible alternative	Information collected by IOBC and companies on voluntary basis	Redefine acceptability according to SETAC/ ESCORT, or abandon aim to protect within-crop non-targets	Redefine acceptability, or abandon aim to protect off-crop non-targets

diverse: the costs and work load for the IPM scenario are huge, but the information produced is meagre and probably too limited and too late to be really useful. For both of the environmental scenarios, the work load and costs may be kept within reasonable limits; however, the within-crop scenario will very seriously disable crop protection by prohibiting practically all pesticides with insecticidal or acaricidal effects. In return, it will protect the arthropods within-crop – a place that strictly does not even belong to the 'environment' but to the 'agricultural production' compartment. This is not applicable to the off-crop scenario, where non-target arthropods in the true 'environment' are protected at the cost of low impediments to crop protection, i.e. the measures necessary to avoid or reduce emission of pesticides towards the environment.

An obvious alternative for each of these scenarios is to abandon the aim of regulatory protection of the non-target arthropods. For the IPM scenario, this means that IPM would need a continuation of the actual route of collecting information on side-effects on beneficial arthropods, i.e. voluntary testing on behalf of IOBC or of interested institutes or companies producing natural enemies. It is clear that until now this non-regulatory source of information has been an important and relatively well functioning support for IPM. This could remain the case in the future.

The alternative of abandoning the aim of protecting non-target arthropods within-crop and off-crop would continue the actual (pre-Uniform Principles) situation, which has not thus far been considered to be unacceptable. However, it would imply abandoning the aim of improving the condition of the environment in so far as terrestrial non-target arthropods are concerned. In a way, this is a strange and undesirable contrast to the accepted protection that exists for aquatic non-target organisms.

A second alternative for the environmental aims is to define acceptability of effects according to the SETAC/ESCORT recommendations (Barrett *et al.*, 1994). These indicate that effects on within-crop non-target arthropods are acceptable if population recovery occurs within a reasonable time, e.g. one season.

The IPM and the within-crop scenarios bring comparatively high costs and limited benefits. For both, the indicated alternatives are preferable. This is in contrast to the off-crop scenario, where the balance of costs and benefits is far more favourable for implementing the indicated risk management aim.

CONCLUSION

Regulatory risk management aimed at improving IPM through protection of within-crop natural enemies is not capable of generating the necessary information for making IPM work. Therefore it is not likely to improve the voluntary approach greatly. Hence it is not worthwhile and would best be abandoned. Here the second requirement of the EU Uniform Principles comes in. Claims for selectivity or suitability for IPM should be substantiated separately by appropriate data and risk assessments. Suitable data are those gathered according to the relevant EPPO schemes (e.g. EPPO, 1989) or the IOBC approach (which has yet to be formalized and validated). These data should be made available to IPM practitioners by regulatory authorities without the possibility that the risk assessment results might prevent registration.

Regulatory risk management aimed at protecting within-crop non-target arthropods by not authorizing harmful products (as prescribed by the EU Uniform Principles) will be most detrimental to crop protection, at least as long as limited effects in laboratory tests alone are considered as unacceptable. At the same time the aim is disputable, as it is not commonly accepted that the actual crop belongs to the 'environment' rather than to the 'agricultural production' compartment. Hence this aim is not worthwhile either – unless the crop is accepted as belonging to the 'environment' and acceptability is defined according to the SETAC/ESCORT recommendations, i.e. in terms of population recovery within a reasonable time (e.g. one season).

In contrast, regulatory risk management aimed at protecting off-crop non-target arthropods is feasible at reasonable cost; it is also justifiable and therefore worthwhile. The EU Uniform Principles should be elaborated to recognize this environmental aim of risk management.

The view of the aim of risk management as described above is fully in line with the SETAC/ESCORT recommendations (Barrett *et al.*, 1994). It enables a simple, scientifically sound, feasible and timely implementation of the Uniform Principles. It would solve the problem of different and conflicting risk management aims. The EPPO/CoE Working Group on Natural Enemies has taken up the elaboration of the existing risk assessment scheme to make the Uniform Principles workable according to these aims. This way real progress can be made by making Directive 91/414/EEC work, and giving the natural enemies their due protection in Europe.

REFERENCES

Barrett, K.L., Grandy, N., Harrison, E.G. *et al.* (1994) *Guidance Document on Regulatory Testing Procedures for Pesticides with Non Target Arthropods. From the ESCORT workshop, Wageningen, Netherlands, March 1994*, SETAC-Europe, Brussels.

CoE (1991) Council Directive 91/414/EEC of 15 July 1991 Concerning the Placing of Plant Protection Products on the Market. *Official Journal of the European Communities*, **L230**, 1–32.

CoE (1994) Uniform Principles for Evaluation and Authorization of Plant Protection Products. Council Directive 94/43/EC of 27 July 1994, establishing Annex VI to Directive 91/414/EEC concerning the placing of plant protection products on the market. *Official Journal of the European Communities*, **L227**, 31–55.

EPPO (1989) Guideline for the evaluation of side effects of plant protection products. Guideline No. 142: *Encarsia formosa*. *Bulletin OEPP/EPPO*, **19**, 355–372.

EPPO (1994) Decision-making schemes for the environmental risk assessment of plant protection products – arthropod natural enemies. *Bulletin OEPP/EPPO*, **24**(1), 17–35.

23

Effects of plant protection products on beneficial organisms: the current authorization procedure in Germany

Rolf Forster

INTRODUCTION

This chapter gives an overview on how risk assessment, evaluation of effects and labelling of risk are implemented within the German authorization procedure for plant protection products (PPP). It gives an analysis of whether the current testing strategy fulfils the objectives as laid down in Council Directive 91/414/EEC, and finally looks at how risk assessment and risk management could be improved with regard to non-target terrestrial arthropods.

CURRENT AUTHORIZATION PROCEDURE IN GERMANY

Legal background

As laid down in section 15 of the German Plant Protection Act of 15 September 1986 (Pflanzenschutzgesetz – PflSchG), authorization for the placing of a PPP on the market is granted:

> if it is established, in the light of current scientific knowledge, that the PPP has no unjustifiable effects, especially on the environment,

Ecotoxicology: Pesticides and beneficial organisms.
Edited by P.T. Haskell and P. McEwen. Published in 1998 by Chapman & Hall, London. ISBN 0 412 81290 8.

having regard to the conditions under which it shall be used and to the consequences of its use.

Further, as laid down in section 6(1):

farmers shall not use a PPP if harmful effects, especially for the environment, are anticipated.

The strategy established supports two main objectives:

- the introduction and promotion of integrated pest management (IPM);
- the protection of the environment.

Since the early 1970s the German Working Group for Integrated Plant Protection and later the Pesticides and Beneficial Arthropods working group of the International Organization for Biological Control of Noxious Animals and Plants/West Palaearctic Regional Section (IOBC/WPRS) has elaborated testing methods for standard beneficials to identify suitable PPP for specific use in IPM systems. A tiered system has been implemented in the testing strategy using laboratory tests, semi-field tests and field tests.

Studies on the effects of PPP have been required since 1987 for intended use in vines using *Typhlodromus pyri* (Acarina: Phytoseiidae) and since December 1989 for other intended uses and other species relevant to the crop (Brasse, 1990).

Today, testing is based on guidelines published by the German Federal Biological Research Centre (BBA) and the IOBC/WPRS. According to the requirements of the BBA the effects of PPP have to be tested on at least two species, but usually not more than six, using the highest field rate as specified in the intended uses (single-dose test design), if exposure of beneficial arthropods is likely.

Key factors considered to make up a realistic worse case in basic laboratory tests are:

- the intended uses of the PPP (to identify the possibility of exposure and species relevant to the crop);
- The intended time of application (to identify the possibility of exposure of a relevant species and its developmental stage);
- the intended maximum field rate (to produce a worst-case scenario);
- the application technique (e.g. spray vs. granules and seed treatments) (to identify the possibility of exposure and species relevant to the use);
- the mode of action of an active substance (e.g. insect growth regulators – IGRs) (to identify relevant developmental stages of the species tested).

Evaluation of risk and labelling

For the time being the IOBC classification proposed by Hassan (1992) is adapted for the evaluation, because no species-specific criteria are available today. Classification is generally based on the results of laboratory tests; testing in higher tiers (semi-field and field tests) is optional. PPP have been labelled since 1987 for use in vines and since 1993 for all uses. Three classes have been established to classify the effects found for the species tested (Forster, 1995):

- The PPP is classified as not harmful for populations of [name of species tested] (= low risk), for effects < 30% in laboratory test and < 25% in semi-field and field tests.
- The PPP is classified as slightly harmful for populations of [name of species tested] (= medium risk), for effects 30% to 79% in laboratory test and 25% to 49% in semi-field and field tests.
- The PPP is classified as harmful for populations of [name of species tested] (= high risk), for effects > 80% in laboratory test and > 50% in semi-field and field tests.

There are two general labels:

- The PPP is classified as not harmful for populations of relevant species because exposure is not likely (= negligible risk).
- The PPP is classified as harmful for populations of relevant species (= high risk) because harmfulness is established for a range of species or is anticipated.

On 10 May 1996, 960 PPP were authorized for placing on the German market. Of these PPP, 62% were labelled and 1441 labels were applied: 95.6% of the labelled PPP give information about the effects on up to six species tested; 4.4% give information for more than six species (maximum 11); most often, two labels are granted (40.6% of PPP).

Figure 23.1 gives an overview of the species tested and shown on the label for PPP registration – in total, 26 species. The six species most often named on the label, which make up 70.9% of all labels, are:

- *Poecilus cupreus* (Coleoptera: Carabidae)
- *Aleochara bilineata* (Coleoptera: Staphylinidae)
- *Chrysoperla carnea* (Neuroptera: Chrysopidae)
- *Typhlodromus pyri* (Acarina: Phytoseiidae)
- *Coccinella septempunctata* (Coleoptera: Coccinellidae)
- *Trichogramma cacoeciae* (Hymenoptera: Trichogrammatidae).

Figure 23.2 demonstrates the frequencies of the three risk classes within the specified groups of PPP.

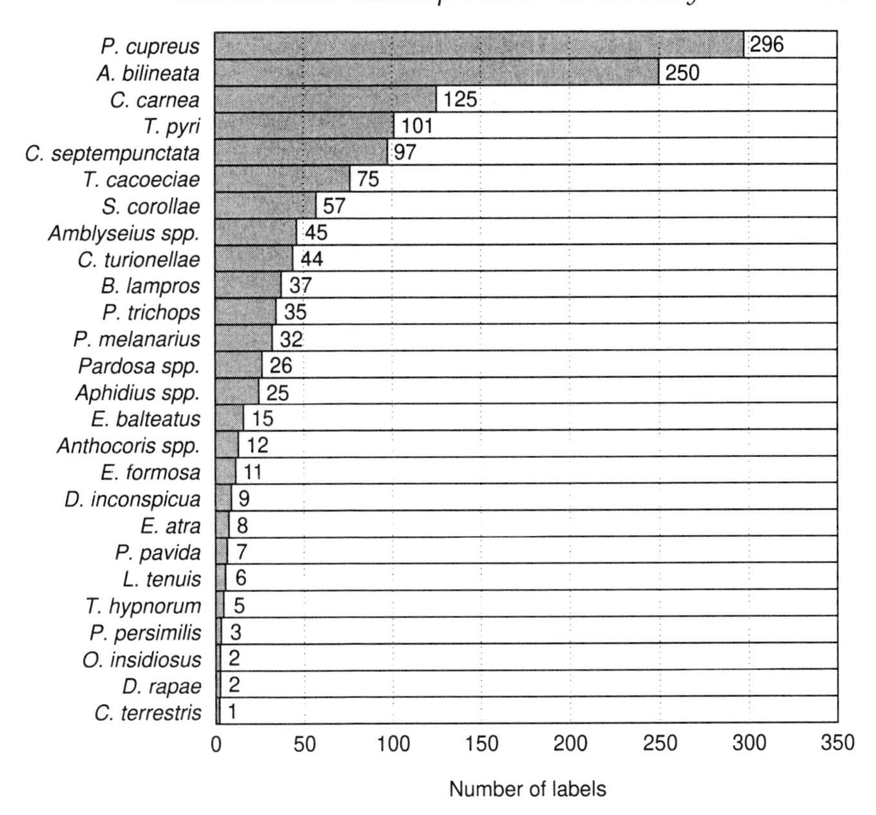

Figure 23.1 Total number of labels for each species (1326 labels, 26 species; 10 May 1996).

- Of the products analysed, 7.6% are acaricides/insecticides, 17.1% are insecticides, 32.8% are fungicides and 42.5% are herbicides.
- Harmfulness for the species tested (high risk) is indicated for 68.9% of the labels granted for acaricides/insecticides, 62.8% for insecticides, 11.6% for fungicides and 4.8% for herbicides.
- In contrast, harmlessness for the species tested (low and negligible risks) is indicated for 90.4% of the labels granted for herbicides, 69.3% for fungicides, 31.8% for insecticides and 19.7% for acaricides/insecticides.

Figure 23.3 illustrates the proportion of PPP labelled in different classes:

- not harmful (low risk and negligible risk): 57%;
- slightly harmful (medium risk) for at least one species tested: 13%;
- harmful (high risk) for at least one species: 30%.

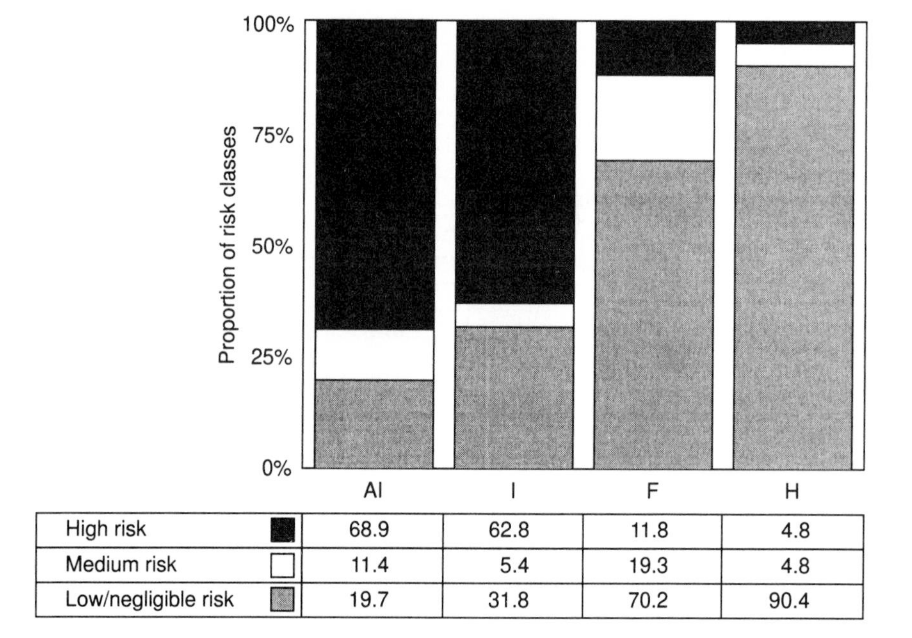

	AI	I	F	H
High risk ■	68.9	62.8	11.8	4.8
Medium risk □	11.4	5.4	19.3	4.8
Low/negligible risk ▒	19.7	31.8	70.2	90.4

Figure 23.2 Proportion of classes in each group of plant protection products (473 PSM; 10 May 1996). **AI**, acaricides/insecticides; **I**, insecticides; **F**, fungicides; **H**, herbicides.

To summarize, the strategy outlined is suggested to support IPM, both with regards to the species tested most often and in the way that labelling allows for a more selective use of PPP.

EUROPEAN LEGISLATION

Legal background

Unlike the German procedure, European legislation is subdivided into the notification of active substances, their inclusion in Annex I as the first step and the authorization of PPP as the second step, as laid down in Directive 91/414/EEC. Annex VI, the 'Uniform Principles', defines the criteria for the placing of PPP on the European market. The following principle is established in Annex VI, C. Decision-Making, 2. Specific principles, Point 2.5.2.4:

> Where there is a possibility of beneficial arthropods other than bees being exposed, no authorization shall be granted if more than 30% of the test organisms are affected in lethal or sublethal laboratory tests conducted at the maximum proposed application rate, unless

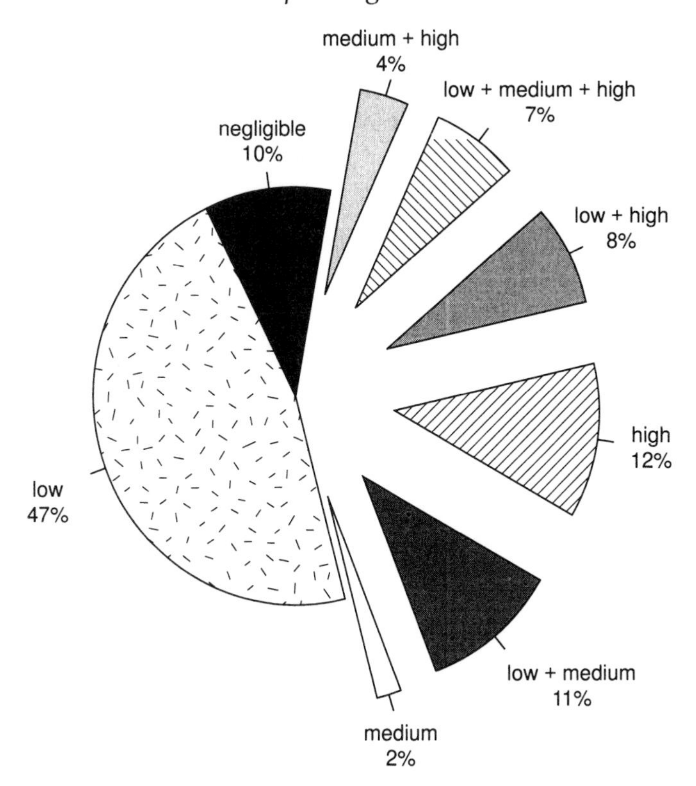

Figure 23.3 Proportion of plant protection products labelled in different risk classes (549 plant protection products; 10 May 1996).

it is clearly established through an appropriate risk assessment that under field conditions there is no unacceptable impact on those organisms after use of the PPP according to the proposed conditions of use. Any claims for selectivity and proposals for use in IPM systems shall be substantiated by appropriate data.

The strategy established is designed to:

- support protection of the environment;
- leave the promotion of IPM in concrete terms as an option.

Cut-off criteria are implemented and clear evidence is required that under field conditions there is no unacceptable impact on the organisms tested. The companies are responsible for proving that effects are not unacceptable before an authorization can be granted and a PPP is placed on the market.

Risk assessment

As laid down in Annexes II and III of Directive 91/414/EEC, information is required except where non-target arthropods are not exposed, such as:

- food storage in enclosed spaces;
- wound sealing and healing treatments;
- rodenticidal baits.

The testing of two standard species and two relevant species is recommended for the evaluation of active substances. Two more tests on relevant species are required if effects were found in previous tests, using a representative formulated product. The standard species proposed are *T. pyri* (Acarina: Phytoseiidae) and *Aphidius rhopalosiphi* (Hymenoptera: Braconidae). These species are suggested to play the role of indicator species. As further laid down in Directive 91/414/EEC, the choice of test species relevant to the intended uses, the testing methods and the principles of testing should follow the SETAC Guidance Document (Barrett *et al.*, 1994). To summarize, Directive 91/414/EEC recommends that studies are conducted using a representative formulation of the active substance on at least four species but usually not more than six (out of 13 species) using the maximum recommended field rate and maximum number of applications. A tiered system is implemented in the testing strategy using laboratory tests, semi-field tests and field tests following the decision-making scheme in EPPO (1994).

Evaluation of risk and labelling

For the evaluation of effects the classification criteria of the IOBC (Hassan, 1992) are implemented within the EPPO/CoE risk assessment scheme (EPPO, 1994). The classes are:

- negligible risk, if exposure is not likely for relevant species;
- low risk, for effects < 30% in laboratory tests and < 25% in semi-field and field tests;
- medium risk and high risk categories, left unspecified.

No labelling requirements are agreed upon for the placing of PPP on the European market with regards to risk classification.

THE DILEMMA OF PROTECTING NON-TARGET SPECIES VERSUS AUTHORIZATION OF PPP

For PPP being classified as medium risk or high risk for at least one non-target arthropod species, the final decision according to the Uniform

Principles within the evaluation of active substances may be one of the following:

- no inclusion of the active substance into Annex I; or
- the inclusion of the active substance into Annex I shall be postponed or be conditional.

For the authorization of PPP this would read:

- the authorization shall not be granted; or
- the authorization shall be conditional.

Obviously, testing and classification according to the current risk assessment strategy for IPM does not fulfil Annex VI requirements for authorization.

It cannot be overemphasized that the final decision for authorization does not aim at risk classification for IPM but at proving acceptability of the effects after use of the PPP according to the proposed conditions. Therefore more meaningful methods are needed to implement appropriate risk assessment and risk mitigation measures to make acceptability of effects provable and thus allow for the authorization of medium risk and high risk PPP. However, in order to preserve stability of agricultural ecosystems it is essential to establish a risk classification for IPM purposes as well.

The value of dose-response data

Dose-response data do facilitate a more differentiated risk assessment which is specifically helpful to evaluate risk for:

- different field rates (e.g. different intended uses, reduced rates for IPM);
- multiple applications (with regards to an accumulation of residues);
- different times of application (with regards to crop cover);
- different types of crops (that may differ in crop cover and initial deposits);
- the extrapolation of effects on species not tested;
- off-crop non-target arthropods (with regards to spray drift).

To summarize, dose-response data allow for a more efficient use of test results (e.g. risk assessment for different intended uses for the evaluation of active substances, for the mutual recognition of authorizations of PPP and for a more differentiated advice for use in IPM systems) and a specification of appropriate risk mitigation strategies in order to fulfil the requirement of acceptability of effects as laid down in Directive 91/414/EEC for authorization of PPPs.

CONCLUSIONS

What are the overall conclusions drawn from the situation described?

Risk assessment and classification of PPP according to the German procedure and Annexes II and III of Directive 91/414/EEC are generally based on similar strategies, as far as IPM is concerned. However, concerning the placing of PPP on the market the implementation of Directive 91/414/EEC will have serious implications. For at least 30% of the PPP authorized in Germany it is not established whether or not the requirements of the Uniform Principles are fulfilled (i.e. high risk PPP).

It is therefore essential to:

- elaborate a new risk assessment scheme aimed at proving that effects are not unacceptable rather than at IPM classification alone;
- include off-crop habitats into risk assessment schemes, because exposure of non-target species is likely, and protection of non-target species within the crop will often be unattainable.

Furthermore, strategies for risk mitigation must be implemented to reduce the effects on non-target species and to allow for the authorization of potentially 'high risk' PPP according to Annex VI.

Risk assessment scheme

A new risk assessment scheme is needed for the assessment of risk for both the within-field exposure (due to the direct application of PPP) and the off-field exposure (due to spray drift of PPP) of non-target terrestrial arthropods. Both exposure scenarios are most frequent (Ganzelmeier *et al.*, 1995), relevant for non-target arthropods (Jepson *et al.*, 1994; Çilgi and Jepson, 1995) and interrelated, since dispersal of non-target arthropods plays a major role for recolonization of sprayed fields (Sherrat and Jepson, 1993). Both scenarios ought to be covered by a risk assessment scheme in order to protect the environment from unacceptable effects.

The principles suggested for a new scheme are in line with the requirements of the SETAC Guidance Document (Barrett *et al.*, 1994):

- Risk labelling is implemented for both in-field risk and off-field risk (where found necessary after expert judgement).
- Risk mitigation strategies are optional for in-field risk but obligatory for off-field risk (where found necessary after expert judgement).
- Effects are not unacceptable if:
  - recovery of in-field populations is possible within reasonable time;
  - pest resurgence is not likely to occur;
  - protection of off-field populations through risk mitigation is possible, i.e. ecologically significant effects are unlikely to occur.

Further recommendations are:

- to elaborate a risk assessment scheme based on dose-response data;
- to intensify research on unacceptable effects;
- to identify habitats at risk and species at risk in off-crop habitats.

Risk mitigation strategies

In order to make sure that unacceptable effects must not be anticipated under the conditions of use, suitable risk mitigation strategies should be implemented.

It is recommended that the aims of risk mitigation should be defined and strategies should be implemented, i.e.:

- restrict the time of application;
- reduce the maximum number of applications;
- restrict the use to application techniques that reduce spray drift;
- restrict the use to strip application;
- require buffer strips to adjacent habitats.

REFERENCES

Barrett, K.L., Grandy, N., Harrison, E.G. *et al.* (1994) *Guidance Document on Regulatory Testing Procedures for Pesticides with Non Target Arthropods. From the ESCORT workshop, Wageningen, Netherlands, March 1994*, SETAC-Europe, Brussels.

Brasse, D. (1990) Einführung der obligatorischen Prüfung der Auswirkung von Pflanzenschutzmitteln auf Nutzorganismen in das Zulassungsverfahren. *Nachrichtenblatt des Deutsche Pflanzenschutzdienstes*, **42**(6), 81–89.

Çilgi, T. and Jepson, P. (1995) The risks posed by deltamethrin drift to hedgerow butterflies. *Environmental Pollution*, **87**, 1–9.

EPPO (1994) Decision-making scheme for the environmental risk assessment of plant protection products. *EPPO Bulletin*, **24**(1), 1–87.

Forster,R. (1995) Auswirkungen von Pflanzenschutzmitteln auf Nutzorganismen – Kennzeichnung in Rahmen des Zulassungsverfahrens. *Nachrichtenblatt des Deutschen Pflanzenschutzdienstes*, **47**(9), 233–236.

Ganzelmeir, H., Rautmann, D., Spangenberg, R. *et al.* (1995) Studies on the spray drift of plant protection products – results of a test program carried out throughout the Federal Republic of Germany. *Mitteilungen aus der Biologischen Bundestandt für Land- und Forstwirtschaft Berlin-Dahlem*, **305**, 111.

Hassan, S.A. (1992) Guidelines for testing the effects of pesticides on beneficial organisms: description of test methods, in *International Organization for Biological and Integrated Control of Noxious Animals and Plants/Working Group 'Pesticides and Beneficial Organisms', University of Southampton, UK, September 1991. IOBC/WPRS Bulletin*, **XV**(3), 1–3.

Jepson, P.C., Çilgi, T. and Cuthbertson, P.A. (1994) Pesticide spray drift into field boundaries and hedgerows: direct measurements of drift deposition. *Environmental Pollution* (preprint).

Sherratt, T.N. and Jepson, P.C. (1993) A metapopulation approach to modelling the long-term impact of pesticides on invertebrates. *Journal of Applied Ecology*, **30**, 696–705.

24

Labelling and risk management strategies for pesticides and terrestrial non-target arthropods: a UK proposal

Peter J. Campbell

INTRODUCTION

UK pesticide registration requirements for non-target arthropods

In the early 1990s in the UK, registration requirements for investigating the effect of pesticides on non-target arthropods were restricted to those insecticides used on summer cereals for which there was a field test guideline (Anon., 1986). However, in 1993 the Advisory Committee on Pesticides (ACP) endorsed a Position Document SC 9204 (Aldridge and Carter, 1992) prepared by the Pesticides Safety Directorate (PSD in conjunction with the Environmental Panel of the ACP, which outlined a more comprehensive regulatory testing strategy for non-target arthropods. This paper proposed a tiered testing strategy for all pesticides based on the European and Mediterranean Plant Protection Organization (EPPO) and the Council of Europe (CoE) 'Arthropod Natural Enemies' risk assessment scheme (EPPO/CoE, 1994). This testing strategy started with laboratory testing of a selection of indicator species (two to six species, depending on use), followed by semi-field and field testing where significant effects (as measured using EPPO threshold criteria) were reported in the laboratory tests. PSD subsequently used

Ecotoxicology: Pesticides and beneficial organisms.
Edited by P.T. Haskell and P. McEwen. Published in 1998 by Chapman & Hall, London. ISBN 0 412 81290 8.

this testing strategy in conjunction with the EPPO 'Arthropod Natural Enemies risk assessment scheme as a basis for assessing the risk to non-target arthropods for both pesticide new active ingredient applications and routine reviews.

EC pesticide registration requirements for non-target arthropods

In 1994 it was recognized by the 'Ecotoxicology Expert Working Group' responsible for drafting the ecotoxicology data requirements for Annexes II and III of the new EC Plant Protection Product Directive 91/414/EEC (Council Directive, 1991; Anon., 1995) that more guidance was required in order to draft non-target arthropod data requirements. Consequently, the EC helped to fund a non-target arthropod specialist workshop called ESCORT (European Standard Characteristics of beneficials Regulatory Testing). The proceedings arising from the workshop have since been published as a *Guidance Document on Regulatory Testing Procedures for Pesticides with Non Target Arthropods,* published under the auspices of the Society for Environmental Toxicology and Chemistry (SETAC) (Barrett *et al.*, 1994). The recommendations from this workshop have also been adopted into the relevant section of Annexes II and III of 91/414/EEC. This EC proposal was similar to the UK position document, in that it involved a tiered testing strategy, which was to be applied to all pesticides and was based on the EPPO/CoE (1994) 'Arthropod Natural Enemies' risk assessment scheme.

Integrated pest management (IPM) beneficial arthropod testing requirements

The ESCORT workshop recognized that there is a considerable amount of overlap between integrated pest management (IPM) and non-target arthropod effect testing. In particular, most of the recommended non-target arthropod indicator species are in fact beneficial species; also the main test methods available are International Organization for Biological and Integrated Control of Noxious Animals and Plants (IOBC) methods, specifically designed for IPM assessment. It was pointed out that IPM requires specific testing to support individual positive IPM label claims, whereas non-target effects testing is aimed at assessing and minimizing the risk to non-target arthropods in general. Therefore it was accepted at ESCORT that the regulatory requirements for both IPM testing and non-target effect testing should be kept separate. It was recognized, however, that some of the data generated for beneficial and non-target arthropods could be used to support both requirements.

Non-target arthropods labelling and risk management strategies for pesticides

Annexes IV and V of 91/414/EEC, which are to detail the standard label risk and safety phrases, have not yet been prepared. While these Annexes are being drafted, the Commission recommends that member states use 'national rules' for labelling products. In the UK, there are currently no non-target arthropod risk label phrases (apart from those specifically relating to honey bees) and the only risk management action taken in this area has been restrictions on the use of certain insecticides on cereals in the summer (Anon., 1986). This use was singled out due to the high biological activity of these insecticides, the wide scale of usage involved and summer being the season of highest insect activity. In response to initial concerns, there was a complete moratorium on the use of synthetic pyrethroid insecticides on UK cereals in summer, because of the potentially wide scale of use. However, after assessing field data on certain insecticides, the ACP agreed to a lifting of this moratorium towards the end of 1989 and instead endorsed a statutory 6 m unsprayed buffer zone around cereal crops for all synthetic pyrethroid products approved for such use (SC 8598; Aldridge, 1990). Following the routine review of dimethoate in 1992, such a buffer zone restriction was also extended to products containing dimethoate used on cereals in the summer (SC 9189; Anon., 1992). Such statutory buffer zones have been the only risk management actions taken to protect non-target arthropods in the UK and have only been applied to the use of certain insecticides on cereals in the summer.

Aim

As a result of current UK and EC pesticide regulatory requirements, a substantial amount of data are now required and are being generated on the effects of pesticides on non-target arthropods. In contrast to the currently available UK risk management options for non-target arthropods, such data requirements are not restricted to insecticides used on cereals in the summer. Consequently, both the UK non-target arthropod position document (SC 9204; Aldridge and Carter, 1992) and the ESCORT workshop identified that further work was required in order to develop appropriate labelling and risk management strategies for non-target arthropods. To this end, a meeting of non-target arthropod specialists from research establishments, industry and regulatory authorities throughout the UK was held (Third UK Forum on Non-Target Arthropods). The aim of this meeting was to help to draft a UK labelling and risk management proposal for non-target arthropods, which may in turn also help in the development of non-target arthropod label risk management guidelines for 91/414/EEC.

PROPOSAL FOR A UK NON-TARGET ARTHROPOD LABELLING AND RISK MANAGEMENT STRATEGY

The following risk management proposal is based on agreements reached at the Third UK Forum on Non-Target Arthropods, held at Chesterford Park on 29–30 March 1995, and has since been endorsed by the ACP and implemented by PSD.

General recommendations

(i) This labelling and risk management proposal is based on an ecotoxicological risk assessment for effects on non-target arthropods and is not restricted to 'beneficial' arthropods. IPM suitability assessments and associated labelling requirements or positive claims are considered to be a separate issue and not therefore incorporated into this labelling and risk management proposal.

(ii) The ESCORT Guidance Document together with its associated EC proposal for Ecotoxicological Annexes II and III data requirements for non-target arthropods will form the basis of the regulatory testing strategy on which this labelling and risk management proposal is based.

(iii) Risk labelling and risk management measures for non-target arthropods, where required, should be extended to all classes of pesticides and not only insecticides.

Implementation strategy for risk labelling and management

(i) The threshold values (> 30% effects), as recommended by the EPPO/CoE Arthropod Natural Enemies risk assessment scheme, should be used to trigger higher tier testing, until such time as species-specific trigger values are validated.

(ii) From the initial laboratory data, it should be possible to identify those pesticide products that are harmless (< 30% effects) and that do not need to be labelled.

(iii) Interim risk labelling and risk management recommendations should initially be applied to all pesticide products based on the results of the laboratory studies, whilst higher tier extended laboratory, semi-field or field data are being generated. For such higher studies (extended laboratory/semi-field/field), the 'acceptable' threshold trigger value of < 25% effects, as recommended by the EPPO/CoE Arthropod Natural Enemies risk assessment scheme, should apply.

(iv) For pesticide products attracting a risk label phrase and an appropriate risk management restriction based on initial laboratory data, additional higher tier testing is not required if the risk labelling and risk management restriction are accepted by the company. However, additional higher tier data may be submitted to support a case for removing such risk labelling and risk management restrictions from the product, or to reduce the risk management buffer zone restriction from 'statutory' to 'advisory'.

(v) Cases submitted to remove risk label and risk management phrases from product labels, which are based on field data, should not be based on demonstrating 'harmlessness' but on demonstrating an 'acceptable effect'. This is in order to distinguish between broad-spectrum and other more selective pesticides.

(vi) Key factors that should be taken into consideration when determining the acceptability of effects on non-target arthropods in field studies include:
- acreage of susceptible crop;
- specificity of effects;
- proposed time of application;
- method of application;
- number of applications;
- severity of pest/disease problem;
- intensity of effects;
- duration of effects;
- comparison of effects with soft and toxic references as well as appropriate marketed products;
- economic value of the damage caused by the pest/disease to be controlled.

Risk label proposal

(i) **For arable and tractor-mounted spray boom applications**: where data indicate a risk to non-target arthropods, expert judgement should be used to decide which of the following two categories of risk label should apply.

HIGH RISK TO NON-TARGET INSECTS OR OTHER ARTHROPODS

RISK TO NON-TARGET INSECTS OF OTHER ARTHROPODS

The 'HIGH RISK.. . .' label should only be used for perceived high-risk use situations for non-target arthropods – for example, use on summer cereals in the UK. In other European countries different crop uses may be perceived as high-risk situations for non-target arthropods.

The expert judgement used to decide which of the above risk labels should apply should take into consideration several factors, including:

- level of risk to non-target arthropods;
- crop;
- economic viability;
- application type/timing;
- acreage;
- agronomic implications.

(ii) **For all other pesticide application situations**: where data indicate a risk to non-target arthropods the following risk label should apply.

RISK TO NON-TARGET INSECTS OR OTHER ARTHROPODS

Risk management proposal

(i) For arable and tractor mounted spray boom applications, the non-target arthropod risk management restriction, which will accompany the risk label phrase, is to be a 6 m untreated buffer zone, which is based on the current UK summer cereal insecticide restriction. In other European countries with different cropping regimes, appropriate buffer zones or other risk management options may be recommended. The level of implementation of this restriction, i.e. 'advisory' or 'statutory', depends on which of the above risk label phrases is recommended.

In the case of pesticide products with highly specific effects on certain non-target arthropods, management through timing or method of application (as recommended in some IPM situations) may be considered as well as, or instead of, a buffer zone restriction.

(a) For pesticides classified as 'High risk to non-target insects or other arthropods', **Statutory Restriction** applies. Where a 'statutory' restriction applies the following risk label phrase and its associated buffer zone restriction should appear in the 'Statutory box' as well as the 'Precautions' section of the product label:

HIGH RISK TO NON-TARGET INSECTS OR OTHER ARTHROPODS. DO NOT SPRAY [crop] within 6 m of the field boundary.

This statutory 'buffer zone'' should only be used for perceived high-risk use situations for non-target arthropods – for example, use on summer cereals in the UK. In other European countries different crop uses may be perceived as high-risk situations for non-target arthropods.

(b) For pesticides classified as 'Risk to non-target insects or other arthropods', **Advisory Restriction** applies. Where an 'advisory' restriction is recommended the following risk label phrase should appear in the 'Precautions' section of the product label:

RISK TO NON-TARGET INSECTS OR OTHER ARTHROPODS. See Directions for use.

The following associated advisory buffer zone restriction phrase should then appear in the appropriate 'Directions for use'section of the label:

Avoid spraying within 6 m of the field boundary to reduce effects on non-target insects or other arthropods.

(ii) For broadcast air-assisted spray applications, such as those made in orchards, top fruit, hops, grapes and some soft fruits, it is recognized that a buffer zone to protect non-target arthropods is not a practical restriction and that IPM is more active. In such use situations, in order to reduce any potential impact of pesticides on non-target arthropods occupying field margin areas, the following risk label phrase should appear in the 'Precautions' section of the product label.

RISK TO NON-TARGET INSECTS OR OTHER ARTHROPODS. See Directions for use.

The following associated advisory phrase should then appear in the appropriate 'Directions for use' section of the label:

The best available application technique, which minimizes off-target drift, should be used to reduce effects on non-target insects or other arthropods.

(iii) For hand-held spray applications that attract risk labelling, it is considered that as this is a localized type of application, which is generally not made to large expanses of crop, it should not present a major risk to non-target arthropods. Therefore, in such cases the following risk label phrase should appear in the 'Precautions' section of the product label.

RISK TO NON-TARGET INSECTS OR OTHER ARTHROPODS. See Directions for use.

The following associated advisory phrase should then appear in the appropriate 'Directions for use' section of the label:

Minimize spray drift away from target area to reduce effects on non-target insects or other arthropods.

(iv) For solid-based products such as pellets and granules that attract risk labelling, it is considered that the main risk would be only to surface-active and soil-active polyphagous predators that may consume the seed or granule. Therefore in such cases, in order to protect non-target arthropods in field margin areas, the following risk label phrase should appear in the 'Precautions' section of the product label.

RISK TO NON-TARGET INSECTS OR OTHER ARTHROPODS. See Directions for use.

The following associated advisory phrase should then appear in the appropriate 'Directions for use' section of the label.

Avoid application within 6 m of field boundary to reduce effects on non-target insects or other arthropods.

RECOMMENDATIONS

It is recommended that a two-tiered risk labelling approach for effects on non-target arthropods be adopted, where necessary, for all pesticide products in order to inform the user. Such risk labelling should be accompanied by either an advisory or statutory risk management restriction, which protects field margins and in some cases also crop margins. The highest category of risk and the associated 'statutory' restriction should only be considered for very high-risk pesticide uses such as on summer cereals in the UK. It should be noted that although the recommended 'advisory' risk management restrictions essentially have concentrated on protecting non-target arthropods in field margin areas, the associated risk phrase ('Risk to non-target insects . . .') applies to non-target arthropods within the crop as well as within field margin areas.

ACKNOWLEDGEMENTS

I would like to acknowledge the helpful contributions of the following, who attended the 3rd UK Forum on Non-Target Arthropods, held at Chesterford Park in March 1995: Katie Barrett and Anna Waltersdorfer (AgrEvo); Kevin Brown (Ecotox); Alistair Burn (English Nature); Arnold Cooke (English Nature); David Cooper (MAFF Chief Scientists Group); Lynda Farelly (Zeneca); Tony Hardy (MAFF Central Science Laboratory); Simon Hoy and David Richardson (Pesticides Safety Directorate); Lawrence King (Rhône-Poulenc); Gavin Lewis (Corning Hazleton Europe); Wilf Powell (IARC, Rothamsted).

REFERENCES

Aldridge, C.A. (1990) Use of synthetic pyrethroids in cereals during the summer for aphid control. (SC8598, unpublished.)

Aldridge, C.A. and Carter, N. (1992) Principles of risk assessment for non-target terrestrial arthropods in agroecosystems: a UK registration perspective on current European initiatives. (SC2904, unpublished.)

Anon. (1986) *Data requirements for approval under the Control of Pesticides Regulations 1986. Working document 7/7 'Guideline to study the within-season effects of insecticides on beneficial arthropods in cereals in summer'.*

Anon. (1992) Joint report on Dimethoate and Omethoate. (SC9189, unpublished.)
Anon. (1995) Draft Annexes II and III (doc 1642/VI/95): Ecotoxicological data requirements of Directive 91/414/EEC.
Barrett, K.L., Grandy, N., Harrison, E.G. *et al.* (1994) *Guidance Document on Regulatory Testing Procedures for Pesticides with Non Target Arthropods. From the ESCORT workshop, Wageningen, Netherlands, March 1994*, SETAC-Europe, Brussels.
Council Directive (1991) Council Directive 15 July 1991 concerning the placing of plant protection products on the market (91/414/EEC). *Official Journal of the European Communities*, **L230**, 1–32 (August 1991).
EPPO/CoE (1994) Decision making scheme for the environmental risk assessment of plant protection products – arthropod natural enemies, *EPPO Bulletin*, **24**, 17–35.

25

Pesticides and beneficial arthropods: an industry perspective

Stephen W. Shires

INTRODUCTION

With few exceptions (e.g. Germany, UK), testing the effects of plant protection products on non-target terrestrial arthropods is a new regulatory requirement introduced through the recently adopted European Union registration harmonization directive (Council of the European Union, 1991). Unfortunately, basic parameters such as clearly defined target objectives, evaluation schemes and testing methodologies are not yet as well developed as in other more established disciplines such as aquatic ecotoxicology. As a consequence, there is much concern that the regulatory requirements may become out of step with the practical needs of environmental risk assessment, mitigation or management.

In particular, the regulatory system should not be regarded as a vehicle for developing integrated pest management (IPM) recommendations: individual test results cannot be applied globally, because crop, pest and agronomic conditions are highly specific to local situations. Nevertheless, it is recognized that results from regulatory testing may help to identify those products that are likely to merit further evaluation in some localities where IPM practices are being followed or developed.

Regulatory requirements for non-target terrestrial arthropods must also take into account that agricultural ecosystems are dissimilar from natural habitats, in that they are accustomed to continuous perturbation

Ecotoxicology: Pesticides and beneficial organisms.
Edited by P.T. Haskell and P. McEwen. Published in 1998 by Chapman & Hall, London. ISBN 0 412 81290 8.

due to normal cultivation practices, e.g. soil preparation, drilling, rotating monocultures, hedge cutting, harvesting, etc. Large fluctuations in arthropod populations are therefore commonplace, indicating that most are well adapted to unstable environmental influences. Indeed, no control practice (chemical or otherwise) can be applied without indirectly affecting a wide range of arthropod species, through such factors as removal of food supplies and depletion of habitats. In view of the above, it is strongly recommended that regulatory testing must add value through conserving arthropod populations and not waste resources by trying to stabilize the temporal abundance of single species.

This chapter briefly examines what the focus of regulatory testing should be for non-target terrestrial arthropods, how the testing might be approached and, more importantly, how the resultant information can be used in environmental risk assessment, management (mitigation) and product labelling.

IDENTIFYING OBJECTIVES

The first goal of any new regulatory requirement must be to define:

- what data need to be provided;
- what effects can be considered as unacceptable.

On the first issue, there has been substantial debate about the relative environmental significance of within-crop and off-crop effects. Given the unnatural and unstable characteristics of within-crop environments, most workers now recognize that the emphasis should be given to protecting the off-crop fauna. This habitat is much more stable and therefore can often be an important reservoir for arthropods to colonize the crop. Moreover, effects (direct or indirect) on within-crop arthropods are inevitable, whereas they can be largely avoided in most off-crop habitats.

As regards the acceptability of effects, again there needs to be a clear difference between within-crop and off-crop situations. For example, impacts of more than an agreed percentage reduction in different groups of arthropods could be an effective way of protecting off-crop habitats. In contrast, such small changes would often be meaningless for within-crop arthropods, where a more appropriate measure could be related to the ability of the treated area to recover by the following cropping season. This would then ensure the effective regulation of persistent products, whose use could result in the long-term deterioration of agricultural land.

TESTING FRAMEWORK

In other areas of regulatory ecotoxicology, there is a well established basic sequential testing framework that ensures a logical progression from simple laboratory acute tests, through more detailed (e.g. chronic) tests to field trials or monitoring studies. The fundamentals of such a sequential testing programme for non-target terrestrial arthropods were proposed at a meeting of experts in Wageningen (Barrett *et al.*, 1994). In these recommendations, the first stage (Tier I) involved single dose limit tests with highly susceptible species, exposed on artificial substrate. Assuming that unacceptable effects were observed in these tests, more detailed evaluations (Tier II) were proposed using extended laboratory tests on natural substrate and/or semi-field tests (as appropriate). Where there was strong evidence available that adverse effects would be inevitable on artificial substrate (e.g. with broad-spectrum insecticides), the programme provided sufficient flexibility to allow direct progression to the more realistic Tier II tests. In this way, resources could be efficiently channelled to performing meaningful tests that provided relevant information for realistic risk assessments.

INDICATOR SPECIES

Established areas of ecotoxicological regulatory testing involve only a limited number of indicator species, e.g. two birds, two fish, one earthworm, one or two aquatic invertebrates. In contrast, many more terrestrial arthropod species are listed in the new EU registration data requirements (Council of the European Union, 1996):

- Annex II – two susceptible plus two crop-relevant species;
- Annex III – two more relevant species.

It is difficult to understand the rationale for requiring such a wide range of species, particularly when early screening data on pest insects are also available for most active substances. Data will also be presented on topical and oral toxicity for honey bees, another arthropod indicator species.

In order to use resources effectively and to bring terrestrial arthropods in line with other areas of ecotoxicology, it is recommended that efforts be made to limit testing to a few sensitive indicator species carefully selected from the key functional groups: epigeal predators, predatory foliar insects, predatory mites and insect parasitoids.

TRIGGER VALUES

Unfortunately, an arbitrary trigger of 30% effects for higher tier testing or risk management/mitigation is currently adopted in the EU. This figure, which is unsupported by reliable data, is based on a value recommended in the Uniform Principles of the EU registration harmonization directive (Council of the European Union, 1994). However, if rigid triggers are to be used for regulatory risk assessment, they should be based on values that are both experimentally and biologically meaningful, i.e. those that are:

- larger than the natural variability within a test system;
- larger than the normal variation between tests;
- appropriate to the unstable nature of the agricultural ecosystem.

Trigger values must be experimentally validated, and large enough to be relevant for use in agricultural systems.

MULTI-RATE TESTING

The use of multiple dose rates in ecotoxicological tests is clearly only cost effective on a small number of indicator species. At present, therefore, it is inappropriate for terrestrial non-target arthropods, where results from many species are required at the first tier. However, it can play a valuable role at Tier II, where studies on only one or two sensitive species could help to quantify likely effects resulting from different levels of spray drift and to identify differences between medium and high risk products.

At Tier II it is essential that tests provide results that can be used to extrapolate likely effects to the field situation. Proposals for multi-rate testing should therefore incorporate the following features.

- Dose rates that reflect the range between maximum recommended field rate and minimum spray deposit at a realistic distance from the field margin.
- Use of a relevant natural surface (e.g. leaf, soil) in preference to an artificial substrate (e.g. glass, sand).
- Simulation of realistic spray drift through the application of appropriately reduced volumes, rather than lower concentrations at a fixed volume. This is important, since spray coverage will play a key role in determining effects.
- Focus on acute endpoints and not complex chronic effects (e.g. behaviour, fecundity, etc.) that are difficult to standardize and very time consuming to perform at multiple dose rates.

RISK ASSESSMENTS

Risk assessments on terrestrial non-target arthropods should take into account both agronomic and ecological factors. In particular, consideration should be given to:

- ecology of the group or species at risk (e.g. reproduction and dispersal capability, distribution frequency and abundance);

- environmental fate of the plant protection product under field conditions (e.g. persistence and movement in soil);
- method, timing and number of applications (e.g. broadcast spray, seed treatment, granule, spring or autumn treatments);
- national and local agronomic practices (e.g. field size, total treated area, quality of application equipment);
- economic value and area of the treated crop.

In addition, great care should be taken with the interpretation of Tier I tests, which are conducted under worst-case conditions of exposure due to:

- greater bioavailability from artificial substrates like glass or sand;
- higher spray deposits than with commercial application equipment;
- limited dissipation through biodegradation and/or hydrolysis.

Tier II tests are generally more realistic but can also significantly exaggerate exposures compared with the field, due to reduced dissipation processes and higher spray deposits.

Finally, it is also recommended that realistic estimates of exposure (predicted environmental conditions – PECs), based whenever possible on field data, are used in all final risk assessments.

RISK MANAGEMENT/MITIGATION

The expenditure of significant resources to evaluate the effects of plant protection products on non-target terrestrial arthropods is only justified when the information generated can be used to provide a sound risk management or mitigation recommendation. Some of the proposals of risk mitigation could include:

- prescription of untreated buffer zones around the crop perimeter;
- restriction of certain application techniques;
- reduction of application frequency and/or dose rates.

However, all risk mitigation measures should be consistent with the physical and chemical properties of the plant protection product and relevant to the ecological parameters of the target area. In addition, they need to be compatible with agricultural objectives, otherwise they will be very difficult and costly to enforce at the farmer level.

PRODUCT LABELLING

A basic scheme for evaluating effects, conducting risk assessments and labelling plant protection products for their effects on non-target terrestrial arthropods is summarized in Figure 25.1. In this scheme, three

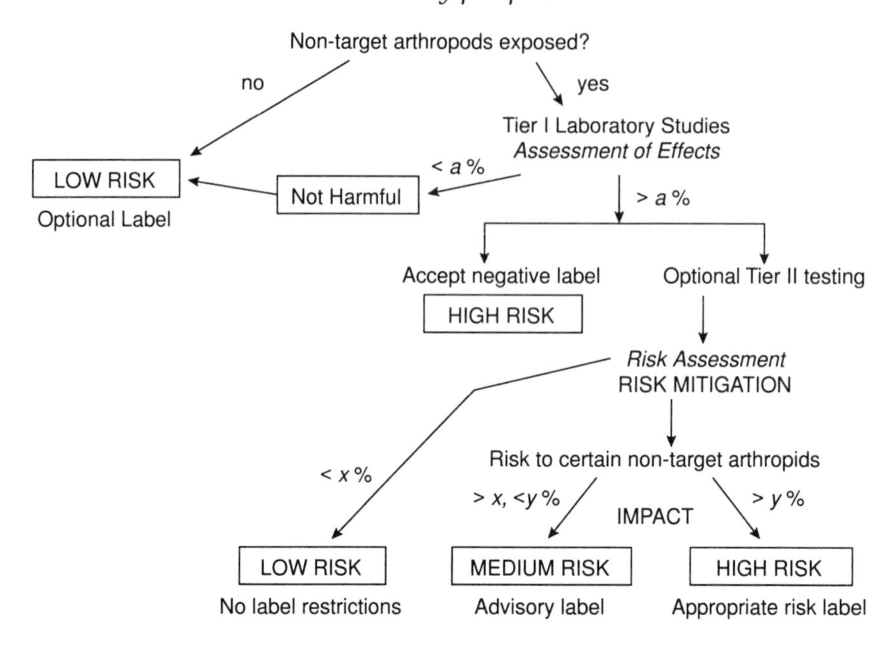

Figure 25.1 Proposed labelling strategy for non-target arthropods. The level of effects indicated by **a**, **x** and **y** needs to be developed for separate functional groups (see section on trigger values).

categories of risk are proposed, based on the level of effects observed. Each of these categories then leads to a separate label requirement appropriate to the type of action necessary to mitigate the risk.

It is also recommended that label statements refer to functional groups, rather than individual species, since the latter are largely unknown by farmers, dealers and many advisers. Furthermore, all labels should be clear and simple in order to ensure that they are both understood and adopted by farmers. Without the latter, the whole process of regulatory testing is of little value to anyone.

The details of labelling are likely to vary between national authorities, though there would be an obvious benefit for harmonization – especially at the EU member state level. Actual phrases used will also differ, but in all cases should aim to remain simple and practical. For example, products in the high risk category requiring special risk mitigation measures could be labelled as follows:

> This product could be harmful to non-target arthropods. Avoid spray drift on to adjacent untreated natural areas by not spraying in windy conditions, by using appropriate application techniques and/or by leaving an untreated buffer zone.

CONCLUSIONS AND RECOMMENDATIONS

- Distinguish between regulatory and IPM testing.
- Consider the normal, frequent perturbations in agricultural ecosystems.
- Define clear objectives in terms of within-crop or off-crop fauna.
- Maintain a logical testing framework with simple data requirements.
- Use experimentally validated and ecologically meaningful trigger values.
- Use multi-rate tests when appropriate to help to extrapolate results from Tier II studies on natural substrates to field situations.
- Incorporate agronomic and ecological factors in risk assessment.
- Focus risk management/mitigation recommendations on practical conservation measures.
- Communicate (label?) risk management requirements in a simple and concise way.

ACKNOWLEDGEMENTS

The author wishes to thank colleagues in the Beneficial Arthropods Regulatory Testing (BART) group and the Environmental Expert Group of the European Crop Protection Association (ECPA) for their valuable contributions.

REFERENCES

Barrett, K.L., Grandy, N., Harrison, E.G. *et al.* (1994) *Guidance Document on Regulatory Testing Procedures for Pesticides with Non Target Arthropods. From the ESCORT workshop, Wageningen, Netherlands, March 1994*, SETAC-Europe, Brussels.

Council of the European Union (1991) Council Directive of 15 July 1991 concerning the placing of plant protection products on the market. *Official Journal of the European Communities*, **L230**, 1–32.

Council of the European Union (1994) Uniform Principles for evaluation and authorisation of plant protection products. Council Directive 94/43/EC of 27 July 1994, establishing Annex VI to Directive 91/414/EEC concerning the placing of plant protection products on the market. *Official Journal of the European Communities*, **L227**, 33–55.

Council of the European Union (1996) Commission Directive 96/12/EC of 8 March 1996 amending Council Directive 91/414/EEC concerning the placing of plant protection products on the market. *Official Journal of the European Communities*, **L65**, 20–37.

26

Honey bees in Europe: lessons for other terrestrial non-target arthropods

Gavin B. Lewis, J.H. Stevenson and P.A. Oomen

INTRODUCTION

In terms of regulatory risk assessment, there are two main groups of non-target arthropods. One of these, the pollinating insects, has been the subject of regulatory requirements in many countries for about 20 years. The other group, represented by the arthropod natural enemies, is relatively new in regulatory terms, with only a few countries having developed statutory or *ad hoc* requirements over the last few years. However, this group has now been included as part of the standard ecotoxicological data requirements that are needed for the registration of pesticides within the European Union (EU) (Council Directive 91/414/EEC as amended by Commission Directive 96/12/EC).

The pollinating insects are mainly represented by the honey bees (*Apis mellifera*). In Europe a technical forum, the Bee Protection Group of the International Commission for Plant-Bee Relationships (ICPBR, formerly the International Commission for Bee Botany) has been meeting every 2–3 years since its first meeting at Wageningen, The Netherlands, in 1980. This group, comprising experts from government institutes, academia and industry, has been developing harmonized methodology for testing the effects of pesticides on honey bees. These methods have been incorporated into a sequential testing scheme by the European Plant Protection Organization/Council of Europe (EPPO/CoE) honey bee

Ecotoxicology: Pesticides and beneficial organisms.
Edited by P.T. Haskell and P. McEwen. Published in 1998 by Chapman & Hall, London. ISBN 0 412 81290 8.

subgroup and now form the basis of the regulatory requirements of the European Commission. However, the development process is ongoing and the ICPBR Bee Protection Group met in Branschweig, German (September, 1996) in order to continue evaluation of the existing scheme and to consider new developments.

Given the recent regulatory developments within the EU, there is now much debate concerning how the data requirements for arthropod natural enemies should be met in terms of the scientific methodology and the overall risk assessment. The aim of this chapter is to compare the risk assessment and management of the pollinating insects with the other regulatory group of terrestrial non-target arthropods: the natural enemies. It will consider whether there are lessons that can be learnt from experiences with the honey bee risk assessment scheme which can be used in the development of the scheme for this new group.

SPECIES SELECTION

For the purposes of pesticide risk assessment, it has been considered that the pollinating insects can be satisfactorily represented by a single indicator species, the honey bee. More recently this assumption has been questioned – with particular reference to bumble bees, which have become more closely associated with pesticide exposure as a result of their use for commercial pollination of glasshouse crops. A subgroup has been set up within the Bee Protection Group to develop methodology to assess the effects of pesticides on bumble bees and to make a comparison of the toxicity values generated for a range of compounds with the corresponding values for honey bees. This is to verify the hypothesis that bumble bees are equally or less sensitive to pesticides than honey bees and that honey bee risk assessment consequently would include the necessary assessment of risk for bumble bees as well. In this way, the degree of extrapolation between species can be assessed with confidence and the need for additional regulatory data more clearly defined.

The arthropod natural enemies are a more diverse group, both taxonomically and functionally, compared with the pollinating insects. For this reason it has been considered necessary from the outset to have more than one indicator species, though it has been difficult to reach a consensus on the number and identity. At the SETAC-ESCORT workshop (SETAC-Europe, 1994), 13 species were identified for regulatory purposes, based on various taxonomic/functional and crop groupings. It would perhaps be useful if more information were generated similar to that for the honey and bumble bees. It would allow a more informed selection of the appropriate indicator species within and between different groups of non-target arthropods, which could result in a reduction in the overall number of test methods that need to be developed and validated as well as a reduction in the data requirements for each pesticide as a result of the need for fewer species to be tested. Another advantage of working with fewer species overall is that a comparative database can be rapidly built up – which is very useful in the assessment process, as has been found with the honey bee.

HONEY BEE RISK ASSESSMENT SCHEME

Laboratory testing

The sequential testing scheme that has been developed for the honey bees is risk based, taking into account both toxicity and exposure as appropriate throughout. It starts with a simple consideration of exposure: where this is clearly very unlikely (e.g. for products used in food storage or as seed treatments), then no further work is necessary. In all other cases, an initial assessment of toxicity is made using a standard dose-response test for both the contact and the oral routes of administration. These represent the two main routes of exposure under field conditions: direct overspray of bees foraging on a crop at the time of application, or uptake of contaminated nectar and pollen from the treated plants.

The toxicity of pesticides to honey bees used to be seen as a satisfactory way of indicating their hazard, or effects, in the field. However, with the advent of new classes of pesticides, such as the pyrethroids, it was realized that this approach had serious limitations (Smart and Stevenson, 1982). While the toxicity of pyrethroid insecticides is similar to other compounds known to present a serious hazard to honey bees, their typical application rates are much lower. This is manifested in the ratio of the application rate (g a.i./ha) and the LD_{50} (μg a.i./bee), which gives the number of LD_{50} doses per hectare $\times 10^{-6}$. A comparison of values for a range of compounds showed that those for pyrethroids were intermediate between those for insecticides known to be hazardous to bees (high toxicity and high application rates) and those considered to present a low risk (low toxicity and high application rates). In these circumstances, laboratory toxicity data will not be a sufficient indicator of field hazard as other factors will have an influence, including the application rate as well as the behaviour of the chemical and of the honey bees themselves.

This approach was formally adopted in the risk assessment scheme proposed by the ICPBR Bee Protection Group (Felton *et al.*, 1986) and subsequently incorporated into the EPPO/CoE (1993) scheme. The hazard ratio or quotient has become a figure without any specific units and is assessed on a comparative basis using pre-set threshold values. These values were proposed on the basis of a comparison of the hazard ratios of pesticides in use in The Netherlands with the existing bee hazard classification, which was based on extensive practical experience (Oomen, 1986). In addition, it has now been recognized that an initial dose-limit test can be introduced before the full dose-response test, so that if the LD_{50} is shown to be greater than a maximum value, no further testing is required.

Realistic worst-case exposure under field conditions for honey bees is thus seen as a relatively simple process of direct overspray (i.e. bees foraging on a crop as it is being sprayed), and so can be given directly by the maximum application rate. In the case of the arthropod natural enemies, field exposure is much more complex. Direct exposure is possible but will be a relatively less important route for populations living within the crop canopy, in which case a more important route will be indirect or residual contact as a result of their activity after the initial application. The degree of exposure will thus be affected by a number of factors – most immediately by the position of the arthropod species within the crop, which will determine the degree of spray interception that has occurred. Thus, species predominantly active on the ground will experience lower levels of residues compared with those living on the foliage. In addition, a proportion of the locally available population for a given species may be off-crop (i.e. outside the immediate deposition zone) and exposure here will be as a result of spray drift, which will involve much lower residue levels.

Currently, arthropod natural enemy testing is conducted at single rates according to standard International Organization for Biological Control of Noxious Animals and Plants (IOBC) methodology. The pesticide is applied at the maximum recommended application rate and the test organisms are exposed to the fresh, dried residues, within about an hour of application. This approach thus adopts the residual contact route of exposure but otherwise is very much a worst-case scenario. The exposure is very severe as the test conditions involve application to glass plates or similar inert substrates, as appropriate, which are unstructured and thus maximize contact. The interpretation of the tests is inflexible as each one is conducted at a single rate and so cannot be easily related to any other conditions, e.g. for different rates, different positions either within or off the crop, etc. An exposure factor has been incorporated into the testing with application rates reduced by a factor dependent on the conditions appropriate to the proposed application conditions – for example, × 0.4 for species living on the ground under the target trees in orchards, but × 1 for all arable species. This approach is a very basic attempt to accommodate different field exposure levels and is still limited to a test being required for each proposed application scenario.

In the case of the arthropod natural enemies, then, exposure is incorporated into the test, with application at the maximum recommended rate or at a proportionately reduced rate, as appropriate. This is in contrast to the honey bees, where there is an initial absolute assessment of toxicity with the dose-response test, incorporation of exposure being carried out in the subsequent assessment using the hazard ratio. This has the advantage that with the result from one test a much more flexible risk assessment is possible at this early laboratory stage in the testing scheme. Different exposure scenarios can be compared with the toxicity value produced, as with the hazard ratio, in order to see what effect this may have on the risk assessment. For honey bees, only limited use of this potential flexibility is appropriate: relevant changes in exposure only relate to the application rate. For arthropod natural enemies, other factors can also be taken into account. For example, the development of mathematical models allows exposure estimates to be made over time or as a result of multiple applications. Different levels of exposure within the crop can also be incorporated into the assessment, with reduction factors as currently used in the testing. Off-crop exposure can also be considered, with drift factors being used to estimate exposure (e.g. using the values of Ganzelmeier, 1993).

The use of the hazard ratio in the honey bee risk assessment scheme allows a very flexible risk assessment using the minimal amount of data generated at the laboratory stage. This not only reduces the amount of testing necessary at this Tier I level but also minimizes the amount of higher tier testing subsequently triggered. The incorporation of a limit test reduces unnecessary testing for compounds with no anticipated toxic effects. While the situation is more complex for arthropod natural enemies, this approach would have similar benefits. An important element in this would be the setting of appropriate threshold values. Given the complexity of field exposure and the ecology of the systems involved, this could be based on comparison with compounds of known field hazard, as for honey bees.

Additional laboratory testing

In the honey bee testing scheme, additional laboratory testing beyond the initial acute toxicity test may be considered necessary. One test that is optional is for residual toxicity, in which the bees are exposed to pesticide residues that have experienced various periods of ageing. Where a compound has been found to be toxic in the initial testing, this provides additional information that can be useful in assessing the likely risk under field conditions, i.e. by incorporating the duration of the toxicity. This allows a number of specific management options, such as the possibility of an evening application where the duration is very short or, where it is longer, determining the re-entry period for honey bee colonies into treated crops. These options are possible because of the control that can be exercised over bee colonies, by shutting up hives or moving them into and out of specific areas. Clearly this control is not possible with arthropod natural enemies and so information on the duration of toxicity will be of less value. An exception to this is in the case of biological control agents, where natural enemies are introduced into the crop, and here persistence information would be of value.

Another honey bee test that is an obligatory requirement under the appropriate conditions is the bee brood test. Where a compound is a known insect growth regulator (IGR), then a test must be conducted to assess its effects on bee brood. This is of particular concern to honey bees as the brood, located within the hive, is not normally directly exposed to pesticides and the primary risk occurs through adult exposure. However, the adults may not be a sufficient indicator of risk to the brood in the case of IGR compounds, which may well not have an effect on them but can be transferred into the hive on food and thus expose the young bees and affect their development. This approach is straightforward because of the clear separation of the bee brood from the area of application. It is not so simple in the case of arthropod natural enemies, as the adult and juvenile stages for many species are found in the same area of exposure. This is reflected in the test guidelines, which require that the tests are conducted (initially at least) on the most vulnerable life stage, taking into account susceptibility and exposure. On this basis, it may be important that the IGR status of the test compound is taken into account in selecting the life stage to be tested.

Higher tier testing

Where the hazard ratio threshold is exceeded, honey bee testing follows a simple path. Semi-field testing may be conducted, in which free-flying colonies of bees are confined to plots of crop in mesh-covered cages. Where no effects are seen, a low risk category can be assigned without further testing. If significant effects are seen in the semi-field test (or directly from the hazard ratio assessment), field testing is necessary to avoid higher risk classification. In this test the bees are exposed under normal agricultural conditions. These tests incorporate the more realistic conditions of field exposure but, as a consequence, the information derived is more complex and difficult to interpret than the simple laboratory data. Risk assessment involving these higher tier results has thus emphasized 'expert judgement', with no simple thresholds being appropriate. One tool that has been found to be useful in this process has been the inclusion of reference products for comparison – either ones that are known to be hazardous to bees, such as dimethoate or parathion, or ones that have a low risk (small but acceptable effects), such as phosalone. The hazardous compounds can also be included as toxic standards, to demonstrate that exposure has occurred.

The arthropod natural enemy testing scheme (EPPO, 1994) follows a similar path to that for the honey bees. Beyond the laboratory tests there are assorted semi-field tests, depending on the particular species concerned, which are conducted under field conditions but with appropriate containment of the test animals. At the final stage are field tests, conducted using the crop type of concern and incorporating all the species present in the natural system. The results from these trials are even more complex than those for the bee field trials, incorporating several species, each with their own population patterns depending on life history strategies, local population dynamics, trial conditions and, possibly, the treatments themselves. This makes interpretation very difficult, particularly given our current limited knowledge of the ecology of these systems. Accordingly, the approach that has been used with honey bees might also be helpful in this case, with the use of reference products of known risk in the appropriate systems against which the effects of the test compound can be compared.

IN-USE MONITORING

The registration of a pesticide with regard to honey bees, involving appropriate recommendations for use, is based on the implementation of the risk assessment scheme as described here. This involves a finite amount of test data, as required, and a certain amount of judgement. The outcome of the scheme can be validated by consideration of in-use monitoring that gathers data relating to incidents involving pesticides and honey bees. Such a scheme, the Wildlife Incident Monitoring Scheme, has been operating in the UK for over 15 years in its current form. In the case of honey bees, this has involved the investigation of incidents reported by beekeepers or others, culminating (where appropriate) in the identification of the pesticide to which it can be attributed. This has resulted in a database which has allowed validation of the risk assessment scheme by applying it to approved products and comparing the outcome with identified incidents over a period form 1983 to 1990 (Aldridge and Hart, 1993). It shows that all of the identified incidents relate to crop-product combinations for which the hazard ratio is either in the high risk category (> 2500) or at the upper end of the intermediate category (50–2500). This indicates that the risk assessment scheme successfully identifies pesticides as low risk in that they have not been implicated in any poisoning incidents. It further suggests that the hazard ratio threshold above which further testing is required could be increased while still maintaining a satisfactory safety margin.

Monitoring data can thus play a very useful role in validating risk assessment schemes and possibly also improving their efficiency by reducing the amount of unnecessary testing. A similar scheme to the

UK one currently operates in Germany and at the recent ICPBR Bee Protection Group meeting in Braunschweig it was recommended that such schemes should be implemented throughout Europe, if possible. However, such schemes are clearly easier to implement for honey bees than they would be for other beneficial arthropods. Honey bees have easily located populations which are housed in hives and also have a network of expert observers, the beekeepers, who readily report any observed incidents. These conditions do not apply to other non-target arthropods and so any incident scheme would have to be more proactive in order to obtain sufficient information. The value of any such programme would have to be clearly measured against the effort that would be involved.

CONCLUSIONS

The ICPBR Bee Protection Group has brought together experts from government, academia and industry to discuss methodology for assessing the effects of pesticides on honey bees since 1980, resulting in a harmonized risk assessment scheme for pollinating insects. This has set the example in environmental risk assessment and management and is now used for pesticide registration in the EU. A number of factors have contributed to this successful development. The group is represented by a single 'farmed' species, the honey bee, which has a relatively long regulatory history so that an extensive database of test results and in-use monitoring information has been accumulated. Consequently, the honey bee risk assessment scheme is the first and only environmental scheme that has been validated with the experiences of practical use.

The other regulatory group of non-target arthropods, represented by the natural enemies is, in contrast, a recent addition. It is also a more complex group in terms of its taxonomy and their ecology. Taking this into account, useful lessons are identified from a consideration of the development and implementation of the honey bee risk assessment scheme which might be used in the development of the comparable scheme for this new group. These include approaches for the selection of species for testing, the advantages of multiple rate testing compared with single rate tests, aspects of higher tier (semi-field and field) methodology and the validation and refinement of risk assessment schemes with in-use monitoring data.

REFERENCES

Aldridge, C.A. and Hart, A.D.M. (1993) Validation of the EPPO/CoE risk assessment scheme for honey bees, in *Proceedings, 5th International Symposium on the Hazards of Pesticides to Bees, 26–28 October 1993*, Appendix 5, pp. 37–42.

Barrett, K.L., Grandy, N., Harrison, E.G. *et al.* (1994) *Guidance Document on Regulatory Testing Procedures for Pesticides with Non Target Arthropods. From the ESCORT workshop, Wageningen, Netherlands, March 1994*, SETAC-Europe, Brussels.

EPPO (1993) Decision-making schemes for the environmental risk assessment of plant protection products – honeybees. *Bulletin OEPP/EPPO*, **23**, 151–165.

EPPO (1994) Decision-making schemes for the environmental risk assessment of plant protection products – arthropod natural enemies. *Bulletin OEPP/EPPO*, **24**, 17–35.

Felton, J.C., Oomen, P.A. and Stevenson, J.H. (1986) Toxicity and hazard of pesticides to honeybees: harmonization of test methods. *Bee World*, **67**, 114–124.

Ganzelmeier, H. (1993) Drift of plant protection products in field crops, vineyards, orchards and hops, in *Proceedings of 2nd International Symposium on Pesticides Application Techniques, Strasbourg, France, 22–24 September 1993*, **1**, 125–132.

Oomen, P.A. (1986) A sequential scheme for evaluating the hazard of pesticides to bees, *Apis mellifera. Mededelingen Faculteit Landbouwwetenschappen Rijksuniversiteit Gent*, **51**, 1205–1213.

SETAC-Europe (1994) Guidance document on regulatory testing procedures for pesticides and non-target arthropods. (Eds K. Barrett, N. Grandy, L. Harrison, S. Hassan and P. Oomen.)

Smart, L.E. and Stevenson, J.H. (1982) Laboratory estimation of toxicity of pyrethroid insecticides to honey bees: relevance to hazard in the field. *Bee World*, **63**, 150–152.

27

Risk assessment and management: is it working? An NGO perspective

Peter Beaumont

NGOS: WHAT ARE THEY?

NGOs are non-government organizations. This in theory includes the associations of agrochemical producers such as the British Agrochemicals Association (BAA) and the European Crop Protection Association (ECPA). For the purposes of this chapter the acronym NGOs refers to not-for-profit organizations that have no connections with either government or industry.

It is difficult to point to a single constituency of identifiable NGOs or a typical NGO view. NGOs are no longer primarily concerned just with 'banning' particular pesticides – except perhaps lindane, methyl bromide, or organophosphorous sheep-dips. Replacing one chemical with another is not necessarily safer or better.

NGOs are sometimes categorized as 'environmentalists' but that is not often the case. Many groups are involved in pesticides issues but not from a point of view that is primarily environmental: they may be consumers or trade unionists or even farmers and users.

The common ground shared between most of the NGOs working on pesticide issues is this: all are concerned with the externalities of pesticide use. These are the non-market impacts of pesticides on health and safety, environmental media, impacts on non-target and beneficial species and overall sustainability. Most of us would rather pay for

Ecotoxicology: Pesticides and beneficial organisms.
Edited by P.T. Haskell and P. McEwen. Published in 1998 by Chapman & Hall, London. ISBN 0 412 81290 8.

farmers to farm using less pesticides than pay for pesticides to be removed from water.

CURRENT POLICY DEVELOPMENTS IN RISK MANAGEMENT AND ASSESSMENT

Developing countries

In 1994 global pesticide sales reached US$25 000 billion, with increasing sales in Asia, Latin America and to a lesser extent in Africa. Developing countries may account for 35% of all pesticides sales by 2000, compared with only 19% in 1988. Yet many of the products sold in developing countries are older, cheaper pesticides which are more persistent in the environment or more hazardous to the health of the user. In areas of regular pesticide use the environment is deteriorating, resulting in contaminated water supplies, loss of plant diversity, increased insect resistance and cattle deaths. A survey for the 10th anniversary of FAO International Code of Conduct on the Distribution and Use of Pesticides (FAO, 1995) concluded that, although there has been an increase in government legislation and regulatory powers relating to pesticides in developing countries: 'Health hazards remain a major preoccupation and improvement on this point generally appeared to be limited,' and: 'The effect of pesticides on the environment . . . was substantially worse than in 1986, which may reflect the increased awareness of the importance of the subject.'

The increased use of pesticides by those least able to manage and assess risks is fuelled by international policies. Structural adjustment has encouraged intensive agricultural production. New rules under the World Trade Organization (WTO) also encourage agricultural exports, and the pesticide industry will take advantage of these trends to expand its markets. The argument is generally put that an increase in world population needs more intensive production to feed the new mouths. The opposite view is that it is poverty and lack of access to food that causes hunger.

In contrast to the pressures to increase use of pesticides, there are areas where progress is being made in the management and assessment of risk. New developments in the field of integrated pest management (IPM) offer the hope that realistic alternative solutions to pest control problems can be found. After some years of debate about the possibilities and practicalities of IPM and a series of Working Group meetings, the new Global IPM Facility is being established after careful consideration by a task force of donors, agencies and practitioners and will focus on the issue of implementation. It will concentrate on the approaches of farmer participation and policy support that have been successful in parts of Africa and Asia.

Another promising avenue has been the impetus that arose from the 1992 Earth Summit in Rio, when governments and agencies pledged themselves to improve the management of toxic chemicals. A Prior Informed Consent convention is being drafted, and the International Organization for the Management of Chemicals is promoting a programme to strengthen the chemical management of developing countries – which will help to link the procurement of safer pesticides with better chemical management capacity.

Europe

Increasingly there is a recognition of common ground between agriculture and the environment. In Europe this is acknowledged in the Authorizations Directive 91/414/EEC (temporarily annulled) which refers in its preamble to the need for pesticide use within the context of IPM. The process of rapprochement between the Agriculture and Environment Directorates of the European Commission that has led to the Sustainable Use of Plant Protection Products programme, which may resolve some of the perceived contradictions between high input farming and the environmental and other costs that result. Until that happens, policy is pulling farmers in different directions at once.

It is clearly in everyone's interest that registration should be harmonized – provided that this is at the highest level of health and environmental protection, not at the lowest. But that process is only likely to apply initially to the newest active ingredients that achieve Europe-wide registration, and it may take between 12 and 20 years or more for the other 600 or so active ingredients to be similarly reviewed. Whilst maize or sugar beet herbicides and other products with a large market share may go through the system, what about minor and not-so-minor uses? Will fruit and vegetable growers be using orphan products? Will local authorities be able to use safer products and will domestic home and garden products still be on sale?

There is also recognition of an overuse of pesticides in Europe. Some countries have introduced pesticide reduction policies – which have often been wrongly characterized as simple percentage reductions of the weight of active ingredient used. (This is an issue to which further reference is made later.) It seems unlikely that much progress can be made in reducing environmental and non-target risks until progress is made with Common Agricultural Policy (CAP) reform. Only some 3% of the CAP budget is currently devoted to what might be termed environmentally friendly agriculture, and little will change until that figure is increased considerably and farmers can be paid for stewardship of resources instead of volume of production.

UK pesticide minimization

In the UK the concept of pesticide minimization holds sway. This says that the amount of pesticide used should be the minimum necessary to protect crops, health and the environment. The aims of pesticide minimization are perfectly acceptable but what a minimization policy seems to lack is any clear long- or short-term strategy and goals. What are the means to achieve minimization, and how shall we know when we have achieved it? Pesticide reduction policies, as explained below, seem clearer about where they are going and why. Perhaps the label of minimization or reduction does not matter as much as the goals and strategies that enable a policy, whatever it is called, to be implemented and monitored.

In order to try to flesh out pesticide minimization, the Ministry of Agriculture, Fisheries and Food (MAFF) and Department of Environment (DoE) have joined together at the behest of ministers to create the Pesticides Forum (MAFF, 1996a), the task of which is to square the circle by suggesting an Action Plan by which minimization can be progressed.

PRACTICAL PROBLEMS OF RISK ASSESSMENT IN THE FIELD

Pesticides are, of course, perfectly safe if used according to label directions. Another view is expressed by Waage (1996):

> In terms of insect pests, it is difficult to think of a serious pest problem in the developing world that has not been created by pesticide use. This is true of all the major crops: rice, cotton, cereals and vegetables. The second major cause has been the movement of pests around the world through the trade in crops which introduce pests to areas where no natural enemies are able to keep the balance.

The situation in the developed world may be more complicated. How can risk assessment and management take account of beneficial organisms? Farmers – who cannot be experts in 600 active ingredients and several thousand formulations – have little independent advice to enable them to choose safer chemicals.

We are encouraged to think that the trend in pesticide usage is to newer active ingredients, effective at lower dose rates and with specific modes of action that are better targeted and less broad spectrum. However, there are countervailing trends. There are increasing pressures on the market for active ingredients and formulations. There is concern that as products reach the end of their patent protection they will return to the market as generics. Some of these are older, more toxic and more broad spectrum in their field of activity. There are also concerns about the production processes. Production in some parts of the world may not be to an acceptable standard of quality: impurities can be dangerous not only to producers but also to users.

All will be well if farmers can be properly trained and advised. The process of user certification has been one of the best features of UK law. The training in the use of pesticides that the law requires focuses on the control of pesticides, not on the control of pests. The message is clearly that pesticides should be used safely – which is right – but little attention is paid to other means of control, non-target effects or the role of beneficials. The Control of Substances Hazardous to Health regulations which prescribe the decision-making process for pesticide use are aimed at human health, not environmental health.

The role of advisers is changing. In the UK, Agricultural Development Advisory Service (ADAS) is shrinking and has been privatized. Farmers are turning increasingly to crop consultants, some of whom are tied to distributors. Agrochemical companies now sell services as much as products. Nevertheless, increasingly independent crop consultants are making decisions over pesticide use – it is rumoured that some seven people in Berkshire decide on pesticide use for 70% of the county. Commentators have observed that a communications network which is rapidly becoming dominated by the agricultural supply industry is unlikely to be in the longer-term interests of the farming community as a whole (Fearne, 1991).

Many farmers now say that as they pay for their advice, they have no incentive to share best practice with their neighbour. There is no comparative advantage in good environmental management – it is not a saleable commodity. Whilst independent advice is clearly desirable, neither farmers nor advisers are going to take risks using non-chemical control.

It is also the case that label recommendations specify full rates. Manufacturers have little choice, as the legal obligations of product liability mean that they have to guarantee that the product, if used according to label directions, works. Although there is an increasing body of research to demonstrate the efficacy of pesticides at lower-than-label rates, this requires a greater knowledge and awareness on the part of the user, a greater time input and a greater willingness to take risks. Half-rates and lower are common, but make more demands on the farmer.

All pesticides have to be approved and used according to the conditions of approval. There has already been discussion of how the impact of pesticides on beneficial and non-target organisms can be improved. At present the effects of pesticides are largely conveyed through the label. There is concern that the current requirements for label information are so overwhelming as to dilute the effect of the label. Clarity has been sacrificed for complexity. Although there are a considerable number of risk phrases, they are not prioritized, and some are so general as to be of little help. Other than (perhaps) checking application rates, does anyone know if labels are read and whether they change behaviour?

An ADAS survey was commissioned in 1995 to investigate the level of awareness of farmers and growers about integrated techniques (ADAS, 1996). When selecting a method of pest control, growers rated the most important factors as (in descending order) operator safety, accuracy of application, effect on produce quality, maximizing yields and then minimizing harmful effects on the environment. When selecting a particular pesticide, operator and public safety and pesticide residues were considered the most important aspects. Reduction in spray drift and contamination of ditches and streams were also important, followed by environmental considerations such as persistence in soils and effects on non-target organisms. The survey noted that the rankings were possibly influenced by the availability – or not – of information on these issues at the point of use.

There are proposals from the European Commission (EC) to harmonize the classification, packaging and labelling of chemical active ingredients and formulations – including pesticides. The proposal is to provide hazard warning, instead of risk warning. A hazard warning covers the possible effect of a chemical as opposed to the risk that the effect might occur. Current proposals include hazard warning where there is a hazard to the environment.

What about advice not on the label? The advice on good farming practice contained in a number of MAFF Codes of Practice is not reaching farmers. According to a recent survey (MAFF, 1996b), under half of all the farmers surveyed were aware of any of the Codes. Other findings were that only one farmer in five owned a Code. Farmers who were not conforming with good practice did not feel they needed advice. However, as increasing deregulation puts more and more of the onus for good practice on farmers, it is clear that the voluntary message is not yet getting across.

WHAT CAN BE DONE TO IMPROVE RISK ASSESSMENT AND MANAGEMENT?

There is pressure to reduce the use of pesticides. Current agrochemical practice is fighting a battle that it is not winning, and may be losing, against resistant pest populations. A narrow genetic basis for crop varieties leaves many vulnerable to pests and disease. Rates of resistance are increasing, and rates of pest and disease attack are not diminishing. Some useful answers have been provided in the fields of pesticide reduction and IPM.

Pesticide reduction policies

Pesticide reduction does not mean using less weight of pesticide: the use of newer lower-dose active ingredients, which are correspondingly higher in biological activity, achieve this in any event. Use reduction means lower than label doses, fewer applications and less waste or overuse. This trend is already actively developing in UK agriculture. Scotland has reduced its use of pesticides in arable agriculture without a pesticide reduction policy and by using developments in technology that relate to products and applications.

Reduction of use is only one element in pesticide reduction policies. Reduction of risk is a second element. But then, what is risk? The perception of risk is a difficult area and it is not simply between the public and 'scientists' that there is a difference of risk perception. There is frequently a lack of consensus among scientists – which in turn reflects the difficulty of some of the issues that are considered. Examples of such differences are those relating to the appropriate pesticide parameter in the current Drinking Water Directive, and the question of the criteria by which an animal carcinogen may be judged safe to humans. There is a clear need for an informed debate on 'acceptable' levels of risk.

The third element of reduction is qualitative – reduction of the reliance of agriculture (and horticulture and other sectors) on pesticides. Many crop varieties are vulnerable to particular pests, and depend on one or two active ingredients to achieve control. Non-chemical alternatives are urgently required. This will become a problem of increasing importance as industry concentrates on the largest and most profitable markets, leaving smaller or minor markets without support as off-label uses. Reduction of dependence is unlikely to be achieved without re-focusing CAP support by introducing cross-compliance measures and whole farm management.

Another interesting element of reduction policies has been the discussion of policy with all the stakeholders. The Dutch experience was that use reduction should be preceded by a sector-by-sector evaluation of the practicalities and possibilities. Percentage reduction targets may be practical but will differ in each sector; but percentage reduction targets are not unknown in the UK. Recent MAFF advice has been to reduce the use of isoproturon by 40%; and it is intended that discharges of the Red List pesticides should be reduced by 50% of 1985 levels (a target that has not been met).

The results of research over some years – particularly in the Less Intensive Farming and the Environment/Integrated Farming Systems (LIFE/IFS) in Europe – now indicates that lower input regimes in many situations and rotations achieve comparable or slightly lower yields to conventional higher input regimes; but lower input costs protect margins. The LIFE work and other European comparisons demonstrate consistent results in validated field trials. Lower inputs are achieved across the board for herbicides, insecticides and fungicides.

IPM recommendations

How can ideas such as pesticide reduction and IPM contribute to a better assessment and management of risk? Increasingly both lead to the conclusion that reducing the risks from pesticides is more than just using the right products according to label directions. The use of pesticides as a technical solution to a pest problem is seen not to be working. An international IPM meeting in the Philippines produced the 'IPM Manifesto' (FAO, 1993). This stressed the need to:

- recognize and train farmers as experts in IPM;
- make research and training participatory;
- adopt IPM as a national policy;
- eliminate pesticide subsidies;
- establish the real costs of pesticide use;
- eliminate the use of pesticides in the most acutely toxic World Health Organization (WHO) hazard classes and chlorinated hydrocarbon pesticides;
- promote environmentally friendly non-chemical methods of pest management.

IPM now recognizes that two other elements are crucial: the full participation of farmers and growers in research, training, and practice; and a strong element of government support. For a successful long-term strategy governments must put IPM practices within the reach of farmers through farmer networks, extension services and research institutions. This seems to be the best way to reduce pesticide risks.

CONCLUSIONS

An emphasis on pests, not pesticides

It would also be helpful to re-focus policy, not on controlling pesticides but on pest control and pest management. This would require a comprehensive and strategic view of pest control in agriculture and public health – perhaps an IPM Advisory Committee, except that it should include pesticide users (unlike many advisory committees).

Goals and targets

Goals are now recognized as necessary in all sectors of economic life, whether in education, health or agriculture, in order to measure progress and make comparisons. It is important that goals be appropriate. Too much attention has been paid to percentage use reduction targets for pesticides. Others that might be suggested are:

- farmers farming using IPM techniques by 2000;
- improving independent extension advice provision and ratios of on-farm advisers to farmers;
- 50% of application machinery tested by 2000;
- LIFE/IFS demonstration farms established by 2000;
- a simplified and rationalized environmental and conservation grant structure leading to 50% of farms participating on the basis of whole farm management plans.

Without a more comprehensive approach to strategies and goals or targets, minimization will remain a minimal policy.

Ecotoxicologists must become environmentalists

We cannot presume that everyone agrees on the importance of 'the environment'. It stands a good chance of being important in times of economic well-being, but in times of recession it sinks to the bottom of the stream of consciousness and adsorbs to the sediments of public and political concern – and that of funders.

An example of this is provided by recent changes in the IPM policy of the World Bank. Many of the larger and more influential donor agencies and governments themselves now have policies that are intended to promote the safer management of pesticides. The World Bank was one of the first; and the European Union (EU) is itself developing an approach to safer use and promoting IPM. However, there is a cautionary note to be sounded. The World Bank's policy was never fully implemented, and it has just been replaced (World Bank, 1996) with a much weaker version that fails to accord the same priority to health and the environment. Are we seeing environmental standards sacrificed to freer trade?

If that is the bad news, there is at least good news to follow. Ecotoxicology brought to public notice the possible association of endocrine disrupting chemicals – many of them pesticides – with adverse health effects in wildlife populations, and perhaps humans. This debate will have enormous consequences for the future. What is good for beneficials may now be seen to be good for all of us.

REFERENCES

ADAS (1996) *Awareness of Use of IPM and ICM in Agriculture and Horticulture*, Survey Report by ADAS for Department of the Environment, ADAS Market Research Team, Guildford, Surrey.

FAO (1993) *Summary of the Global IPM Field Study Tour and Meeting*, FAO Inter-country Programme, Philippines, FAO, Rome.

FAO (1995) *Review of the implementation of the International Code of Conduct on the Distribution and Use of Pesticides*, FAO, COAG/95/8, FAO, Rome.

Fearne, A.P. (1991) Communication in agriculture. Results of a farmer survey. *Journal of Agricultural Economics*, **X**, 371–380.

MAFF (1996a) Government announces plans to establish a Pesticides Forum. MAFF News Release 34/96, February 1996. MAFF, London.

MAFF (1996b) *Report on the Evaluation of Codes of Good Agricultural Practice*. Circular, August, 1996, Environmental Protection Division, MAFF, London.

Waage, J.K. (1996) From biological controls to farmer led research [interview]. *Pesticides News*, 3 September 1996.

World Bank (1996) Operational Policies: Pest Management (OP4.09), in *The World Bank Operational Manual*, July 1996, World Bank, Washington DC.

28

Comparative impact of insecticide treatments on beneficial invertebrates in conventional and integrated farming systems

John M. Holland

INTRODUCTION

Integrated farming is seen as the way to reduce pesticide use in arable crops whilst maintaining profitability and protecting the environment. The LINK Integrated Farming Systems project was initiated firstly to develop an integrated system for arable rotations and secondly to compare this system with conventional practice. Located at six sites, a range of climatic zones, soil types and agronomic practices are represented. The project is now in the final year of a five-course rotation of cereals, break crops and set-aside. A minimum of seven pairs of plots (minimum area of 2.5 ha and 70 m width) per site ensured that each course of the rotation was present every year. Economic, agronomic and environmental factors were investigated and invertebrates were used as environmental bio-indicators, because they have been shown to be responsive to many farming practices. Furthermore, their encouragement is an essential component of integrated farming, where invertebrates can be important predators of crop pests.

Ecotoxicology: Pesticides and beneficial organisms.
Edited by P.T. Haskell and P. McEwen. Published in 1998 by Chapman & Hall, London. ISBN 0 412 81290 8.

In the current economic climate of high crop values and arable area aid, there appears to be little incentive for farmers to reduce inputs which directly influence yield or quality. However, other farming system projects have demonstrated that integrated farming can be equally or more profitable than the conventional practice (El Titi, 1991; Jordan and Hutcheon, 1995). If integrated farming is to be widely adopted, it is essential to demonstrate the environmental benefits and to present these with the potential economic risk to the farmer of adopting such a system. With respect to integrated pest management, it may take several years for predatory and parasitic species to reach sufficient levels to supply acceptable levels of biological control. Therefore, ensuring that pesticide inputs are only applied when necessary would appear to be essential.

The impact of insecticides on beneficial invertebrates may be influenced by the spectrum of activity of the active ingredient, the spray distribution, the type and growth stage of the crop, and the invertebrate species and their phenology. In addition, diurnal activity is important in determining exposure and may be used as a management tool. The intensity of insecticide treatments may also prolong the effect by preventing recovery through breeding or re-invasion.

The LINK IFS project provides the opportunity to examine insecticide effects under current spray regimes as in the conventionally farmed plots, as well as allowing comparisons with the lower inputs on the plots farmed using an integrated system. The effect of insecticide treatments can also be examined with respect to other farming practices that may influence invertebrates. Each of the major arable crops will be discussed with respect to their insecticide inputs and their impact on non-target invertebrate species will be appraised using invertebrate data from four of the LINK IFS project sites.

ASSESSING THE RISK

Cereal crops

Insecticides are most commonly applied to cereals in the autumn to prevent the transmission of barley yellow dwarf virus (BYDV) by aphids. Early sown crops are most vulnerable; therefore in the LINK IFS project wheat crops are drilled later, after mid October, to avoid the main aphid immigration period. There is a yield penalty for this practice and a risk for the farmer that adverse weather may prevent drilling or that deteriorating soil conditions may result in poorer crop establishment, consequently increasing vulnerability to slugs. In comparison, insecticide costs are small and often the application cost is low as the insecticide is usually applied with a herbicide (Oakley, 1994). Moreover, fewer invertebrates are active in late autumn compared with the summer (Den Boer and Den Boer-Daanje, 1990), but exposure may be greater because of poor ground cover in the autumn. Consequently, the risk of crop damage from not applying the insecticide to earlier drilled crops is very high, but the benefits in terms of number of species protected are low. In addition, biological control by predators in the autumn has not been demonstrated and is unlikely to be sufficient to prevent virus transmission.

Molluscicides can be toxic to ground-active predatory invertebrates, but again the risk to the crop is very high compared with the benefits of not applying any molluscicide. Cultural measures can often reduce the need for any molluscicide. Other autumn/winter-applied insecticides for the control of pests on grassland or for stem-boring species may be more damaging, as the recommended products are broad spectrum. However, because their use is not so widespread on each farm, recovery from unsprayed areas is more likely.

Despite the availability of the more selective but expensive product, pirimicarb, broad-spectrum organophosphate products such as dimethoate are still commonly used for summer cereal aphid infestations, because of their lower cost (Oakley, 1994). Potentially this is harmful because they are applied when many invertebrate species are active (mid June to early July) in both the field and the field margins. Spray thresholds can help to reduce the number of applications but close monitoring is required, because cereal aphid populations can develop quickly and differ between individual fields. Consequently, prophylactic treatments are a temptation.

Evidence from the LINK IFS project showed that pirimicarb had no effect on numbers of Carabidae and Linyphiidae (Table 28.1). Paired t-tests showed no significant difference (t = 1.65, df = 9, P = 0.066) in numbers of Carabidae for pre- to post-spraying between conventional farm practice (CFP) and integrated farming system (IFS) plots for the occasions when an insecticide was applied to only the CFP plots, though the numbers of Carabidae declined by 2.46 per day per pitfall trap in the CFP plots compared with a pre-/post-spray difference of only +0.22 in the IFS plots. Numbers of Linyphiidae increased from pre- to post-spraying by 8.20 and 4.79 per day per pitfall trap in the CFP an IFS plots, respectively, but the difference was not significant (t = 0.49, df = 8, P = 0.32). Predominantly ground-dwelling species such as Carabidae and Linyphiidae may be protected from insecticide deposition by the cereal canopy at this time of the year (Jepson, 1989) or by their nocturnal habit, but many beneficial species in cereal crops fly or inhabit the upper parts of the crop during the day.

Summer-applied insecticides are also very damaging to gamebird populations because the chicks rely on invertebrates for food during their first three weeks (Potts, 1986). Because chicks forage most commonly in cereal field margins, their prey can best be preserved by leaving the outer

Table 28.1 Effect of insecticides applied to wheat and oilseed rape crops on mean numbers of Carabidae and Linyphiidae per pitfall trap/day in the conventional farm practice (CFP) and integrated farming system (IFS) plots at four of the LINK IFS Project sites during 1993–1995

Crop	*Treatment*				*Date of pre-sample*	*Date of post-sample*	*Numbers of Carabidae*						*Numbers of Linyphiidae*					
	Insecticide		*Date applied*				*In CFP plots*			*In IFS*			*In CFP plots*			*In IFS*		
	CFP	*IFS*	*CFP*	*IFS*			*Pre*	*Post*	*Change*	*Pre*	*Post*	*Change*	*Pre*	*Post*	*Change*	*Pre*	*Post*	*Change*
Wheat																		
	α-cyp	–	17.7.93	–	2.7.93	16.8.93	1.21	0.26	–0.95	1.40	0.54	–0.86	2.83	0.06	–2.77	2.31	0.06	–2.25
	α-cyp	pir	17.7.93	17.7.93	2.7.93	16.8.93	1.64	1.18	–0.46	0.18	0.3	–0.15	0.80	0.20	–0.60	0.30	0.07	–0.23
	pirim	–	28.6.95	–	1.6.95	5.7.95	3.22	2.46	–0.76	3.00	0.90	–0.21	9.70	26.7	+17.0	7.80	39.5	+31.7
	pirim	–	28.6.95	–	1.6.95	5.7.95	8.02	12.7	+4.68	13.5	4.92	–8.58	6.70	22.6	+15.9	13.5	34.1	–20.6
	dimet	–	7.7.94	–	5.6.94	14.7.94	0.08	0.03	–0.05	0.22	0.14	–0.08	0.33	0.04	–0.29	0.86	0.54	–0.32
	dem-S	–	28.6.94	–	5.6.94	14.7.94	3.24	1.40	–1.84	3.68	0.48	–3.20	1.92	0.80	–1.12	2.94	2.02	–0.92
	oxydm	–	23.6.93	–	25.5.93	23.6.93	0.07	0.12	+0.05	0.08	0.32	+0.24	0.02	0.13	+0.11	0.04	0.07	+0.03
Oilseed rape																		
	cyper	cyper	25.3.93	25.3.93	12.3.93	31.3.93	1.27	0.34	–0.93	0.75	0.22	–0.53	2.78	0.04	–2.74	3.97	0.04	–3.93
	–	α-cyp	–	21.5.93	7.5.93	17.6.93	1.30	6.90	+5.60	4.06	4.16	+0.10	0.10	0.71	+0.61	0.10	14.8	+14.7
	cyper	cyper	5.7.94	5.7.94	30.6.94	13.7.94	0.50	0.24	–0.26	0.75	0.29	–0.46	0.53	0.04	–0.49	0.73	0.08	–0.65
	λcyh	–	5.11.93	–	15.10.93	17.11.93	2.14	0.22	–1.92	1.62	0.70	–0.92	0.40	0.06	–0.34	0.26	0.14	–0.12
	α-cyp	–	10.5.94	–	21.4.94	18.5.94	0.72	0.86	+0.14	1.48	1.22	0.26	0.56	0.16	–0.40	2.42	0.84	–1.58

α-cyp = alpha-cypermethrin; **pirim** - pirimicarb; **dimet** = dimethoate; **dem-S** = demeton-*S*-methyl; **oxydm** = oxydemeton-methyl; **cyper** = cypermethrin; **λ-chy** = lambda-cyhalothrin margins of the field unsprayed or by using selective inputs of pesticides (Sotherton, 1991). This policy costs the farmer very little financially and poses little or no risk to the remainder of the crop. It remains to be shown how the recent compulsory implementation of a 6 m unsprayed insecticide buffer zone around cereal fields will help to preserve invertebrates in the field margins.

More recently, the orange wheat blossom midge caused severe crop damage throughout England during 1993 and 1994 (Oakley, 1994). The problem occurred because this is a sporadic pest which is expensive to monitor routinely from soil cores, and difficult to monitor accurately in the crop; consequently prophylactic spraying is common. The currently approved products, chlorpyrifos and triazophos, have a broad spectrum of activity and are potentially damaging to non-target invertebrates as they are applied when beneficial invertebrate activity is high, i.e. end of May.

Oilseed rape

As with cereals, the most common timings for insecticides on oilseed rape are in the autumn or winter for cabbage stem flea beetle and in May/June for pollen beetle and seed weevil. Ground cover is relatively sparse in the autumn but activity of beneficial species is also low. However, both the November and the March insecticide applications resulted in reductions in numbers of Carabidae and Linyphiidae (Table 28.1). The November application appeared to cause a greater reduction of invertebrates in the CFP plot compared with the natural seasonal decline observed in the unsprayed IFS plot.

The risks to ground-dwelling beneficial species from insecticide applied in the summer are lower than in cereal crops because the spray is less likely to penetrate the dense, mature crop canopy of oilseed rape. This was confirmed for the few occasions when insecticides were applied in the summer (Table 28.1). In addition, the spray thresholds are easier to implement and patch or spraying of individual fields is more common, but because of the height needed for the spray boom the risk of insecticide drift outside the crop may be greater.

Peas

Because of the late sowing date and development of ground cover, pea crops are not favoured by many ground-active invertebrates (Booij, 1994). However, the crop may receive insecticide applications when many beneficial species are active, i.e. May-July – first for pea aphid and later for pea moth. Demeton-*S*-methyl (DSM) is widely used for pea aphid control and because it is a broad-spectrum product it can be potentially damaging to non-target invertebrate populations. At two of the sites a broad-spectrum organophosphate insecticide was used in the CFP plots whilst pirimicarb was applied to the IFS plots (Table 28.2). Numbers of Carabidae and Linyphiidae did not appear to differ between the CFP and IFS plots and this was because either the organophosphate insecticides were having no effect on these invertebrates or the pirimicarb was having the same impact. When the differences pre- to post-spraying were compared for all those plots receiving pirimicarb, there was no significant difference in numbers of Carabidae or Linyphiidae.

Pea moth only requires control in pea crops intended for human consumption or for seed. Pheromone traps can be used to assess the risk and to aid timing of insecticide applications. Broad-spectrum products are approved (triazophos and deltamethrin/heptophos) and their use may result in a large impact on beneficial species, but treatment may be unavoidable given the high value of these crops. Pyrethroid insecticides are also approved but lack the persistence of the organophosphate products. Where cypermethrin was used, there appeared to be little effect on numbers of Carabidae and Linyphiidae (Table 28.2).

Potatoes

The very high value of potato crops combined with the characteristics of the pests make the crop one of the most difficult to manage using an integrated system. Nematodes are one of the most damaging pests and control measures are often unavoidable once they are present. Applications of nematicides combined with the intensive soil cultivations associated with potato production may be responsible for the low numbers of invertebrates found at sites growing potatoes. Organophosphate insecticides applied in the summer for aphid control also appeared to reduce numbers of Carabidae (Table 28.2).

Potato viruses transmitted by aphids require control in seed potatoes throughout the growing season. The repeated use of pirimicarb at the site where seed potatoes were grown may have been responsible for the decline in Carabidae, although it is more likely to be a combination of later ground cover and soil cultivations. When comparisons of pre- to post-spraying of pirimicarb were examined there was no significant difference in the numbers of Carabidae (t = 0.22, df = 15, P = 0.41) or Linyphiidae (t = 0.99, df = 15, P = 0.17).

DISCUSSION

The LINK IFS project has shown that considerable variation in invertebrate activity and density occurs between the sites, and between fields within sites, and that these differences are often greater than any caused

Table 28.2 Effect of insecticides applied to pea and potato crops on mean numbers of Carabidae and Linyphiidae per pitfall trap/day in the conventional farm practice (CFP) and integrated farming system (IFS) plots at four of the LINK IFS Project sites during 1993–1995

Crop	Treatment				Date of	Date of	Numbers of Carabidae						Numbers of Linyphiidae					
	Insecticide		Date applied		pre-	post-	In CFP plots			In IFS			In CFP plots			In IFS		
	CFP	IFS	CFP	IFS	sample	sample	Pre	Post	Change	Pre	Post	Change	Pre	Post	Change	Pre	Post	Change
Peas																		
	oxydem	pirim	11.6.93	11.6.93	3.6.93	17.6.93	11.6	9.6	–2.0	11.4	10.6	–0.80	1.06	2.00	+0.94	9.80	2.40	–7.40
	metas	pirim	23.6.94	28.6.94	15.6.94	14.7.94	2.32	0.92	–1.40	5.14	1.54	–3.60	0.74	1.24	+0.50	0.84	0.38	+0.4
	cyp	cyp	1.7.93	1.7.93	17.6.93	16.7.93	9.60	3.40	–6.20	1.05	3.90	+2.85	2.00	1.80	–0.20	2.40	1.80	–0.60
	–	pirim	–	25.6.93	17.6.93	30.6.93	9.60	1.50	–8.10	1.06	2.98	–1.92	2.00	2.40	+0.4	0.20	2.30	+2.1
	pirim	pirim	14.6.95	14.6.95	21.5.95	21.6.95	0.19	0.01	–0.18	0.01	0.04	+0.03	0	0.001	+0.001	0	0	0
	pirim	–	9.6.93	–	19.5.93	18.6.93	0.32	0.20	–0.12	0.13	0.45	+0.32	0.03	0.07	+0.04	0.04	0.09	+0.0
Potatoes																		
	pirim	pirim	7.7.92	7.7.92	24.6.92	9.7.92	0.14	0.06	–0.08	0.11	0.20	+0.09	0.14	1.60	+1.46	0.20	2.38	+2.1
	pirim	pirim	23.7.92	23.7.92	9.7.92	27.7.92	0.06	0.03	–0.03	0.20	0.05	–0.15	1.60	0.41	–1.19	2.38	0.70	–1.68
	pirim	pirim	6+28.6.92	6+28.6.92	1.6.93	2.7.93	0.32	0.06	–0.26	0.85	0.20	–0.65	0.30	0.06	–0.24	0.43	0.36	–0.07
	pirim	pirim	12+22.7.93	12+22.7.93	2.7.93	16.8.93	0.06	0.20	+0.14	0.20	0.63	+0.43	0.06	0.43	+0.37	0.83	0.06	–0.77
	pirim	pirim	13.6.94	13.6.94	24.5.94	21.6.94	0.19	0.25	+0.06	0.16	0.15	–0.01	0.03	0.14	+0.11	0.01	0.03	+0.0
	pirim	pirim	27.6+12.7.94	27.6+12.7.94	21.6.94	19.7.94	0.25	0.46	+0.21	0.15	0.29	+0.14	0.14	0.12	–0.02	0.03	0.05	+0.0
	pirim	pirim	23.7.94	23.7.94	19.7.94	30.8.94	1.64	1.18	–0.46	0.18	0.30	+0.12	0.12	0.80	+0.68	0.05	0.30	+0.2
	oxydem-*n*	dem-*S*	7.7.94	7.7.94	15.6.94	14.7.94	0.23	0.06	–0.17	0.15	0.07	–0.08	0.14	0.21	+0.07	0.13	0.08	–0.05
	pirim	pirim	13.6.93	15.6.93	19.5.93	18.6.93	0.02	0.18	+0.16	0.02	0.22	+0.20	0.11	0.05	–0.06	0.00	0.09	+0.0

oxydem = oxydemeton-methyl; **pirim** = pirimicarb; **metas** = metasystox; **cyp** = cypermethrin; **oxydem-n** = oxydemeto-*n*-methyl; **dem-S** = demeton-*S*-methyl by insecticide treatments (Holland *et al.*, 1996) The variation may be attributed to previous management, ratio of adjacent uncropped areas, soil type, geographical location and climate. Highest numbers were found at the site in Hampshire, which is probably a reflection of the management practices there which aim to preserve chick-food insects for gamebirds. Pesticide inputs were also relatively low (Ogilvy *et al.*, 1994) and conservation headlands (Sotherton, 1991) are established around most cereal fields. Climate may also be important as this is the most southerly of the sites. In contrast, the site in Cambridgeshire had the lowest activity of Carabidae and the highest use of broad-spectrum insecticides in the conventional plots (Holland *et al.*, 1996).

The sites in this study represent farms with low to medium pesticide input for their conventional practice; and reductions beyond this do not appear to be justified, because invertebrates have not yet been enhanced in the integrated plots. This would appear to indicate that insecticide treatments, despite publicity, may not be the most important factors affecting invertebrates in agroecosystems. Other farming practices may have a larger influence (Hance and Gregoire-Wibo, 1987) and these must be examined in the context of pesticide inputs. Few differences in invertebrate numbers between conventional and integrated farming were found in the Less Intensive Farming and the Environment (LIFE) Project (Jordan and Hutcheon, 1995). In the Dutch Nagele Project, however, invertebrate density – but not diversity – was encouraged by integrated farming (Booij, 1994). This may be because pesticide inputs are typically higher in The Netherlands than in the UK. Invertebrate activity-density was also encouraged by integrated farming in a long-term study of agricultural systems in Switzerland (Pfiffner and Niggli, 1996).

The experimental design of the LINK IFS project does not include or allow examination of the effect of field boundaries and non-cropped areas. The higher invertebrate numbers at the Hampshire site might indicate that it is the management and encouragement of field boundaries and margins that is more important than the cropped areas, provided insecticide inputs are only applied when careful monitoring has shown that thresholds have been exceeded and that broad-spectrum products are avoided wherever possible. The importance of field boundaries was demonstrated in the Lautenbach Project in Germany where the planting of hedges decreased a number of insect pests and increased crop yields (El Titi, 1991).

The Sussex study carried out by the Game Conservancy Trust has shown a decline in many invertebrate guilds since the 1970s (Aebischer, 1991) and consequently species density and diversity may already be low at many locations; therefore the species remaining are those that have adapted to the current pesticide and husbandry regimes. This may only be revealed by comparison with organic farms, or farms with no pesticide inputs. Invertebrates are encouraged in some organic systems (Booij, 1994), although greater weediness and lower plant densities may be as influential as no pesticides (Pfiffner and Niggli, 1996). Other ways of encouraging beneficial insects in integrated crop management systems include the development of more selective products, better spray targeting, buffer zones to protect field boundaries, more patch spraying in conjunction with precision farming, and an increase in the ratio of non-crop areas to act as refuges and as sources for migrants. Further information on the relative selectivity of existing products would also assist agronomists and conservation advisers when making a crop protection recommendation.

It may be that differences between the farming systems in this and other long-term farming system experiments are not detected if the sampling methods are not suitable. Pitfall traps may not be the most appropriate when comparing treatments that may alter invertebrate activity and thereby capture rate. Density-dependent methods such as fenced pitfall traps (Chapter 18) and photoelectors offer the best compromise (Sunderland *et al.*, 1995).

ACKNOWLEDGEMENTS

The LINK IFS project is jointly funded by the Ministry of Agriculture, Fisheries and Food, the Scottish Office Agriculture and Fisheries Department, Zeneca Agrochemicals, H-GCA (Cereals and Oilseeds) and the British Agrochemicals Association.

REFERENCES

Aebischer, N.J. (1991) Twenty years of monitoring invertebrates and weeds in cereal fields in Sussex, in *The Ecology of Temperate Cereal Fields*, (eds L.G. Firbank, N. Carter, J.F. Darbyshire and G.R. Potts), Blackwell Scientific Publications, Oxford, pp. 305–332.

Booij, C.J.H. (1994) Diversity patterns in carabid assemblages in relation to crops and farming systems, in *Carabid Beetles: Ecology and Evolution*, (ed. K. Descender), Kluwer Academic Publishers, Dordrecht, pp. 425–431.

Den Boer, J. and Den Boer-Daanje, W. (1990) On life history tactics in carabid beetles: are there only spring and autumn breeders?, in *The Role of Ground Beetles in Ecological and Environmental Studies*, (ed. N.E. Stork), Intercept, Andover, Hants, pp. 247–258.

El Titi, A. (1991) The Lautenbach project 1978–89: integrated wheat production on a commercial arable farm, south-west Germany, in *The Ecology of Temperate Cereal Fields* (eds L.G. Firbank, N. Carter, J.F. Darbyshire and G.R. Potts), Blackwell Scientific Publications, Oxford, pp. 399–411.

Hance, T. and Gregoire-Wibo, C. (1987) Effect of agricultural practices on carabid populations. *Acta Phytopatholica et Entomologica Hungarica*, **22**, 147–160.

Holland, J.M. and Smith, S. (1997) Capture efficiency of fenced pitfall traps for predatory arthropods within a cereal crop, in *New Studies in Ecotoxicology* (eds

P.T. Haskell and P.K. McEwen), The Welsh Pest Management Forum, Cardiff, pp. 34–36.

Holland, J.M., Drysdale, A., Hewitt, M.V. and Turley, D. (1996) The LINK IFS Project – the effect of crop rotations and cropping systems on Carabidae. *Aspects of Applied Biology*, **47** (119–126).

Jepson, P.C. (1989) The temporal and spatial dynamics of pesticide side-effects on non-target invertebrates, in *Pesticides and Non-target Invertebrates* (ed. P.C. Jepson), Intercept, Wimborne, Dorset, pp. 95–128.

Jordan, V.W.L. and Hutcheon, J.A. (1995) Less-intensive farming and the environment: an integrated farming systems approach for UK arable crop production, in *Ecology and Integrated Farming Systems: Proceedings of the 13th Long Ashton International Symposium*, (eds D.M. Glen, M.P. Greaves and H.M. Andersen), John Wiley & Sons, Chichester, pp. 307–318.

Oakley, J.N. (1994) Impact of CAP reforms on decision making for pest control on cereals. *Aspects of Applied Biology*, **40**, 183–192.

Ogilvy, S.E, Turley, D.B., Cook, S.K. *et al.* (1994) Integrated farming – putting together systems for farm use. *Aspects of Applied Biology*, **40**, 53–60.

Pfiffner, L. and Niggli, U. (1996) Effects of bio-dynamic, organic and conventional farming on ground beetles (Col. Carabidae) and other epigaeic arthropods in winter wheat. *Biological Agriculture and Horticulture*, **12**, 353–364.

Potts, G.R. (1986) *The Partridge: Pesticides, Predation and Conservation*, Collins, London.

Sotherton, N.W. (1991) Conservation headlands: a practical combination of intensive cereal farming and conservation, in *The Ecology of Temperate Cereal Fields* (eds L.G. Firbank, N. Carter, J.F. Darbyshire and G.R. Potts), Blackwell Scientific Publications, Oxford, pp. 373–398.

Sunderland, K.D., de Snoo, G.R., Dinter, A. *et al.* (1995) Density estimation for invertebrate predators in agroecosystems. *Acta Jutlandica*, **70**, 133–162.

Part Three: B

Practice

29

Relative toxicity of pesticides to pest and beneficial insects in potato crops in Victoria, Australia

Catherine A. Symington and Paul A. Horne

INTRODUCTION

Potato crops in Victoria may be affected by two primary insect pests: the potato tuber moth, *Phthorimaea operculella* (Zeller) (Lepidoptera: Gelechiidae) (Rothschild, 1986) and the green peach aphid, *Myzus persicae* Sulzer (Hempitera: Aphididae). Insecticides are often used to control these pests (Dillard *et al.*, 1993) but many Victorian growers want to reduce their pesticide input (Henderson, 1993). Integrated pest management (IPM) involving biological control of *P. operculella* by the parasitoids *Orgilus lepidus* Muesebeck (Braconidae), *Apanteles subandinus* Blanchard (Braconidae) and *Copidosoma koehleri* Annecke & Mynhardt (Encyrtidae) can be a successful alternative to reliance on synthetic pesticides (Horne, 1990). Insecticides may still be required in some cases to control insect pests, and fungicides remain necessary for management of fungal pathogens. A complication in integration of biological and chemical control is that predatory and parasitic insects tend to be more susceptible to pesticides than the pests (Lingren *et al.*, 1972; Plapp and Vinson, 1977; Keeratikasikorn and Hooper, 1981; Powell *et al.*, 1986). The effects of insecticides and fungicides on the parasitoids of *P. operculella* were largely unknown. Knowledge of these is necessary for effective integration of synthetic pesticides and biological control.

Ecotoxicology: Pesticides and beneficial organisms.
Edited by P.T. Haskell and P. McEwen. Published in 1998 by Chapman & Hall, London. ISBN 0 412 81290 8.

Pesticides commonly used in potato production, or in tomatoes where *P. operculella* may be present, were assessed in the laboratory. The relative toxicity of pesticides to adult and juvenile *P. operculella* and parasitoids and to adult *M. persicae* were established. Effects on development time, pupation and emergence of parasitized and unparasitized fourth instar *P. operculella* were also assessed.

MATERIALS AND METHODS

Insects were reared in the laboratory using techniques based on Platner and Oatman (1968) for *P. operculella*, Horne (1990) and Horne and Horne (1991) for parasitoids, and Curtis (unpublished) for *M. persicae*. Assays on the relative toxicity of pesticides tested first and fourth instar larvae and adult male *P. operculella*, adult male parasitoids and adult apterous female *M. persicae*, 0–24 hours post-eclosion. All assays involved five replicates of 10 insects per treatment. Each species was exposed to logarithmic serial dilutions of pesticide in distilled water. Calculations were based on concentration of active ingredient. Controls were treated with distilled water. Seven to 10 concentrations over the minimum concentration range that caused 0–100% mortality were then tested in the same manner. Dose-response curves and LC_{50} (50% lethal concentration) for each species were established from these data. The pesticides tested are listed in Table 29.1.

Contact toxicity of pesticides to adults of each species was evaluated by exposing insects to sprayed glass. Aliquots (1 ml) of pesticide solution were sprayed on to the lid and base of glass petri dishes (diameter 80 mm) through a Potter tower (Burkard) (40 kPa, mean droplet diameter 30–40 μm), providing spray coverage of approximately 0.03 ml cm^{-2}.

Table 29.1 Pesticides

Generic name	*Trade name*	*Class*	*Manufacturer*
Endosulfan[a]	Thiodan E.C.®	Organochlorine	Hoechst Australia
Methamidophos[a,b]	Nitofol®	Organo-phosphate	Bayer Australia
Thiodicarb[a]	Larvin 375®	Carbamate	Rhône-Poulenc
Pirimicarb[b]	Pirimor W.P.®	Carbamate	ICI Crop Care
Permethrin[a]	Ambush E.C.®	Synthetic pyrethroid	ICI Crop Care
Imidacloprid[b]	Confidor S.C.®	Nitroguanidine	Bayer Australia
Mancozeb[c]	Dithane M-45®	Polymeric Mn with Zn salt	Rhom and Haas
Difenoconazole[c]	Score	Conazole	Ciba-Geigy

Targets: [a] *P. operculella*
[b] *M. persicae*
[c] Fungal pathogens

Dishes were dried for 5 minutes in a fume hood, leaving a fresh residue. *P. operculella* and parasitoids were lightly anaesthetized with CO_2, transferred to the treated dishes, and incubate at 25 ± 1.5°C (12L : 12D photoperiod). *M. persicae* were individually transferred to the dishes using a fine camel-hair brush. The dish bases were secured over the discs and larvae with elastic bands. Dishes were inverted and incubated as above. Larval mortality was assessed after 24 hours. Moribund larvae were considered dead.

Substrate preparation for exposure of parasitized and unparasitized fourth instar *P. operculella* was as for the adult insects. Treated dishes each contained a small strip of corrugated white plastic cut from the sides of cups (Polar Cup® CB6) under which the larvae pupated. Larvae were assessed for mortality 24 hours after exposure, and dose-response curves ascertained as above. To investigate the effects on development time, pupation and emergence, 100 larvae of each species were exposed to LC_{50} pesticide residue, and 50 of each species to glass treated with distilled water. Larvae were incubated in the treated petri dishes until emergence, and monitored for length of pupal period, percentage pupation and percentage emergence.

The results of relative toxicity to the pest and beneficial species were analysed using Probit-P.C. (LeOra Software, California). Relative effects of each pesticide on the species tested were compared through ratios of LC_{50}. Effects of exposure of parasitized and unparasitized fourth instar *P. operculella* to LC_{50} on development, pupation and emergence were analysed through ANOVA and one-sample *t*-tests after arcsin transformation.

RESULTS

Thiodicarb was the only insecticide for use against *P. operculella* that was less toxic (8.3-fold) to a parasitoid, *O. lepidus*, than to adult *P. operculella* at LC_{50}. The aphicides imidacloprid and pirimicarb were less toxic to the parasitoids of *P. operculella*, and more toxic to adult *P. operculella*, than to *M. persicae*. First instar *P. operculella* larvae were less susceptible to the aphicides than the adult parasitoids. Methamidophos and permethrin were similar in toxicity to *O. lepidus* and *P. operculella*, and more toxic to *A. subandinus* and *C. koehleri* ($P < 0.0.5$). Endosulfan was significantly more toxic to each of the parasitoid species than to adult *P. operculella* ($P < 0.05$). First instar larvae were between two and 40 times more susceptible to the insecticides than were the adult moths. *O. lepidus* was the least susceptible of the parasitoid species. The fungicides mancozeb and difenoconazole had no effect on the mortality of the adult insects. First instar *P. operculella* were susceptible to the fungicides; LC_{50} for mancozeb = 2×10^{-3}, difenoconazole = 10^{-4}. Fourth instar

Table 29.2 50% Lethal concentration (LC_{50}) (g a.i./ml) and toxicity to parasitoids and pests relative to adult *Phthorimaea operculella*

Pesticide	*Measure*	*Phthorimaea operculella*	*Orgilus lepidus*	*Copidosoma koehleri*	*Apanteles subandinus*	*1st instar P. operculella*	*4th instar*	*Myzus persicae*
Insecticides								
Thiodicarb	LC_{50}	9.2×10^{-4}	7.8×10^{-3}	4.0×10^{-6}	4.5×10^{-4}	3×10^{-5}	8×10^{-5}	–
	Ratio*	N/A	0.12	230	2.0	30.6	11.5	–
Permethrin	LC_{50}	7.8×10^{-6}	5.9×10^{-6}	2.7×10^{-6}	8.8×10^{-6}	9×10^{-7}	10^{-6}	–
	Ratio*	N/A	1.32	2.89	0.88	8.67	7.8	–
Endosulfan	LC_{50}	7.5×10^{-5}	9.0×10^{-6}	4.1×10^{-5}	1.8×10^{-5}	2×10^{-6}	–	–
	Ratio*	N/A	8.32	1.83	4.28	37.5	–	–
Methamidophos	LC_{50}	3.4×10^{-5}	4.4×10^{-5}	8.0×10^{-6}	4.1×10^{-6}	2×10^{-5}	5×10^{-5}	7.2×10^{-5}
	Ratio*	N/A	0.77	4.23	8.24	1.7	0.68	0.47
Aphicides								
Pirimcarb	LC_{50}	3.9×10^{-3}	1.8×10^{-4}	1.0×10^{-4}	9.0×10^{-5}	2×10^{-4}	–	9.8×10^{-6}
	Ratio*	N/A	21.5	38.7	43	19.5	–	
Imidacloprid	LC_{50}	$>10^{-2}$	5.0×10^{-5}	4.8×10^{-5}	5.3×10^{-4}	2×10^{-4}	5×10^{-4}	7.3×10^{-6}
	Ratio*	N/A	N/A	N/A	N/A	N/A	N/A	N/A

*Ratio = relative susceptibility; LC_{50} (adult *P. operculella*)/LC_{50} (parasitoid, or other species)
N/A = not applicable

P. operculella were intermediate between first instar and adult *P. operculella* in susceptibility (Table 29.2).

Imidacloprid reduced pupation, and methamidophos and difenoconazole reduced emergence of *P. operculella* exposed as fourth instars. Thiodicarb reduced pupation and emergence, and methamidophos and imidacloprid reduced emergence of *O. lepidus*. Exposure of fourth instar larvae to pesticide had no effect on the duration of the pupal stage (Table 29.3).

DISCUSSION

Thiodicarb was the only insecticide effective against *P. operculella* which was also less toxic to adults of one parasitoid species, *O. epidus* (8.3-fold difference) than to *P. operculella* at LC $_{50}$. Comparison of parasitoid susceptibility with first instar *P. operculella* increases this differential as the larvae are approximately 30 times more susceptible than adult *P. operculella*. These results suggest that thiodicarb may be useful in IPM strategies to reduce moth numbers without affecting the parasitoids. Thiodicarb is not currently recommended for use against *P. operculella* in Victoria. In Queensland (northern Australia) thiodicarb is recommended for *Helicoverpa* spp. (Lepidoptera: Noctuidae) in tomato crops,

Table 29.3 Proportion of successfully pupated and emerged *P. operculella* and *O. lepidus* after exposure at the fourth instar stage to Lc_{50} of pesticide (arcsin transformation of proportion pupated and emerged – mean and standard deviation; ranked from least to most detrimental to *O. Lepidus*)

Treatment	*P. operculella*		*O. lepidus*	
	Mean	*SD*	*Mean*	*SD*
Pupation				
Control	0.942	0.194	0.551	0.044
Difencoconazole	0.725[a]	0.099	0.506[a]	0.015
Midacloprid	0.513[b]	0.006	0.380[a]	0.008
Methamidophos	0.647[a]	0.170	0.367[a]	0.037
Thiodicarb	0.775[a]	0.256	0.240[b]	0.035
Emergence				
Control	0.571	0.070	0.387	0.008
Difenoconazole	0.107[b]	0.010	0.277[a]	0.019
Midacloprid	0.372[a]	0.008	0.253[b]	0.001
Methamidophos	0.383[b]	0.101	0.227[b]	0.030
Thiodicarb	0.478[a]	0.069	0.129[b]	0.003

[a] = not significantly different from control ($P \leq 0.05$)
[b] = significantly different from control ($P \leq 0.05$).

where it is concurrently effective against *P. operculella* (Kay, 1993). A parasitoid of *Helicoverpa* spp., *Microplitis croceipes* (Hymenoptera: Braconidae), was found to be tolerant of thiodicarb (Powell and Scott, 1985). Hence thiodicarb may be appropriate for use in IPM where *O. lepidus* or *M. croceipes* is active against these pests. *O. lepidus* can be mass reared and released to augment natural populations (Horne, 1993). If field tests indicate that thiodicarb is still less toxic towards *O. lepidus* than *P. operculella*, and verify the laboratory findings that *O. lepidus* is the least susceptible of the three species tested, then mass release of *O. lepidus* can be more strongly advocated as part of IPM in potato crops. However, thiodicarb is 230 times more toxic to *C. koehleri* than to *P. operculella*, suggesting that thiodicarb may not always be appropriate for use.

Parasitoids are more susceptible than *P. operculella* to the aphicides pirimicarb and imidacloprid, and less susceptible than *M. persicae*. Imidacloprid reduced pupation and emergence of *P. operculella* and *O. lepidus*. The concentrations required for aphid control may be sufficiently low to cause insignificant interference with the parasitization of *P. operculella*. As such, these insecticides may be suitable for incorporation into IPM in potatoes. Methamidophos is also used as an aphicide (Dillard *et al.*, 1993). Being more toxic to the parasitoids than to either *M. persicae* or *P. operculella*, this insecticide is unlikely to be compatible with parasitoid activity in IPM.

The fact that endosulfan, methamidophos and permethrin are 0–10-fold greater in toxicity to the adult parasitoids than to *P. operculella* ($P < 0.05$) is consistent with results presented by Keeratikasikorn and Hooper (1981). The reduction of emergence of *P. operculella* and *O. lepidus* from pupae after exposure to methamidophos shows potential interference with both pest and beneficial populations. Targeting the more susceptible first instars may favour the parasitoids. However, the broad spectrum of activity and high toxicity of these insecticides may lead to disruption of an IPM system where these species are active.

The effect of fungicides on insects is highly variable (Hare and Moore, 1988) but neither mancozeb nor difenoconazole affected the mortality of adult *P. operculella* or the parasitoids in the laboratory assays. First instar *P. operculella* were susceptible to both fungicides. Pupation and emergence of fourth instar *P. operculella* is reduced, while that of *O. lepidus* is unaffected. As such, it appears that neither of these fungicides would be detrimental to biological control of the potato tuber moth by these parasitoid species.

Effective integration of pesticides and biological control is not simple. Both laboratory and field trials are necessary to establish the effects of pesticides on populations of parasitoids and *P. operculella*. Future work evaluating a greater range of pesticides against more insects, including predatory Coleoptera (Carabidae and Coccinelidae) and Neuroptera, and establishing details of sublethal effects including development, fertility, fecundity and host location, will provide information for a more detailed database from which informed decisions on pesticide application within an IPM system can be made.

ACKNOWLEDGEMENTS

This work was part of a PhD project undertaken by Catherine Symington at the Institute for Horticultural Development, co-supervised by Dr T.R. New, and generously funded by the Horticultural Research and Development Corporation. Dr R. deBoer and J. Curtis provided useful comments towards the manuscript. Ms Symington's presentation at the conference was facilitated by Ann Milankovic and Isabel and Clive Carter.

REFERENCES

Dillard, H.R., Wicks, T.J. and Philip, B. (1993) A grower survey of diseases, invertebrate pests, and pesticide use on potatoes grown in South Australia. *Australian Journal of Experimental Agriculture*, **33**(5), 653–661.

Hare, J.D. and Moore, R.E.B. (1988) Impact and management of late season populations of the Colorado potato beetle (Coleoptera: Chrysomelidae) on potato in Connecticut. *Journal of Economic Entomology*, **81**(3), 914–921.

Henderson, A.P. (1993) Adoption of IPM for potato moth control: attitudes and awareness, in *Proceedings of the Seventh National Potato Research Workshop, Ulverstone, Tasmania, Australia*, pp. 31–36.

Horne, P.A. (1990) The influence of introduced parasitoids on the potato moth, *Phthorimaea operculella* (Lepidoptera: Gelechiidae) in Victoria, Australia. *Bulletin of Entomological Research*, **80**, 159–163.

Horne, P.A. (1993) The potential of inundative releases of parasitoids as part of an IMP strategy, in *Pest Control and Sustainable Agriculture* (eds S.A. Corey, D.J. Dall and W.M. Milne), CSIRO, Melbourne, Australia, pp. 118–120.

Horne, P.A. and Horne, J.A. (1991) The effects of host density on the development and survival of *Copidosoma koehleri*. *Entomologia Experimentalis et Applicata*, **39**, 289–292.

Kay, I.R. (1993) Insecticidal control of *Helicoverpa* spp. and *Phthorimaea operculella* on tomatoes, in *Pest Control and Sustainable Agriculture* (eds S.A. Corey, D.J. Dall and W.M. Milne), CSIRO, Melbourne, Australia, pp. 154–157.

Keeratikasikorn, M. and Hooper, G.H.S. (1981) The comparative toxicity of some insecticides to the potato moth *Phthorimaea operculella* (Zeller) (Lepidoptera: Gelechiidae) and two of its parasites *Orgilus lepidus* Muesebeck and *Copidosoma desantisi* Annecke and Mynhardt. *Journal of the Australian Entomological Society*, **20**, 309–311.

Lingren, P.D., Wolfenbarger, D.A., Nosky, J.B. and Diaz, M. Jr (1972) Response of *Campoletis marginiventris* to insecticides. *Journal of Economic Entomology*, **65**, 1295–1299.

Plapp, F.W. Jr and Vinson, S.B. (1977) Comparative toxicities of some insecticides to the tobacco budworm and its ichneumonid parasite *Campoletis sonorensis*. *Environmental Entomology*, **6**, 381–384.

Platner, G. and Oatman, E.R. (1968) An improved technique for producing potato tuberworm eggs for mass production of natural enemies. *Journal of Economic Entomology*, **61**, 1054–1057.

Powell, J.E. and Scott, W.P. (1985) Effect of insecticide residues on survival of *Microplitis croceipes* adults (Hymenoptera: Braconidae) on cotton. *Florida Entomologist*, **68**, 692–693.

Powell, J.E., King, E.G. Jr and Jany, C.S. (1986) Toxicity of insecticides to adult *Microplitis croceipes* (Hymenoptera: Braconidae). *Journal of Economic Entomology*, **79**, 1343–1346.

Rothschild, G.H.L. (1986) The potato moth – an adaptable pest of short-term cropping systems, in *The Ecology of Exotic Animals and Plants*, (ed. R.L. Kipling), J. Wiley, Brisbane, pp. 144–162.

30

The cumulative effect on populations of linyphid spiders and carabid beetles of insecticide sprays applied to two consecutive winter wheat crops

K.F.A. Walters, D. Morgan, A. Lane and S.A. Ellis

INTRODUCTION

Assessment of pesticides for regulatory purposes must consider potential environmental side-effects including possible impact on beneficial invertebrates. The European and Mediterranean Plant Protection Organization (EPPO) and the Council of Europe have formed an expert panel to develop a framework for the assessment of risks associated with the effects of plant protection products, including those for arthropod natural enemies. In addition, the European chemical industry has formed a Beneficial Arthropod Regulatory Testing Group (BART), in order to develop a testing procedure to assess the side-effects of pesticides on natural enemies. In both cases, the testing procedure being considered is a conditional stepwise process in which the decision of whether to move to the next stage depends on the values obtained in the previous one. In most field trials associated with these procedures only short-term (within-season) effects are measured and it may prove difficult to extrapolate these results to long-term consequences for natural enemy populations.

Ecotoxicology: Pesticides and beneficial organisms.
Edited by P.T. Haskell and P. McEwen. Published in 1998 by Chapman & Hall, London. ISBN 0 412 81290 8.

An understanding of such long-term effects is essential if we are to make informed risk assessments or to design practical strategies for managing or reducing environmental impact. This chapter reports preliminary results from a study of the effect of typical insecticide application schedules to winter wheat crops grown for two consecutive years, and discusses their importance in the development of stepwise testing procedures.

MATERIALS AND METHODS

A series of 10 two-year experiments are planned in this research programme. The first experiment to be completed was conducted at Agricultural Development Advisory Service (ADAS), High Mowthorpe. Two fields that shared a suitable common boundary (Italian ryegrass) were selected and both were sown with winter wheat (cv. Avalon) in each year of the experiment.

In each field a rectangular grid of 16 pitfall traps was established, the grids being equidistant from the common boundary of the two fields. In addition, a row of five pitfall traps was set on either side of the boundary. Trapping commenced in April of 1992 and continued until harvest, and then from September until harvest in 1993. Between the sowing date and the end of April, pitfall traps were operated for a period of 7 days every third week. From May until harvest, traps were operated continually and emptied weekly. All polyphagous predators were identified to species.

In one field a single autumn spray of deltamethrin was applied at each site, and an application of the same product in both years at growth stage (GS) 61 (Zadoks *et al.*, 1974). No insecticides were applied to the second field in either year.

Five species of predators (*Erigone atra, Nebria brevicollis, Pterostichus madidus, Pterostichus melanarius* and *Trechus quadristriatus*), which had been caught in large numbers during the experiment, were selected and the results were analysed using a two-level nested analysis of variance with unequal sample sizes.

RESULTS

The initial analysis made within-field comparisons of total trap catches from the different sampling areas (field boundary and within-field trapping grid) and, using data from the trapping grids only, between-field comparisons of the effect of the treatment regimes on the numbers caught. No significant within-field differences were found between the number of invertebrates caught in the trapping grids and the boundary traps for any of the species (Table 30.1). The between-field comparisons of

Table 30.1 Comparison of total number of insects caught in pitfall traps in the centre and edge of cereal fields (within-fields) and in fields receiving different insecticide treatment regimes (between-fields)

Predator species	*Between-field comparison*		*Within-field comparison*	
	F	*P*	*F*	*P*
Erigone atra	21.69	< 0.05	0.41	NS
Nebria brevicollis	0.23	NS	1.66	NS
Pterostichus madidus	0.32	NS	2.53	NS
P. melanarius	2.00	NS	2.53	NS
Trechus quadristriatus	0.08	NS	2.45	NS

NS = not significant

total catches showed that the pesticides significantly reduced ($P < 0.05$) numbers of *E. atra* trapped, although no differences were recorded for any of the carabid species.

A more detailed analysis of the change in invertebrate numbers with time was undertaken using data for *P. melanarius*, the most numerous species. The results showed that in both fields no significant within-field differences between numbers caught in grid and boundary traps were detected in any of the sampling periods (Table 30.2). Similarly, no significant differences in the numbers recorded in fields receiving the different pesticide regimes were detected prior to June 1993. Thereafter, significant reductions in the numbers caught in the trapping grid were recorded following the application of deltamethrin at GS 61. The magnitude of the difference between the treated and untreated field declined rapidly and no significant differences occurred after trapping period 6–12 July 1993, two weeks after the spray was applied (Table 30.2).

DISCUSSION

The ability to predict long-term, cumulative effects on non-target species of multiple applications of insecticides is an important consideration when developing testing procedures for regulatory purposes. The current stepwise procedures often rely on laboratory experiments and short-term, within-season field trials to predict the likely effects of repeated applications in cereal rotations. It is important that more lengthy experiments investigating cumulative effects are undertaken when developing testing procedures to underpin results from smaller-scale techniques. The initial analysis of data from the first of a series of 10 two-year field experiments investigating the cumulative effects of insecticide applications on beneficial predators in cereal fields illustrates

Table 30.2 Comparison of the number of *Pterostichus melanarius* caught in pitfall traps in the centre and edge of cereal fields (within-field) and in fields receiving different insecticide treatment regimes (between-field) in selected trapping periods

Dates of sample	*Between-field comparison*		*Within-field comparison*	
	F	*P*	*F*	*P*
7–14/6/93	0.08	NS	0.30	NS
15–21/6/93	7.56	NS	1.04	NS
22–28/6/93[a]	253.93	< 0.05	0.01	NS
29/6–5/7/93	92.36	< 0.05	0.18	NS
6–12/7/93	0.87	NS	2.55	NS
13–20/7/93	2.85	NS	1.59	NS

[a] Deltamethrin applied on 22/6/93

the urgent need for such work, and for careful data analysis. Different levels of statistical analysis resulted in different amounts of interpretable information being yielded by the data. Further work is under way to refine and optimize the analytical approach.

The results indicate that no significant differences in numbers of *P. melanarius* caught in treated and untreated fields were recorded after the GS 61 sprays in 1992 or the autumn 1993 sprays. This species is partially autumn active and field overwintering (Thiele, 1977), and was present in only low numbers when the sprays were applied. However, for two weeks after the spray at GS 61 in 1993, numbers caught were lower in the treated field. Determination of whether this was the result of a cumulative pesticide application effect on predator populations, the higher numbers of *P. melanarius* present in the fields in summer 1993, or of other factors, will require a comparative analysis of data from all the other sites. Thus the multi-site aspect is a vital component of the design of the full research programme.

The complete data sets from this programme will ultimately be compared with the results of the within-season experiments which form testing procedures for pesticide registration. This will improve our ability to predict the longer-term effects of multiple spray applications to land under cereal rotations, increasing confidence in the interpretation of the results of such short-term tests, and avoiding the need to conduct uneconomic, long-term field experiments as part of normal pre-registration testing.

ACKNOWLEDGEMENTS

We thank ADAS and CSL colleagues for technical assistance and MAFF Pesticides Safety Directorate for funding.

REFERENCES

Thiele, H.U. (1977) *Carabid Beetles and Their Environments*, Springer-Verlag, Berlin, 369 pp.

Zadoks, J.C., Chang, T.T. and Konzak, C.F. (1974) A decimal code for the growth stages of cereals. *Weed Research*, **14**, 415–421.

31

The MAFF SCARAB project: seven years of pesticide side-effects research on arthropods

Geoff Frampton

INTRODUCTION

One of the key results from the Ministry of Agriculture, Fisheries and Food (MAFF) Boxworth project (1981–1988), a farm-scale comparison of pesticide input systems in winter wheat (Greig-Smith *et al.*, 1992), was the vulnerability of some arthropod taxa to repeated and prophylactic use of pesticides under a high input 'full insurance' regime (Burn, 1992; Vickerman, 1992). Catches of small Carabidae such as *Bembidion* and *Notiophilus* spp. and Collembola such as *Sminthurus viridis* L. were markedly and persistently depleted in the high input regime soon after it was initiated. However, the siting of the study on one cereal farm precluded extrapolation of the results to other arable scenarios at different geographical locations. This chapter presents some results from the MAFF SCARAB (Seeking Confirmation About Results At Boxworth) study (1990–1996), which used a different experimental approach to investigate the relevance of the pesticide effects seen at Boxworth to a wider variety of arable farming situations with different pesticide demands.

Ecotoxicology: Pesticides and beneficial organisms.
Edited by P.T. Haskell and P. McEwen. Published in 1998 by Chapman & Hall, London. ISBN 0 412 81290 8.

MATERIAL AND METHODS

Sites

The SCARAB project comprised seven study fields located at three MAFF Agricultural Development and Advisory Service (ADAS) Research Centres in central and northern England (Frampton and Çilgi, 1996). Results from three of the study fields are given to permit comparisons of data from each site. The fields were: Field 5 (8 ha, Drayton, Warwickshire, 52.2°N 1.8°W); Near Kingston (8 ha, Gleadthorpe, Nottinghamshire, 53.2°N 1.1°W); and Bugdale (19 ha, High Mowthorpe, North Yorkshire, 54.1°N 0.6°W). Respective soil types were calcareous clay, stony sand and calcareous loam. The arable rotation of each Research Centre typified the farming practice of the locality, with cropping as shown later in Figures 31.2 and 31.3. All fields received a conventional pesticide regime during 1990 ('current farm practice', CFP). Then, from autumn 1990 to autumn 1996, pesticides were applied as half-field treatments, one half of each field receiving CFP inputs and the other a 'reduced input approach' (RIA) of pesticide use.

Pesticide regimes

MAFF pesticide usage surveys (e.g. Garthwaite *et al.*, 1995) were used to ensure that the CFP pesticide regime mimicked conventional practice for each crop. The RIA regime contrasted strongly with the CFP by avoiding use of insecticides in all crops and years. Fungicide and herbicide use was also reduced in RIA compared with CFP where possible, though this was more easily achieved in cereals and grass than in root break crops (Frampton and Çilgi, 1996). Management of the two halves of each study field differed only in their pesticide regimes, all other husbandry activities (tillage, harvesting and fertilization) being perfomed on a whole-field basis.

Arthropod sampling

Abundance and species richness of arthropods was estimated using suction sampling (D-vac) (Dietrick, 1961). Four D-vac samples (total 1.84 m^2) were taken at matched locations in each field-half, between 25 m and 125 m from a common field boundary as described in Frampton (1997b). Except during adverse weather and periods of cultivation, samples were collected in all months and over 150 taxa were identified and recorded. For brevity, results are presented here for predatory Coleoptera (Carabidae, Staphylinidae) Coccinellidae and Cantharidae and epigeic Collembola on three summer sampling occasions in each year.

RESULTS

Pesticides

During the period 1990–1996, on average the RIA inputs of herbicides, fungicides and insecticides were, respectively, 48%, 53% and 100% lower than CFP inputs. A summary of the overall pesticide inputs in each field is shown in Figure 31.1. The high use of herbicides in Near Kingston field mainly reflects the high herbicide demands of a sugar beet crop grown in 1996.

Arthropods

Summer catches of predatory Coleoptera (Figure 31.2) and epigeic (surface-dwelling) Collembola (Figure 31.3) varied considerably between sites and years. Differences in the number of predatory Coleoptera species trapped in each of the pesticide regimes changed markedly on consecutive sampling occasions in some years (Figure 31.2) whereas the spatial variation in Collembola species was generally more stable within a season, except after some insecticide applications (Figure 31.3). Apart from the winter cereals crops of 1994 and 1995, diversity of epigeic Collembola was relatively low at the Gleadthorpe site (Figure 31.3b) but no other consistent effects of cropping or site were evident. Following sprayed organophosphorus insecticide applications (chlorpyrifos, dimethoate, triazophos) the number of predatory Coleoptera species trapped was lower in the sprayed (CFP) regime on 8 of the 10 occasions when sampling was preceded by an application in the same season (Figure 31.2). Catches of Collembola species showed a similar pattern but the lower numbers of species trapped in the CFP regime after use of chlorpyrifos at Gleadthorpe and Drayton persisted throughout the

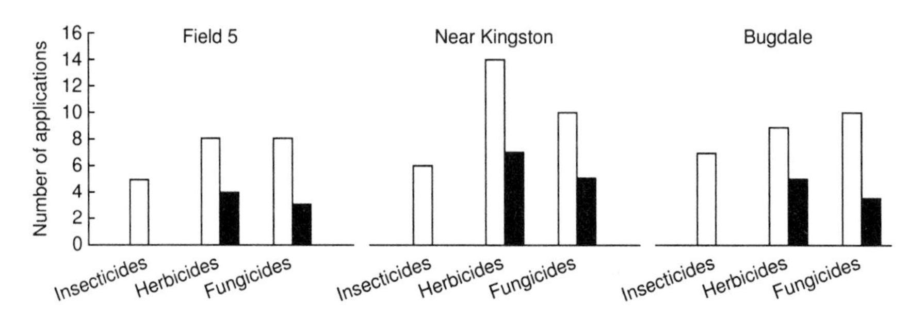

Figure 31.1 Number of current farm practice (■) and reduced input approach (□) full-rate pesticide applications made on three sites between autumn 1990 and autumn 1996.

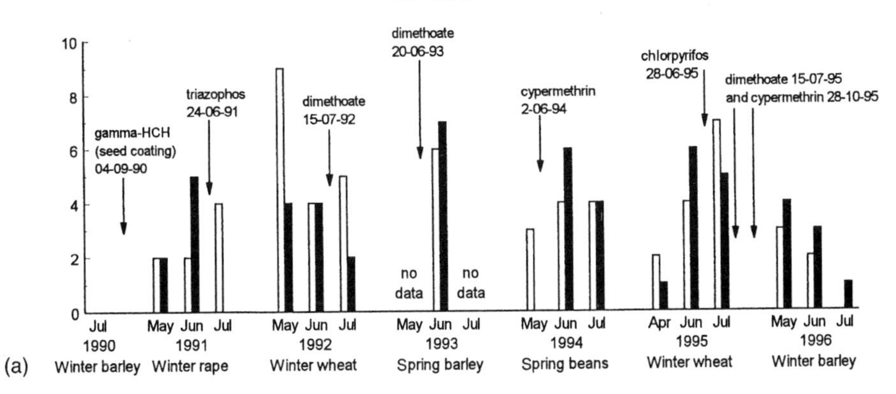

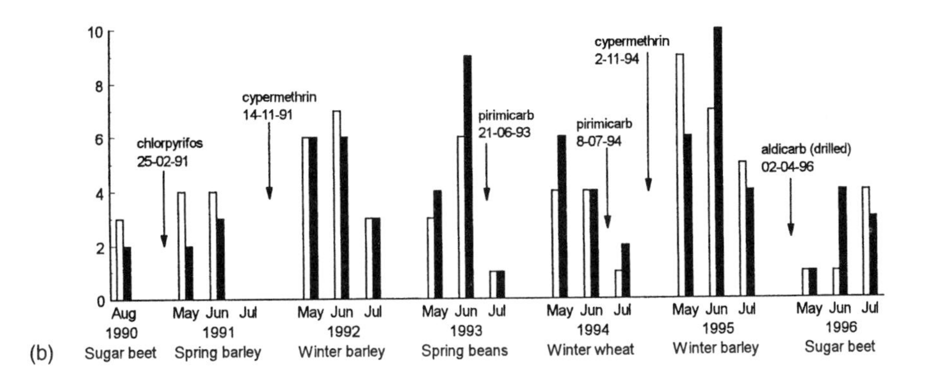

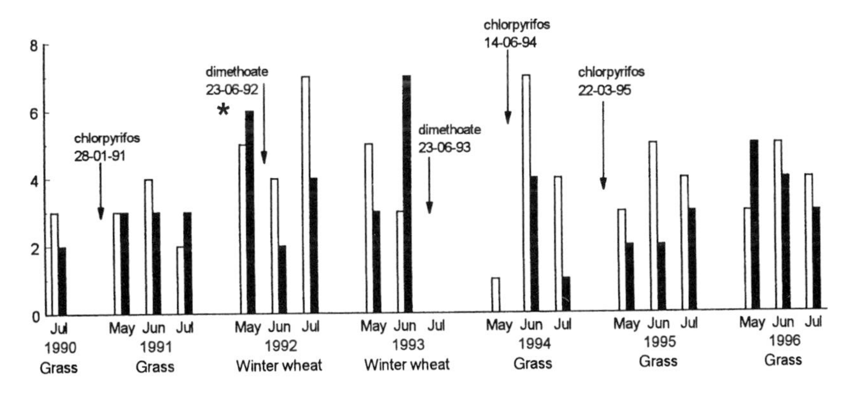

Figure 31.2 Number of predatory Coleoptera species in four suction samples (1.84 m^2) from current farm practice (CFP) (■) and reduced input approach (RIA) (□) pesticide regimes of three rotations. Insecticides were applied only under the CFP regime. Minimum time interval between sampling and a previous insecticide application = 5 days, except * (= 1 day). **(a)** Bugdale field, High Mowthorpe; **(b)** Near Kingston field, Gleadthorpe; **(c)** Field 5, Drayton.

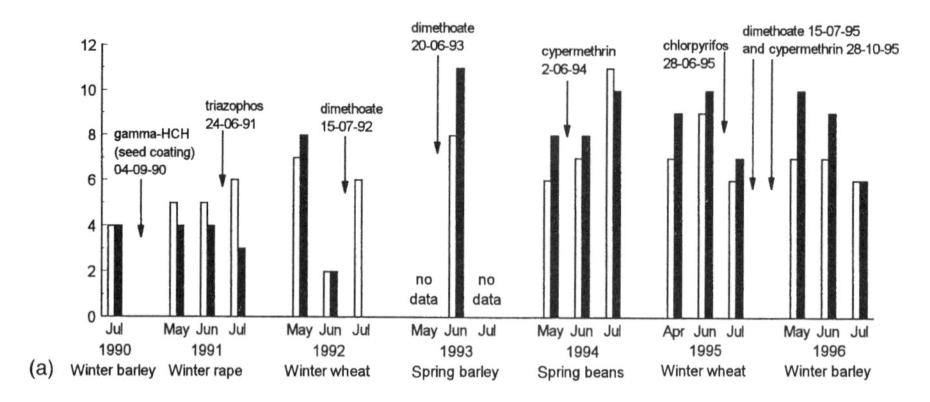

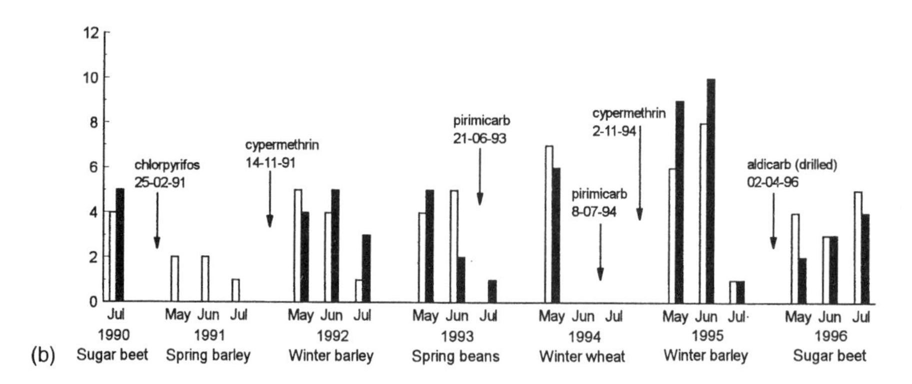

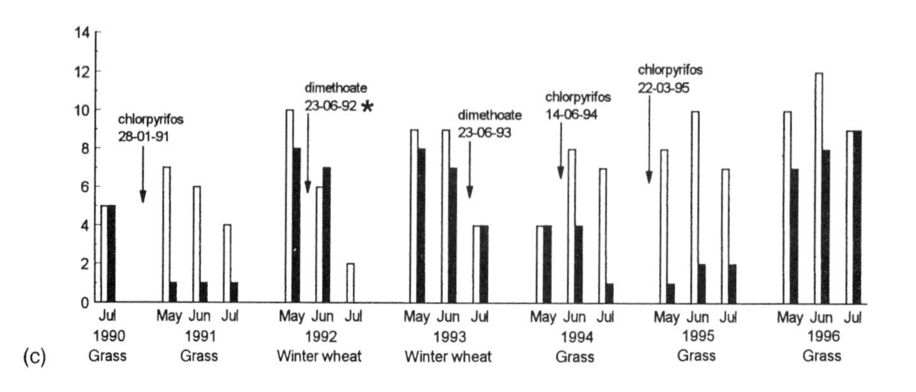

Figure 31.3 Number of abundant ($n > 3$) Collembola species in four suction samples (1.84 m^2) from current farm practice (CFP) (■) and reduced input approach (RIA) (□) pesticide regimes of three rotations. Minimum time interval between sampling and a previous insecticide application = 5 days, except *(= 1 day). **(a)** Bugdale field, High Mowthorpe; **(b)** Near Kingston field, Gleadthorpe; **(c)** Field 5, Drayton.

summer (Figure 31.3). There were no consistent patterns in the total numbers of Collembola and predatory Coleoptera species trapped after use of the synthetic pyrethroid, cypermethrin.

DISCUSSION

The results presented here focus on predatory Coleoptera, Collembola and insecticides in three fields and so represent a subset of the full SCARAB project data. The design of SCARAB, lacking orthodox replication, is amenable to multivariate methods of data analysis (e.g. as in Siepel and Van de Bund, 1988; Sanderson, 1994). For present purposes, confidence that changes in species richness shown in Figures 31.2 and 31.3 reflect actual effects of insecticides is based on: the timing of the changes in relation to pesticide applications; the magnitude of the changes; their persistence; and, for Collembola, the relative parity of CFP and RIA catches in the 'pre-treatment' year (1990). Other farming system studies (examples in Frampton, 1997b) have used similar criteria to infer negative or positive pesticide effects in unreplicated experimental designs.

The primary aim of SCARAB was to investigate the overall environmental consequences of using conventional and reduced-input regimes rather than effects of individual pesticides, but results consistent with negative effects of particular broad-spectrum insecticides have occurred. Some of the spatial variation in the arthropod catches shown in Figures 31.2 and 31.3 perhaps also reflects effects of fungicide and herbicide use, which differed between the regimes (Figure 31.1). For instance, the lower CFP number of Collembola species trapped in spring beans in 1993 (Figure 31.3b) would be consistent with a negative effect of the fungicide benomyl (e.g. Krogh, 1991) which was applied in May as a greater quantity of active ingredient under CFP than RIA. Effects of some pesticides would have been too subtle to detect against the perturbations caused by broad-spectrum insecticides, particularly if they were indirect or cumulative and thereby not coinciding clearly with specific applications. The environmental importance of such use of pesticides should be clarified when the full data set, which includes other fields with less intensive use of broad-spectrum insecticides, is analysed. Confidence in the interpretation of individual non-replicated observations, such as the use of benomyl described above, should also be improved as the full analysis will include a greater number of representative pesticide-field-crop combinations. Present results indicate that the most obvious negative effects of pesticide use were related to organophosphorus insecticides, but even within this class of pesticide there was spatial and temporal variation, with effects of chlorpyrifos on numbers of Collembola species more pronounced than those of dimethoate at two of the sites (Figure 31.3b,c).

There were a number of instances where more species were trapped in the CFP (sprayed) regime than the RIA but with no consistent pattern to give clear support to any positive effects of pesticide use. Statistically significant increases in the abundance of some Collembola species occurred after cypermethrin was sprayed in a replicated-field study in wheat (Frampton, 1997a) but so far no evidence of an effect of cypermethrin has been detected in the SCARAB fields. Given the known broad-spectrum properties of dimethoate, it was unexpected that more species of Collembola and predatory Coleoptera would be trapped in the current farm practice regime after use of dimethoate in spring barley in June 1993 (Figure 31.3a). Abundance of Collembola in CFP catches was 26% higher than in RIA catches and this pattern was also observed for Staphylinidae (36% more abundant in CFP samples), Lathridiidae (53%), Aphididae (26%) and Hymenoptera (33%). These results indicate that a broad-spectrum effect of dimethoate was lacking on this occasion, emphasizing the need for pesticide studies to include a representative range of agricultural, geographical and meteorological conditions for extrapolation of results to be acceptable. In 1992, use of dimethoate in winter wheat was followed by a lower catch of Coleoptera species (Figure 31.2c) but without a decrease in Collembola species (Figure 31.3c). The most likely reason for this result, in which sampling took place one day after the application, is that foliage-dwelling species of predator were exposed to the dimethoate sooner than ground-dwelling Collembola. Another broad-spectrum insecticide application which did not clearly decrease the richness of Collembola species was the use of chlorpyrifos in winter wheat in Bugdale field in 1995 (Figure 31.3a). However, on this occasion both the number of predator species (Figure 31.2a) and the total collembolan abundance, which was 78% lower in CFP than RIA catches in July, were consistent with a negative effect of the insecticide, indicating that species richness alone may be insensitive to some effects of broad-spectrum insecticides.

Despite the limitations of species richness data, it is desirable – when predicting the overall risk to an arthropod community – to ascertain the spectrum of susceptible species and their relative ecological importance. Preliminary results from SCARAB show that there were some specific crop-field-pesticide combinations in which all of the suction sampled (i.e. diurnally active) species of predatory Coleoptera or abundant Collembola species were eliminated, e.g. after use of triazophos in 1991 (Figure 31.2a), chlorpyrifos in 1991 (Figure 31.3b) or dimethoate in 1992 (Figure 31.3a). When determining the range of susceptible species it is important to consider any bias inherent in the sampling method. Pitfall trap catches in grass in 1991 indicated that several species of Carabidae were susceptible to chlorpyrifos (Frampton and Çilgi, 1994) whereas these effects were underestimated by suction samples taken at the same time (Figure 31.2c) because Carabidae formed a very small proportion of the predator catch. In the three SCARAB fields considered here, total elimination of diurnally active predatory Coleoptera or Collembola occurred relatively infrequently but recovery times of some collembolan species subsequently took several years (Frampton, 1997a).

Although the results discussed here are not complete, they illustrate some key points for consideration when conducting pesticide side-effects studies with arthropods:

- Collembola may be suitable for detecting negative effects of organophosphorus pesticide use, as several susceptible species occur in arable crops; currently Coleoptera, but not Collembola, are recommended in field testing procedures with pesticides (e.g. Barrett *et al.*, 1994).
- Extrapolation of pesticide effects from one taxon, geographical location, agricultural or meteorological scenario could be misleading.
- Coleoptera and Collembola are not suitable as indicators of all CFP side-effects, since neither group exhibited consistent negative responses to the synthetic pyrethroid insecticide cypermethrin, which is harmful to other arthropods such as Araneae (e.g. Pullen *et al.*, 1992).

As the occurrence of susceptible species varies spatially and temporally (Frampton, 1997a,b), long-term multi-site studies of pesticide effects are needed to aid interpretation of the results from the many pesticide studies conducted in single fields.

ACKNOWLEDGEMENTS

Funding of the SCARAB project by the Ministry of Agriculture, Fisheries and Food and the assistance of ADAS staff in the collection of arthropod samples at ADAS Drayton, Gleadthorpe and High Mowthorpe Research Centres is gratefully acknowledged.

REFERENCES

Barrett, K.L., Grandy, N., Harrison, E.G. *et al.* (1994) *Guidance Document on Regulatory Testing Procedures for Pesticides with Non Target Arthropods. From the ESCORT workshop, Wageningen, Netherlands, March 1994*, SETAC-Europe, Brussels.

Burn, A.J. (1992) Interactions between cereal pests and their predators and parasites, in *Pesticides, Cereal Farming and the Environment*, (eds P.W. Greig-Smith, G.W. Frampton and A.R. Hardy), HMSO, London, pp. 110–131.

Dietrick, E.J. (1961) An improved back pack motor fan for improved suction sampling of insect populations. *Journal of Economic Entomology*, **54**, 394–395.

Frampton, G.K. (1997a) Species spectrum, severity and persistence of pesticide side-effects on UK arable springtail populations, in *Proceedings, ANPP Fourth International Conference on Pests in Agriculture, Montpellier, January 1997* **1**, pp. 129–136.

Frampton, G.K. (1997b) The potential of Collembola as indicators of pesticide usage: evidence and methods from the UK arable ecosystem. *Pedobiologia*, **41**, 34–39.

Frampton, G.K. and Çilgi, T. (1994) Long-term effects of pesticides on Carabidae in UK farmland: some initial results from the 'SCARAB' Project, in *Carabid Beetles: Ecology and Evolution*, (ed. K. Desender), Kluwer Academic Publishers, Rotterdam, pp. 433–438.

Frampton, G.K. and Çilgi, T. (1996) How do arable rotations influence pesticide side-effects on arthropods? *Aspects of Applied Biology*, **47**, 127–135.

Garthwaite, D.G., Thomas, M.R. and Hart, M. (1995) *Pesticide Usage Survey Report 127: Arable Farm Crops in Great Britain 1994*, MAFF Publications, London.

Greig-Smith, P.W., Frampton, G.K. and Hardy, A.R. (1992) *Pesticides, Cereal Farming and the Environment – the Boxworth Project*, HMSO, London.

Krogh, P.H. (1991) Perturbation of the soil microarthropod community with the pesticides benomyl and isofenphos. I. Population changes. *Pedobiologia*, **35**, 71–88.

Pullen, A.J., Jepson, P.C. and Sotherton, N.W. (1992) Terrestrial non-target arthropods and the autumn application of synthetic pyrethroid insecticides: experimental methodology and the trade-off between replication and plot size. *Archives of Environmental Contamination and Toxicology*, **23**, 246–258.

Sanderson, R.A. (1994) Carabidae and cereals: a multivariate approach, in *Carabid Beetles: Ecology and Evolution*, (ed. K. Desender), Kluwer Academic Publishers, Rotterdam, pp. 433–438.

Siepel, H. and Van de Bund, C.F. (1988) The influence of management practices on the microarthropod community of grassland. *Pedobiologia*, **31**, 339–354.

Vickerman, G.P. (1992) The effects of different pesticide regimes on the invertebrate fauna of winter wheat, in *Pesticides, Cereal Farming and the Environment* (eds P.W. Greig-Smith, G.K. Frampton and A.R. Hardy), HMSO, London, pp. 82–108.

32

A dynamic risk assessment for beneficial insects in cereal canopies

Steen Gyldenkaerne

INTRODUCTION

Pesticide application in the field may pose a risk of unwanted side-effects. To give farmers and other users of pesticides the opportunity to choose pesticides from an environmental point of view and to rank them according to environmental risk, the Danish Decision Support System – PC Plant Protection (Secher, 1991; Rydahl, 1993; Secher *et al.*, 1995), will be provided with a dynamic environmental risk assessment (RA). PC Plant Protection today has more than 3000 users. When using PC Plant Protection the farmer gets a list of suitable pesticides that can help to solve the problem. For each pesticide, PC Plant Protection calculates the minimum dose that can control the weed, disease or pest (between 5% and 100% of normal application rate, depending on the problem). PC Plant Protection will be provided with environmental information based on risk quotients (RQ) or hazard quotients for different compartments (bees, birds, mammals, earthworms, aquatic organisms and beneficials). RQ is calculated as a predicted environmental concentration (PEC) divided by the toxicity of the pesticide. Leaching, degradation and worker exposure will be included in the environmental information (Gyldenkaerne, 1996). Council Directive 91/414/EEC Annex VI (Council Directive, 1994) is the general guideline for applying the RA system, but for beneficials a new approach is developed to meet the need for a dynamic RA. In future

Ecotoxicology: Pesticides and beneficial organisms.
Edited by P.T. Haskell and P. McEwen. Published in 1998 by Chapman & Hall, London. ISBN 0 412 81290 8.

toxicity data will come from Council Directive 91/414/EEC Annex II (Council Directive, 1991). Until Annex II is established, toxicity data comes from the Danish Environmental Protection Agency and other sources.

The ongoing re-evaluation of pesticides in Denmark has led to a reduction of the available number of active ingredients from 214 to 92 in 1996. Despite this, there has been an increase in ecotoxic load (amount of a.i./LD_{50}) for aquatic organisms and only a slight decrease for bees and birds (Secher and Gyldenkaerne, 1996), which has increased the need for environmental information.

To meet the need for precise information and to avoid any unreliability about the RA information, such ranking systems have to be clearly rule-based – for example, EPPO decision schemes (EPPO, 1993, 1994).

Choosing pesticides from the environmental angle may not have priority for the farmer, and so to improve the usefulness to the farmer some demands have to be fulfilled by the RA system. It should be:

- as close to field conditions as possible, by taking into account as many factors as necessary;
- reliable;
- easy to understand and use;
- educational;
- able to rank pesticides from an environmental point of view.

The system can also be used for book-keeping on the individual farm and as an indicator of environmental impact. An overview of PC Plant Protection Danish and the RA module are shown in Figure 32.1.

Beneficials

The category 'beneficials insects' is incorporated in the RA system. However, today there is no consensus on which organisms such a category should include (ESCORT 1994) and toxicity data (e.g. LD_{50}, NOEC) on these insects is not yet available so it is not possible to carry out a dynamic RA. Testing of beneficial insects is based on worst-case scenarios with one dose (full field rate).

When reviewing the literature on pesticide impact on beneficial insects in the field, no coherent trends in the results can be seen, for various reasons. Some factors may be as follows.

- The dose reaching the site where the beneficial insects are living is not measured and varies with the field conditions.
- The occurrence and activity of the beneficial insects during and after the pesticide application are partly unknown.
- The bioavailability of the pesticide after reaching the plant surface or soil surface is partly unknown.

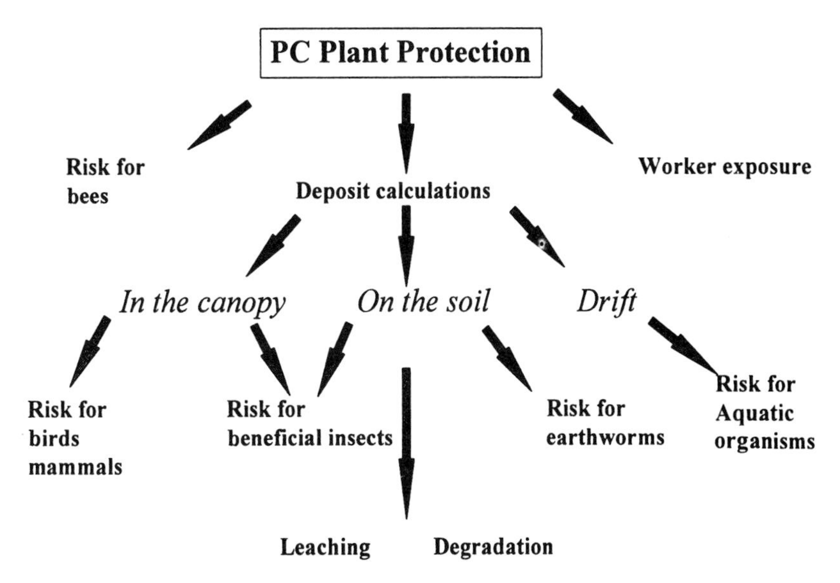

Figure 32.1 Overview of the environmental information to be implemented in PC Plant Protection.

- The efficacy of the pesticide depends on climatic conditions.
- The methods of measuring the impact on a field scale are not precise.

The Danish Institute of Plant and Soil Science has therefore decided to develop a new RA approach for beneficial insects, taking into account the exposure/toxicity ratio. The methods used are based on scientific principles. This includes a precise description of how the ranking is done and a validation of the system. The problems especially addressed will be the actual application rate and the bioavailable pesticide in the soil. Ranking of the pesticides should be based on important predator groups in the actual crop, and the RA should take into account the most sensitive predator at a given time so that maximum attention is directed to the predators. We hope that this may reduce the impact on beneficial insects and in the long term reduce pesticide applications, due to increased numbers of aphid predators in the field.

METHODS

The RA has four assumptions to give it realism and to give the actual most effective predators high priority:

- The insects differ from absence to presence in the field.
- The most effective aphid predator is used as an indicator species.

- The most sensitive species of the two in relation to the pesticide is used.
- The insects occur in different strata in the cereal canopy.

Dose rates

The actual application rate is based on the recommendation from PC Plant Protection. The dose that reaches the soil or a given place in the canopy depends on a model for canopy development and mode of pesticide application (Gyldenkaerne *et al.*, submitted).

Model indicators

Two types of insect are used in the calculations. In the spring and autumn the indicator is the small ground-living polyphagous beetle, *Bembidion lampros*, and after the aphids have arrived in the field the indicator species is a parasitoid (*Aphidius* spp.).

An investigation carried out in Denmark by the Danish Institute of Plant and Soil Science has shown that *B. lampros* hibernates in the field boundaries and moves into the fields when day-degree (DD) (> 6°C) exceeds 12 (Petersen, submitted). Before this time, spraying with pesticides in the field involves no risk. After 12 DD the risk becomes real.

The RA changes *B. lampros* for *Aphidius* spp. when either ladybirds (*Coccinella* spp.) or hoverflies (*Syrphus* spp.) are actually seen in the field by the farmers, or when the aphids are expected to have arrived in the field based on a day-degree model for aphid development (Hansen, in preparation). Until ear emergence – growth stage 55, or 450 DD (4°C) in a winter wheat crop development model (Olesen and Plauborg, 1995) – we expect the *Aphidius* spp. to be inside the canopy, partly protected by the canopy. From growth stage 55 to harvest, we assume the aphids as well as the parasitoids to be sitting on the ear. After harvest the RA is carried out on *B. lampros* until the soil temperature is below 4°C (end of November; Petersen, submitted). Figure 32.2 shows the height and the position of the indicator species in the canopy.

Risk assessment

For *B. lampros* the risk depends on the amount of pesticide reaching the soil. The amount of pesticide penetrating the canopy depends on the actual condition of the crop (leaf area index, LAI). LAI depends on growth stage and crop condition. For hydraulic nozzles the ground deposit depends only on LAI and canopy height (Gyldenkaerne *et al.*, submitted), whereas for other application techniques other models are used. When the pesticide reaches the soil, some pesticide is chemically

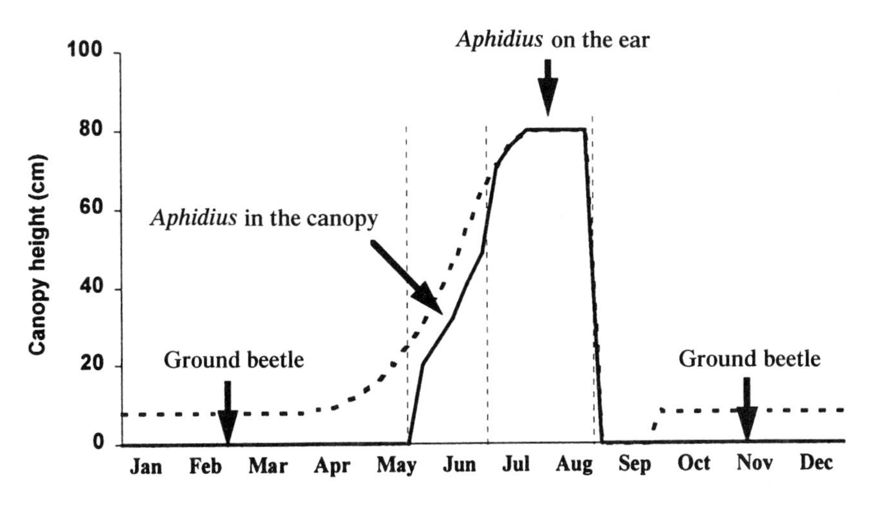

Figure 32.2 Indicator species *Bembidion lampros* and *Aphidius* spp. and their position in the cereal canopy in relation to crop height.

bound and becomes unavailable for the beetle. This process is soil dependent. A basic model for bioavailable pesticides in different soils is developed, to take into consideration the influence of soil type (Gyldenkaerne, 1996). The RA is then carried out by comparing the body burden, which is the calculated integral of pesticide in the soil water, with an estimated LD_{50} value. It is assumed that the mortality of the beetles is caused only by residual uptake from the soil and not from topical application on the beetle. This assumption can partly be verified by comparing toxicity data from Wiles and Jepson (1992, 1994) and Gyldenkaerne and Jepson (in preparation), who showed that the residual uptake is the most important mortality factor for deltamethrin for small beetles compared with the topical mortality. Moreover, only a few beetles can be found on the soil surface, compared with the total number of beetles (Sunderland *et al.*, 1987).

For *Aphidius* spp. we propose a hazard index based on the topical application rate. In the model we assume the *Aphidius* spp. to be partly protected (one LAI) by the plant canopy until growth stage 55, thus reducing the topical application by 40%. The topical dose is calculated from the dorsal size of a standard *Aphidius* female. The received dose is then application rate (g a.i./ha) $\times$ 1^{-10} $\times$ dorsal size (mm^2) $\times$ [1 – (LAI reduction factor)].

At every spraying occasion the risk assessment is carried out by comparing the received dose with an estimated LD_{50} value, inserted in a probit mortality diagram. Figure 32.3 gives an approximate exposure rate in relation to the application rate.

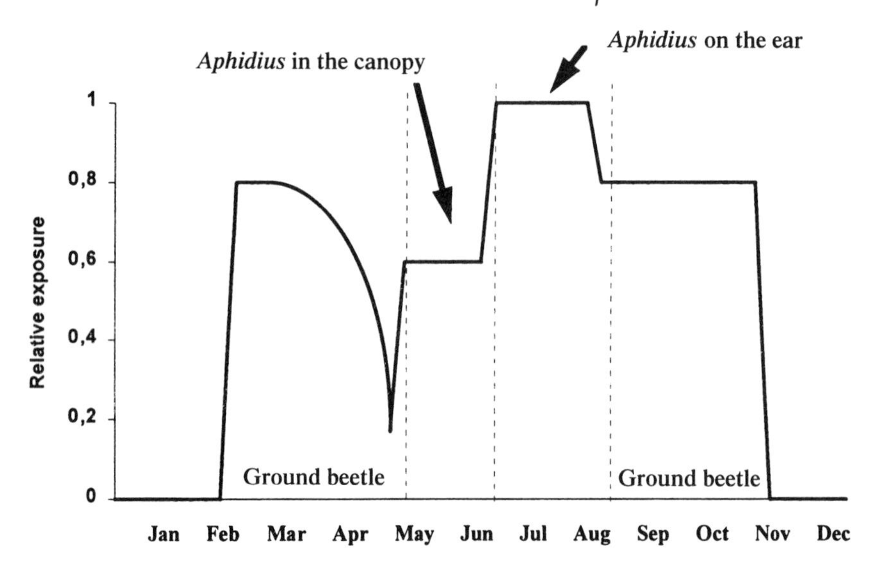

Figure 32.3 Simplified exposure rate for *Bembidion lampros* and *Aphidius* spp. in relation to the actual application rate, based on pesticide penetrating the canopy.

Toxicity data – lack of data – extrapolation

In Council Directive 94/93/EEC there is no demand for testing for LD_{50} values on beneficial insects. In the literature, only limited toxicity data is available on *B. lampros* and *Aphidius* spp. and in particular there are very few dose-response curves for carabids on natural soils. Getting a complete data set seems impossible for the moment. A high degree of extrapolation of toxicity data from other sources is therefore necessary. Until further data are available, we will use a mix of International Organization for Biological Control of Noxious Animals and Plants (IOBC) data supplied with literature data on relevant species and on honey bees to give as much information as possible extrapolated to the size of *B. lampros* and *Aphidius*. This extrapolation assumes that penetration, elimination rates and all possible reaction kinetics of the pesticide inside the beetle and *Aphidius* are identical. Toxicity is normally expressed as amount of active ingredients/mg insect. For topical application of deltamethrin to several species of beetles, in a weight range from 2 to 150 mg, Wiles and Jepson (1992) found an increasing toxicity/mg beetle with increasing body size. This can also be shown for other pesticides. A better way of expressing toxicity (other than mg/beetle) is probably an expression based on surface area, which was shown by Critchley (1972) for thionazin. However, there is a close relationship between the surface area of an insect and its metabolic rate. For insects, an estimate of the metabolic rate can be calculated as body weight (Kayser and Heusner, 1964; Mill, 1985). This suggests that toxicity preferential should be expressed according to metabolic rate, and that the toxic effect is due to a depression of the metabolism.

So far, translating a discrete topical dosage into a continuous uptake from the soil seems difficult. There is therefore a need to make dose-response mortality curves on the bioavailable pesticide in the testing procedure. The model for soil bioavailability (Gyldenkaerne, 1997) offers a good opportunity to extrapolate from one soil type to another.

Risk assessment

In PC Plant Protection the RQ is shown either in a quantitative way or in five classes of RA where the risk is low, low-medium, medium, high or very high. The maximum acceptable mortality for beneficial insects according to Council Directive 94/43/EEC Annex VI is 30%. Thus an RQ giving higher mortality than 30% is classified as very high risk. An RQ less than 0.01 of the maximum acceptable mortality is defined as low risk.

RESULTS

Figure 32.4 gives an example of how the index for beneficials insects differs in a medium-dense wheat field on a sandy loam from deltamethrin, fenvalerate and parathion with normal application rates. Figure 32.4 should only be seen as an example and not as the final version, because further validation is needed. Deltamethrin is the least harmful insecticide, and parathion the most harmful. It can be seen that the predicted mortality decreases in April-June due to an increase in crop cover. In May-June the indicator species is changed from *B. lampros* to *Aphidius*. As a result the predicted mortality increases. After harvest the risk calculations are again based on *B. lampros*. During the winter the predicted mortality is zero because it is presumed that beetles occur in the field.

DISCUSSION

By including the effect of weather, crop growth stages and soil type, a dynamic RA system is developed. The different application recommendations given by PC Plant Protection and the possibilities of using the dynamic of the canopy development can give an advanced RA and an opportunity for the user to see the importance of different management strategies. The technology used in PC Plant Protection gives further possibilities of using dose-response curves instead of a linear RA. The models are based on scientific principles, but a field validation still needs

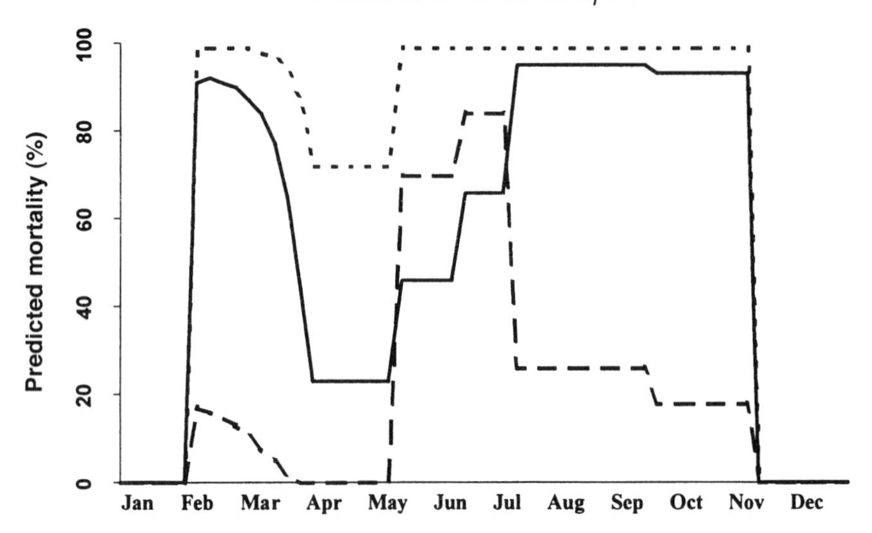

Figure 32.4 Predicted mortality for indicator species by using deltamethrin (– –), fenvalerate (——) and parathion (- - -) throughout the year in proposed risk assessment system in PC Plant Protection.

to be carried out. Not until this has been done will we know if the prediction is made satisfactorily.

The model for bioavailability of pesticides in the soil is based on basic physical and chemical properties of the pesticide and soil properties. The model is not yet fully validated and data to calculate dose-response curves are especially needed.

We have decided to use *B. lampros* as an indicator species for ground-living beetles because we find it an important and common carabid in arable fields. In the registration procedure *Poecilus cupreus* is often used as a test species although *B. lampros* may be more sensible. For an RA system such as this to become a common way of extrapolating to more sensible species, a safety factor could be added or extrapolations could be based on differences in metabolic rates.

Because of the dynamic approach in the RA, dose-response curves based on residual uptake of pesticide from soil would be the best approach for estimating risk for ground-living beetles. Since toxicity data on the indicator species are scarce, and it seems very difficult to extrapolate from standard IOBC tests to real field conditions, we may use data from different sources and of different quality in the RA. We find a need to expand the standard test to comprise dose-response curves on defined soils with known humidity and temperature. In that case, using data for calculating more advanced RA under field conditions would be possible.

REFERENCES

Barrett, K.L., Grandy, N., Harrison, E.G. *et al.* (1994) *Guidance Document on Regulatory Testing Procedures for Pesticides with Non Target Arthropods. From the ESCORT workshop, Wageningen, Netherlands, March 1994*, SETAC-Europe, Brussels.

Chritchley, B.R. (1972) A laboratory study of the effects of some soil-applied organophosphorous pesticides on Carabidae (Coleoptera). *Bulletin of Entomological Research*, **62**, 229–242.

Council Directive (1991) Council Directive 15 July 1991 concerning the placing of plant protection products on the market (91/414/EEC). *Official Journal of the European Communities*, **L230**, 1–32.

Council Directive (1994) Council Directive 94/43/EEC of 27 July 1994, establishing Annex VI to Directive 91/414/EEC concerning the placing of plant protection products on the market. *Official Journal of the European Communities*, **L227**.

EPPO (1993) Decision-making schemes for the environmental risk assessment of plant protection products. *Bulletin OEPP/EPPO*, **23**, 1–165.

EPPO (1994) Decision-making schemes for the environmental risk assessment of plant protection products - arthropod natural enemies. *EPPO Bulletin*, **24**, 17–35.

Gyldenkaerne, S. (1996) Integrating environmental risk assessment in PC Plant Protection, 13, in *Danish Plant Protection Conference 1996, Side Effects of Pesticides*, SP Report 3, Danish Institute of Plant and Soil Science, pp. 85–95.

Gyldenkaerne, S. (1997) Bioavailability of pesticides in different soil, in *New Studies in Ecotoxicology* (eds P.T. Haskell and P.K. McEwen), Welsh Pest Management Forum, Cardiff (pp. 26–29).

Gyldenkaerne, S. and Jepson, P.C. (in preparation) Estimating the effect of deltamethrin on ground living beetles in cereals.

Gyldenkaerne, S., Secher, B. and Nordbo, E. (submitted) Ground deposit of pesticides in cereals in relation of Leaf Area Index.

Hansen, L.M. (in preparation) Prognose for ankomsttidspunkt af havrebladlus i vårbyg.

Kayser, C. and Heusner, A. (1964) Etude comparative du métabolisme énergétique dans la sèries animale. *Journal of Physiology*, **56**, 489–524.

Mill, P.J. (1985) Structure and physiology of the respiratory system, in *Comprehensive Insect Physiology* (eds G.A. Kerut and L.I. Gilbert), Pergamon Press, Oxford, pp. 517–593.

Olesen, J.E. and Plauborg, F. (1995) MVTOOL version 1.10 for developing MARKVAND, in *SP Report, 27*, Danish Institute of Plant and Soil Science, p. 64.

Petersen, M.K. (submitted) Post-winter events for *B. lampros* and *T. hypnorum* - two polyphagous predators in arable fields.

Rydahl, P. (1993) PC Plant Protection: optimizing weed control, in *Workshop on Computer-based DSS on Crop Protection, Parma, Italy, 23–26 November 1993*, SP Report 7, Danish Institute of Plant and Soil Science, pp. 81–87.

Secher, B.J.M. (1991) An information system for plant protection. 2. Recommendation models, structure and performance. *Annals ANPP*, **2**, 153–160.

Secher, B.J.M. and Gyldenkaerne, S. (1996) Regulating pesticide use and environmental impacts, in *Integrated Environmental and Economic Analysis in Agriculture* (eds A. Walter-Joergense and S. Pilegaard), Statens Jordbrugs-og Fiskerioekonomiste Institut Report no. 89, pp. 163–172.

Secher, B.J.M., Joergensen, L.N., Murali, N.S. and Boll, P.S. (1995) Field evaluation of a decision support system for the control of pests and diseases in cereals in Denmark. *Pesticide Science*, **45**, 195–199.

Sunderland, K.D., Hawkes, C., Stevenson, J.H. *et al.* (1987) Accurate estimation of invertebrate density in cereals. *Bulletin SROP/WPRS Bulletin*, **X**(1), 71–81.

Wiles, J. and Jepson, P. (1992) The susceptibility of a cereal aphid pest and its natural enemies to deltamethrin. *Pesticide Science*, **36**, 263–272.

Wiles, J. and Jepson, P. (1994) Substrate-mediated toxicity of deltamethrin residues to beneficial invertebrates: estimation of toxicity factors to aid risk assessment. *Archives of Environmental Contamination and Toxicology*, **27**, 384–391.

Part Four

Tools for Tomorrow

33

Introduction

John Wiles

The recognition of the importance of beneficial arthropods to modern agriculture, and the maintenance of agroecosystem diversity and stability, would seem to have come of age. Certainly, as emphasized a number of times elsewhere in this book, most of the major players involved – such as regulators, policy makers, agrochemical companies, crop management advisers, farmers, growers, produce suppliers and even consumers – are now aware of the importance and value of groups of these organisms. It is clear that research with beneficial arthropods is gaining recognition and expanding into many areas of modern crop protection and production.

The chapters in this section illustrate this expansion and provide insights, food for thought and to some extent an agenda for research to further this development. A common theme of new and expanding collaboration runs through the four chapters. This would seem to indicate that, to some degree, those researching the side-effects of pesticides on beneficial arthropods have been somewhat separated from those working in other areas of crop production. It may now be time to call for closer collaboration between pest biologists, formulation chemists and ecological modellers, to name but a few.

The contributions from Birnie *et al.* (Chapter 34) and Minks and Kirsch (Chapter 36) discuss how pest control measures may either directly or indirectly influence beneficial arthropod performance. In particular they illustrate that a more concerted effort is being made to research tritrophic relationships between host, pest and beneficial organism interactions under the influence of pesticides, where subtle changes in

Ecotoxicology: Pesticides and beneficial organisms.
Edited by P.T. Haskell and P. McEwen. Published in 1998 by Chapman & Hall, London. ISBN 0 412 81290 8.

response may have an unpredictable outcome for beneficial organism populations.

In Chapter 35, Wiles and Barrett explore a range of qualitative and quantitative approaches for risk extrapolation between arthropod species, using case study examples. The time would seem right to evaluate the side-effects data generated to date and to look to establish patterns of species sensitivity, species exposure and chemical bioavailability, so that we may further develop 'rules of thumb' to guide the development of extrapolation methodologies.

It is clear from other sections of this book that those researching the ecotoxicology of pesticides have devised a wealth of methodologies to assess the toxicity to beneficial arthropods, but the theoretical basis for predicting effects at the population level remains somewhat undeveloped. Standard approaches involve direct assessment of mortality or sublethal effects which feed into threshold schemes. Stark *et al.* (Chapter 37) re-evaluate the importance of different toxic endpoints and discuss new approaches, incorporating demographic and toxicological measurements, which may offer considerable potential for improving predictions of the population consequences of observed effects.

If this section of the book gives a snapshot of the new developments in beneficial arthropod research, it would seem that an exciting time lies ahead for those working in this area. In particular the time appears to have arrived for evaluating and testing ecological approaches in order to establish a more solid theoretical base for this field.

34

Implications of insecticide resistance for interactions between pests, natural enemies and pesticides

Linzi Birnie, Barbara Hackett and Ian Denholm

INTRODUCTION

The development of insecticide resistance by insect pests is most commonly perceived merely as a threat to the efficacy or profitability of chemical control agents. Its possible influence on interactions between pests and natural enemies has received relatively little attention in the ecotoxicological literature. However, there is growing evidence that genetic variation in the response of insects to insecticides can exert a significant influence on the survival of natural enemies, and hence their contribution to pest management practices (Furlong and Wright, 1993; Birnie *et al.*, 1996).

This chapter reviews some implications of insecticide resistance for ecotoxicology in both a conceptual and experimental framework. Topics and examples that are considered relate primarily to insect parasitoids, whose life histories depend so intimately on host biology that even minor changes in the latter can conceivably have a significant effect on pest-parasitoid dynamics in the presence of insecticide treatments.

Ecotoxicology: Pesticides and beneficial organisms.
Edited by P.T. Haskell and P. McEwen. Published in 1998 by Chapman & Hall, London. ISBN 0 412 81290 8.

GENERAL CONSIDERATIONS

Some potential interactions between resistance and natural enemy survival are summarized schematically in Figure 34.1. The two dimensions of this represent contrasting perspectives on the outcome of exposing insect pests to insecticides. The situation depicted is a simplistic one entailing a single pest and associated parasitoid, but serves nonetheless to highlight interactions that are likely to assume increasing significance as resistance problems in crop pests continue to expand in breadth and intensity.

Resistance management perspective

The horizontal dimension in Figure 34.1 represents a progressive change in the genetic composition of a host population, whereby alleles conferring resistance increase in frequency under selection with an insecticide. As selection progresses, efficacy of the selecting agent is likely to be increasingly compromised – leading, if unchecked, to failure in controlling one or more life stages of the pest. Although the relationship between resistance and field control can be complex and is often unclear (Denholm *et al.*, 1984), it is self-evident that resistance must confer some protection for the trait to be selected in the first place (Roush and Daly, 1990; Denholm and Rowland, 1992). Increasing resistance *per se* is unquestionably a detrimental process that erodes control options and threatens the sustainability of pest management in numerous cropping

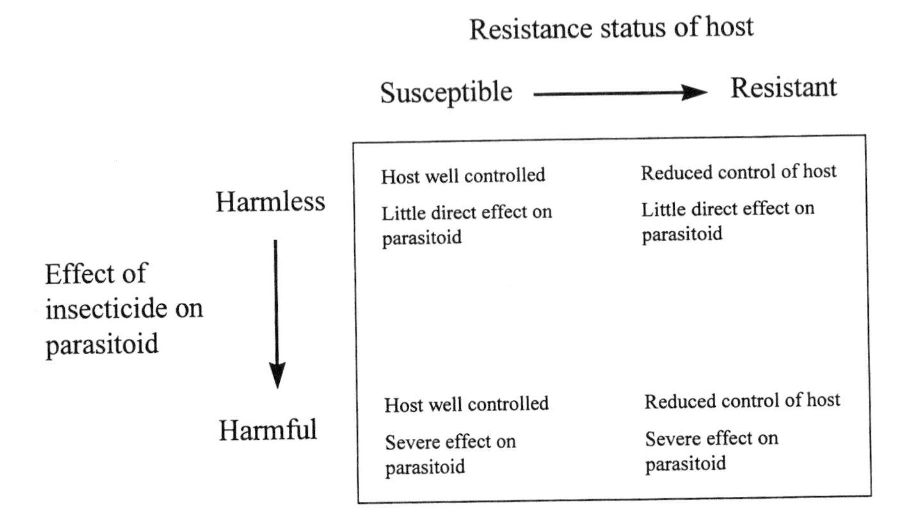

Figure 34.1 Interactions between the resistance status of a host species and the response of an insect parasitoid under treatment with an insecticide.

systems. As a consequence, there has been much effort by scientists, industrialists and policy-makers to identify and implement tactics to delay resistance or reduce its impact on control practices (Denholm and Rowland, 1992; Forrester *et al.*, 1993; Leonard and Perrin, 1994).

Ecotoxicological perspective

The vertical dimension in Figure 34.1 will be more familiar to ecotoxicologists attempting to minimize side-effects of insecticides or enhance the contribution of non-chemical agents, such as insect parasitoids, to pest management. This again reflects a continuum of responses, recording the perceived effect of an insecticide on a parasitoid which, under unsprayed conditions, would be expected to cause appreciable mortality of the pest in question. Methods and approaches for categorizing insecticides in this respect have been subject to intense scrutiny (Hassan, 1983; Hassan *et al.*, 1985), and challenges with extrapolating laboratory data on insecticide selectivity to the field are implicit in several contributions to this volume. However, it is undeniable that insecticides do differ in the extent and persistence of their harmful effects on non-target organisms such as parasitoids. As with resistance management, there has therefore been considerable work by researchers, politicians and the agrochemical industry to coordinate methodology and identify how best to deploy insecticides to conserve indigenous and exotic natural enemies to the greatest extent possible (Council Directive, 1991; Barrett *et al.*, 1994).

INTEGRATION OF PERSPECTIVES

Towards integrated resistance management

The importance of integrating these two perspectives becomes most readily apparent when considering how a knowledge of insecticide selectivity can be used to assist with resistance management and, conversely, how failure to encompass this factor may exacerbate control problems, particularly if resistance is already established. The former underpins a now widely accepted principle of employing all possible non-chemical methods to minimize reliance on insecticides and thereby reduce the selection pressure for resistance genes (Denholm and Rowland, 1992). This is especially critical in environments such as glasshouses where conditions favouring rapid pest build-up, coupled with lack of immigration from unsprayed areas, render resistance an almost inevitable consequence of intensive insecticide applications (Sanderson and Roush, 1995; Denholm *et al.*, in press). The potential drawback of exploiting parasitoids for this purpose, particularly in confined habitats or for species with a narrow range of hosts, is that high mortality by an insecticide to which pests are still largely susceptible (i.e. the left-hand side of Figure 34.1) may preclude the survival of parasitoids irrespective of their response to the insecticide being applied. With more generalist species or under open field conditions, the risk of host depletion is substantially reduced and opportunities for integrating parasitoids with selective insecticides to manage resistance increase accordingly. To date, one of the most successful examples of this approach relates to cotton in Israel, where long-standing problems of combating whitefly (*Bemisia tabaci*) with conventional pyrethroids and organophosphates have been effectively overcome by switching to novel compounds (e.g. buprofezin and pyriproxyfen) active primarily against immature stages of the whitefly, and known to inflict far less direct mortality on parasitoids (*Encarsia* and *Eretmocerus* spp.) attacking this species (Horowitz *et al.*, 1994). Use of pheromones to disrupt the mating and build-up of pink bollworm (*Pectinophora gossypiella*) has further reduced the need to use insecticides in the early season when whitefly parasitoids are present on the cotton crop. As a result, it has been possible to deploy the new chemicals sparingly and in rotation, in accordance with standard resistance management recommendations (Horowitz *et al.*, 1994; Denholm *et al.*, in press).

Causes of pest flaring

The tendency of some pest populations to flare in numbers following insecticide applications is a frequently discussed but still poorly understood phenomenon (Dittrich *et al.*, 1974; Chellia and Heinrichs, 1980; Kerns and Gaylor, 1991; Sundaramurthy, 1992). Causal factors that have been implicated include the removal of competing species, alteration of plant quality, loss of natural enemies, and the reported ability of sublethal exposure to some insecticides to stimulate pest development or fecundity (reviewed by Hardin *et al.*, 1995).

Some of the most compelling data implicating interactions between resistance and natural enemy mortality in flaring comes from work with *B. tabaci* and its aphelinid parasitoid *Eretmocerus mundus* under simulated field conditions in the laboratory (Birnie and Denholm, 1992 and unpublished data). Experiments summarized in Figure 34.2 involved *B. tabaci* strains collected in the Sudan (SUD-R) and Pakistan (PAK–2) that differed markedly in resistance to insecticides. In leaf-dip bioassays, resistance ratios of SUD-R and PAK–2 at LC_{50} (relative to the response of a standard susceptible strain) were 4 and 39, respectively, for the organophosphate profenofos, and 5 and 26, respectively, for the pyrethroid cypermethrin. A release of SUD-R or PAK–2 adults on to cotton plants in large cages was followed 10 days later by adults of *E. mundus* at a density known to cause significant suppression of subsequent whitefly generations. On day

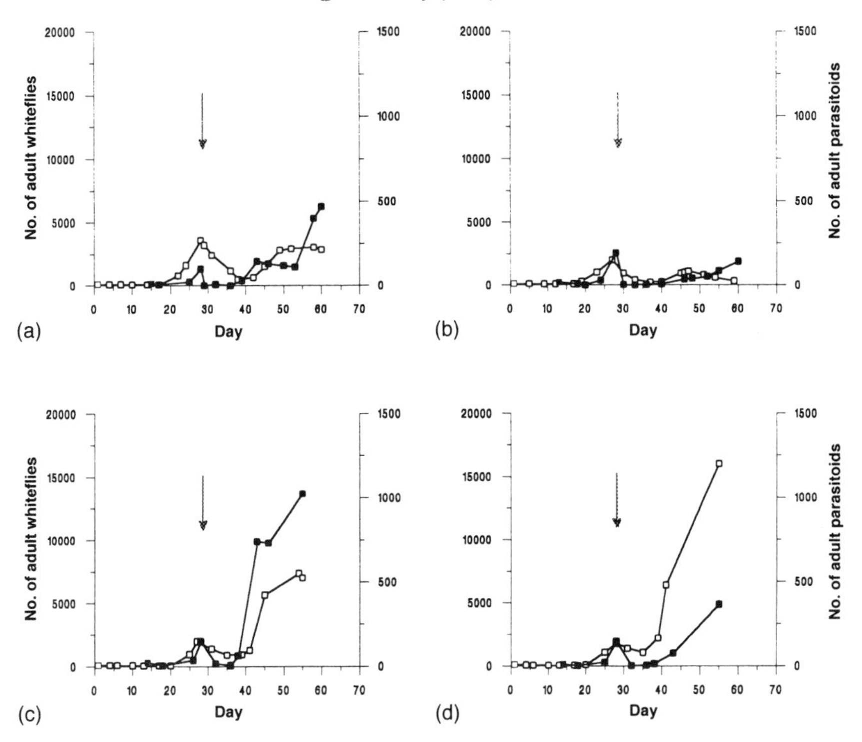

Figure 34.2 Response of weakly (SUD-R) and strongly (PAK-2) insecticide-resistant strains of *Bemisia tabaci* (□) and their parasitoid *Eretmocerus mundus* (■) sprayed with recommended field rates of profenofos and cypermethrin in the laboratory. Note different scales for whitefly (left vertical axis) and parasitoid (right vertical axis). **(a)** SUD-R, profenofos; **(b)** SUD-R, cypermethrin; **(c)** PAK-2, profenofos; **(d)** PAK-2, cypermethrin.

28 of the experiments, cages were sprayed with either profenofos or cypermethrin at their recommended field application rates. Numbers of adult whiteflies and parasitoids were then monitored regularly for a further 4–5-week period.

Results demonstrated substantial differences in the response of the strains to these regimes. In all cases, the sprays caused severe initial mortality of adult parasitoids, although the effects of profenofos were more transient than those of cypermethrin and enabled parasitoid numbers to recover more rapidly in the aftermath of spraying. The most clear-cut difference between SUD-R and PAK–2 occurred with cypermethrin, the more persistent of the two chemicals investigated. As expected, SUD-R proved considerably more susceptible than PAK–2 to this compound applied at the field rate; adult numbers declined gradually but very markedly after spraying and showed little or no tendency to re-build thereafter (Figure 34.2b). PAK–2 numbers were much less affected initially and, freed from regulation by *E. mundus*, built up rapidly to exceed 15 000 per cage by the end of this experiment (Figure 34.2d).

The outcome of such experiments implies that although resistance and parasitoid depletion can each impede pest control in their own right, the combination of the two may be a particularly potent force in promoting pest outbreaks following use of broad-spectrum chemicals such as pyrethroids. As noted by Waage (1992), the situation is analogous to one in which an insecticide directed against a primary pest, and causing a general loss in the natural enemy complex on a crop, may precipitate the flaring of secondary pests better protected from the spray due to inappropriate targeting or their intrinsically greater tolerance of the chemical. The interaction represented in the bottom right-hand corner of Figure 34.1 therefore illustrates the worst possible outcome of insecticide treatment on a pest/parasitoid system and reinforces the importance of obtaining and heeding data on both resistance and insecticide selectivity before implementing integrated pest management tactics.

CAN PEST RESISTANCE FAVOUR THE SURVIVAL OF PARASITOIDS?

The possibility of resistance in host species benefiting parasitoid survival by conferring physical or even biochemical protection from an insecticide spray was most clearly demonstrated with experiments with the diamondback moth, *Plutella xylostella*. Furlong and Wright (1993) investigated the success of an ichneumonid (*Diadegma semiclausum*) and a braconid (*Cotesia plutellae*) parasitoid developing in larvae of *P. xylostella* strains susceptible and resistant to the benzoylphenylurea teflubenzuron. In leaf-dip bioassays with second instar moth larvae, both species of parasitoid proved capable of surviving teflubenzuron concentrations 100-fold to 1000-fold higher when attacking resistant compared with susceptible hosts. Hence the primary factor constraining parasitoid success in the presence of teflubenzuron was the response of the host to the chemical, rather than intrinsic toxicity against *D. semiclausum* and *C. plutellae, per se*.

Laboratory experiments with susceptible and resistant clones of the peach-potato aphid, *Myzus persicae*, provided another example of the response of insect parasitoids being dependent on the genetic composition of a host population (Birnie, unpublished data). In this case, the phenomenon depended in turn on the nature of the insecticide applied, and differed markedly between two chemicals with contrasting spectra of activity. Although spraying oilseed rape plants containing established populations of *M. persicae* and its braconid parasitoid *Diaeretiella rapae* with the recommended field rate of dimethoate yielded, as expected, very distinct degrees of suppression of susceptible and resistant hosts, levels of parasitism observed 10 days after spraying (1–2%) did not differ significantly (Table 34.1). With the more selective aphicide pirimicarb, occurrence of *c.* 11% parasitism among surviving resistant hosts showed that the loss of parasitoids in cages containing susceptible aphids was attributable primarily to host depletion rather than the intrinsic toxicity of pirimicarb against *D. rapae*. The greater survival of resistant aphids to dimethoate compared with pirimicarb was consistent with higher resistance to dimethoate recorded in short-term bioassays for aphids possessing the same esterase-based resistance mechanism (Sawicki and Rice, 1978).

An interesting extension to these arguments is that immature parasitoids could, in addition to benefiting from the continued survival of a resistant host, derive added protection from an insecticide by inadvertently exploiting the mechanism of resistance present in host insects. If this phenomenon occurs at all, it is much more likely to relate to endo- rather than ectoparasitoids, and to mechanisms based on enhanced enzymatic detoxification of an insecticide than a modification to its site of action. Such an effect was claimed for the *P. xylostella* example (Furlong and Wright, 1993), but the data presented are insufficient to preclude the likelihood that parasitoid success in resistant hosts was merely a consequence of the latter's survival at concentrations of teflubenzuron lethal to susceptible *P. xylostella* larvae, particularly since mechanism(s) of benzoylphenylurea resistance had not been adequately resolved at the time of that study. The known occurrence in *M. persicae* of both metabolic and target-site mechanisms of insecticide resistance provides the ideal model for exploring this subject in a more controlled manner, and such experiments are presently under way.

Table 34.1 Response of insecticide susceptible and resistant clones of *Myzus persicae* and its parasitoid *Diaeretiella rapae* to dimethoate and pirimicarb sprayed at recommended field rates in the laboratory

	Susceptible clone		*Resistant clone*	
Insecticide	*% suppression of host*[a]	*% parasitism*	*% suppression of host*[a]	*% parasitism*
Dimethoate	85.8	1	0	1.8
Pirimicarb	100	0	78.6	11
Unsprayed	–	17.7	–	18.4

[a] Expressed relative to unsprayed controls

INSECTICIDE RESISTANCE AS AN ECOLOGICAL TOOL?

On the basis of work outlined above it might be concluded that the occurrence of resistant hosts could, under certain conditions, have a positive influence on integrated pest management as it provides a 'refugium' for parasitoids and facilitates their survival in the presence of insecticide treatments. In practice, opportunities for exploiting this other than on a *de facto* basis are probably very limited, since it could only apply to highly selective insecticides and to cases when the frequency of resistance is sufficiently low not to obviate the cost-effectiveness of applying a chemical in the first place. A potentially more manageable application lies in the area of mainstream ecotoxicology, where the use of resistant strains of insect pests could assist greatly with quantifying the toxicity of compounds against key parasitoid species. Current use of susceptible strains for this purpose must pose a considerable constraint on disentangling direct effects on hosts from unwanted side-effects on immature parasitoids, and is presumably one reason why existing methodology places overwhelming reliance on tests against free-living adults (e.g., Hassan, 1983; Hassan *et al.*, 1994).

CONCLUSIONS

To date, work on insecticide resistance and ecotoxicology appears to have progressed with surprisingly little collaboration or communication between the parties involved. The aim of this chapter has been to emphasize that these two perspectives on insecticide use are complementary, and that a clearer understanding of interactions between resistance and the side-effects on non-target species can assist greatly in anticipating the dynamics of pest and natural enemies in the presence of insecticides, and in the rational design and deployment of integrated pest management strategies. Although the consequences of resistance for crop protection and disease management are generally adverse, there are opportunities for exploiting it to the benefit of ecotoxicological research, and these deserve further scrutiny.

ACKNOWLEDGEMENTS

We thank the Pesticides Safety Directorate of the UK Ministry of Agriculture, Fisheries and Food for support of work at IACR Rothamsted, and W. Powell and M.R. Cahill for advice. IACR receives grant-aided support from the Biotechnology and Biological Sciences Research Council of the United Kingdom.

REFERENCES

Barrett, K.L., Grandy, N., Harrison, E.G. *et al.* (1994) *Guidance Document on Regulatory Testing Procedures for Pesticides with Non Target Arthropods. From the ESCORT workshop, Wageningen, Netherlands, March 1994*, SETAC-Europe, Brussels.

Birnie, L.C. and Denholm, I. (1992) Use of field simulators to investigate integrated chemical and biological control tactics against the cotton whitefly, *Bemisia tabaci*, in *Proceedings Brighton Crop Protection Conference – Pests and Diseases*, Brighton, pp. 1003–1008.

Birnie, L.C., Hackett, B. and Denholm, I. (1996) The impact of resistance in insect pests on interactions with natural enemies, in *Proceedings Brighton Crop Protection Conference – Pests and Diseases*, Brighton, pp. 203–208.

Chellia, S. and Heinrichs, E.A. (1980) Factors affecting insecticide-induced resurgence of the brown planthopper, *Nilparvata lugens* on rice. *Environmental Entomology*, **96**(6), 773–777.

Council Directive (1991) Council Directive 91/414/EEC concerning the placing of plant protection products on the market. *Official Journal of the European Communities*, **L230**, 1–32.

Denholm, I. and Rowland, M.W. (1992) Tactics for managing pesticide resistance in arthropods: theory and practice. *Annual Review of Entomology*, **37**, 91–112.

Denholm, I., Sawicki, R.M. and Farnham, A.W. (1984) The relationship between insecticide resistance and control failure, in *Proceedings Brighton Crop Protection Conference – Pests and Diseases*, Brighton, pp. 527–534.

Denholm, I., Horowitz, A.R., Cahill, M. and Ishaaya, I. (in press) Management of resistance to novel insecticides, in *Novel Insecticides: Mode of Action and Management* (eds I. Ishaaya and D. Degheele), Springer Verlag, Berlin.

Dittrich, V., Streibert, P. and Bathe, P.A. (1974) An old case reopened: mite stimulation by insecticide residues. *Environmental Entomology*, **3**(3), 534–550.

Forrester, N.W., Cahill, M., Bird, L.J. and Layland, J.K. (1993) Management of pyrethroid and endosulphan resistance in *Helicoverpa armigera* in Australia. *Bulletin of Entomological Research Supplement Series*, **1**.

Furlong, M. and Wright, D. (1993) Effects of the acylurea insect growth regulator teflubenzuron on the endo-larval stages of the hymenopteran parasitoids *Cotesia plutellae* and *Diadegma semiclausum* in a susceptible and an acylurea-resistant strain of *Plutella xylostella*. *Pesticide Science*, **39**, 305–312.

Hardin, M.R., Benrey, B., Coll, M. *et al.* (1995) Arthropod pest resurgence: an overview of potential mechanisms. *Crop Protection*, **14**(1), 3–18.

Hassan, S.A. (1983) Procedures for testing the side-effects of pesticides on beneficial arthropods as being considered by the IOBC International Working Group. *Mitteilungen der Deutschen Gesellschaft für allgemeine und angewandte Entomologie*, **4**, 86–88.

Hassan, S.A., Bigler, F., Blaisinger, P. *et al.* (1985) Standard methods to test the side-effects of pesticides on natural enemies of insects and mites developed by the IOBC/WPRS Working Group 'Pesticides and Beneficial Organisms'. *Bulletin OEPP/EPPO*, **15**, 214–255.

Hassan, S.A., Bigler, F., Bogenschütz, H. *et al.* (1994) Results of the sixth joint pesticide testing programme of the IOBC/WPRS – Working Group 'Pesticides and Beneficial Organisms'. *Entomophaga*, **39**, 107–119.

Horowitz, A.R., Forer, G. and Ishaaya, I. (1994) Managing resistance in *Bemisia tabaci* in Israel with emphasis on cotton. *Pesticide Science*, **42**, 113–122.

Kerns, D.L. and Gaylor, M.J. (1991) Induction of cotton aphid outbreaks by the insecticide sulprofos, in *Proceedings 44th Cotton Insect Research and Control Conference*.

Leonard, P.K. and Perrin, R.M. (1994) Resistance management – making it happen, in *Proceedings Brighton Crop Protection Conference – Pest and Diseases*, Brighton, pp. 969–974.

Roush, R.T. and Daly, J.C. (1990) The role of population genetics in resistance research and management, in *Pesticide Resistance in Arthropods* (eds R.T. Roush and B.E. Tabashnik), Chapman & Hall,

Sanderson, J.P. and Roush, R.T. (1995) Management of insecticide resistance in the greenhouse, in *Proceedings 11th Conference Insect Disease Management* (eds A. Bishop, M. Hansbeck and R. Lindquist), Ornman, Fort Myers, Florida, pp. 18–20.

Sawicki, R.M. and Rice, A.D. (1978) Response of susceptible and resistant peach-potato aphids *Myzus persicae* to insecticides in leaf-dip bioassays. *Pesticide Science*, **9**, 513–516.

Sundaramurthy, V.T. (1992) Upsurgence of whitefly *Bemisia tabaci* in the cotton ecosystem in India. *Outlook on Agriculture*, **21**(2).

Waage, J.K. (1992) Biological control in the year 2000, in *Pest Management and the Environment in 2000* (eds A. Abdul, S.A. Kadir and H.S. Barlow), CAB International, Oxon.

35

Approaches for extrapolating pesticide side-effects between arthropod species: how much do we know?

John A. Wiles and Katie L. Barrett

INTRODUCTION

One of the paradigms of ecotoxicology is the practical need to select a limited number of species for evaluating the environmental effects of chemicals, whilst wishing to apply or extrapolate the findings to other (related) species or groups of organisms which are at risk of exposure within the same environmental compartment. This is true for those working with organisms inhabiting the marine, freshwater, soil and crop environments.

Within the field of ecological risk assessment of pesticides to terrestrial invertebrates over recent years, the need for approaches to address the problem of interspecific extrapolation has been highlighted on many occasions (e.g. Aldridge and Carter, 1992; Greig-Smith, 1992a,b; De Snoo *et al.*, 1994). In this contribution to the discussion we present case study data sets to explore some qualitative and quantitative approaches for interspecific risk extrapolation.

Ecotoxicology: Pesticides and beneficial organisms.
Edited by P.T. Haskell and P. McEwen. Published in 1998 by Chapman & Hall, London. ISBN 0 412 81290 8.

THE SCALE OF THE EXTRAPOLATION PROBLEM IN TERRESTRIAL ECOTOXICOLOGY: ECOLOGICAL AND REGULATORY PERSPECTIVES

Scientists wishing to make risk extrapolations between species groups or types for regulatory or other scientific purposes face a number of frustrations. Firstly, they often have only a rather small number of data sets to make judgements from; secondly, general 'rules of thumb' that would enable extrapolation based upon susceptibility and exposure patterns have yet to be formally established; and thirdly, 'ecological tools' for use in extrapolation procedures remain underdeveloped as the incorporation of ecological theory has yet to be fully explored and utilized in ecotoxicology (e.g. Baird *et al.*, 1996, and papers therein; also see Chapter 37).

The need for extrapolation stems from the fact that a large number of arthropod species inhabit within-crop and off-crop environments. This may put them at risk from suffering direct or indirect effects from pesticide applications. To put this into perspective using the UK as an example, inventory studies by Beard and Mauremootoo (1994) identified 128 species of Coleoptera overwintering in arable field margins in Hampshire alone, and more than 630 species of arthropods have been identified in the cereal ecosystem in Sussex (Potts, 1991). Expanding this further, we find that most of the 25 000 terrestrial arthropod species in the UK have been found to exist in farmland (Aldridge and Carter, 1992). Of course it is not reasonable to test every species and therefore the most practical approach is to extrapolate the findings of studies on a limited set of species (such as predators, parasites and pollinators) to other 'important' species or groups of organisms inhabiting the same environment. These species include, for example, organisms that have an important role in the decomposition process, organisms that are important to the feeding ecology of invertebrate and vertebrate farmland fauna and species considered to have high conservation merit, e.g. butterflies. From a scientific and political point of view the central goal remains maintaining and managing species richness, ecosystem productivity and sustainability.

The need for approaches to extrapolation is explicitly highlighted in two of the key regulatory documents concerning pesticide risk assessment for arthropod natural enemies. The EPPO risk assessment scheme (EPPO, 1994) states:

> This sub-scheme is concerned with assessing the potential risks to natural enemies of pests species from the use of plant protection products. This group comprises species from a wide taxonomic and ecological range, whose composition will vary spatially, between different crops, and temporally. It is, of course, not possible to have

all natural enemy species tested, and results will, therefore have to be extrapolated from the few species tested to groups.

The more recently published *SETAC Guidance Document on Regulatory Testing Procedures with Non Target Arthropods* (Barrett *et al.*, 1995) also addresses the problem, stating:

> Effects on beneficials may give an indication about the potential effects of a product on non-target (neutral) species of the same taxonomic group and the same trophic level.

Current European regulatory requirements for placing plant protection products on the market require ecotoxicity data for four species of non-target arthropod natural enemies, selected to cover a range of taxonomic groups and crop exposure categories relevant to the particular proposed usage pattern of the chemical product. In addition studies are conducted with a pollinator species, most commonly the honeybee *Apis mellifera* L. (Hymenoptera: Apidae). Using this data set, risk assessors are explicitly tasked with determining the acceptability of any effects observed on crop-inhabiting pollinators or natural enemies, and implicitly tasked with judging the likelihood of any effects on other species.

TOWARDS PROVIDING 'TOOLS' FOR EXTRAPOLATION: QUALITATIVE AND QUANTITATIVE APPROACHES

The spectrum of approaches available for use as extrapolation procedures can vary from relatively simple box scoring up to more precise quantification of susceptibility and exposure mechanisms. The level of resolution required may well differ according to requirements of end-users (e.g. regulators, environmental managers or research scientists) with trade-offs between precision and robustness and risk management needs requiring careful consideration (also see Chapter 22).

In the following sections we have used a number of case study data sets, for the synthetic pyrethroid insecticide deltamethrin, to illustrate some qualitative and quantitative approaches that may aid the extrapolation of risk between species. Our aim is not to promote these as definitive approaches for extrapolating risk but to examine the information they yield and how comprehensive our current knowledge base is, and to determine their potential for future use or development.

Characterization of species susceptibility relationships

The profiling of susceptibility relationships has proved useful in other areas of ecotoxicology, e.g. aquatic ecotoxicology (Hoekstra *et al.*, 1992), although pertinent differences must be appreciated between aquatic and

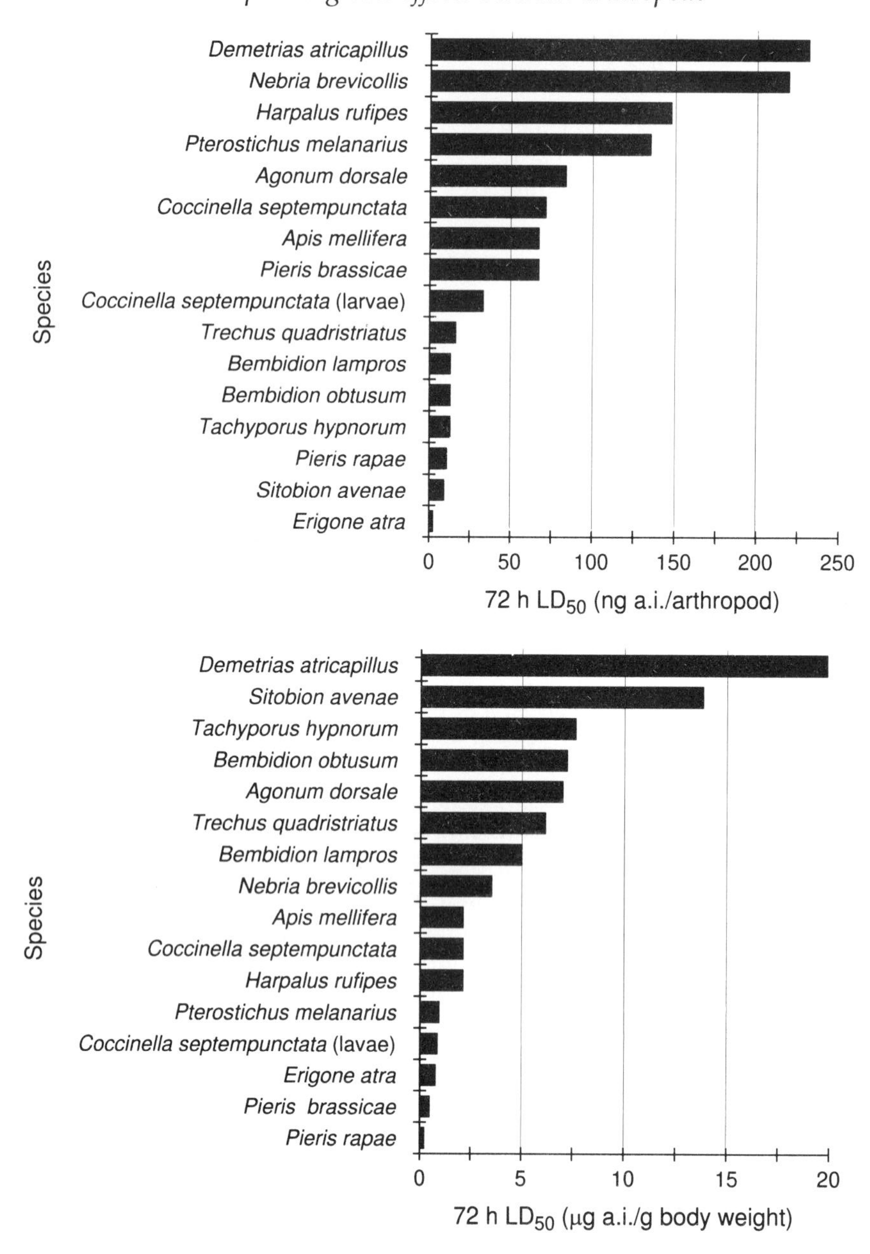

Figure 35.1 Susceptibility profile for some target and non-target arthropods to the synthetic pyrethroid insecticide deltamethrin. (Data for *A. mellifera* from Aitkins, 1981; data for *P. rapae* and *P. brassicae* from Çilgi and Jepson, 1996; all other data from Wiles and Jepson, 1992.) The 72 h LD_{50} value for *E. atra* was 0.8 μg a.i./spider (female spiders tested only) and for *D. atricapillus* 66.2 μg a.i./g

terrestrial environments. To date few serious attempts have been made to construct such profiles for a range of above-ground terrestrial arthropod species. If successful such an approach may allow the inference of patterns of susceptibility between representative species from a range of functional, taxonomic, morphological and size groups.

One factor that has clearly limited this approach is that, even though laboratory bioassays with arthropod species have now been conducted for several decades, comparable data sets enabling profiling of susceptibility relationships between species remain rare. Few papers compare more than three or four species, and relatively few papers have studied the same species or the same products. Even when this is the case, comparison of results is often confounded by methodological differences in the bioassays, or the use of different routes of exposure. In particular many bioassays have examined the residual route of exposure; however, attempts to produce useful susceptibility profiles based upon these data are complicated by many unknowns regarding exposure and uptake.

Data from topical (direct application) bioassays may provide the most useful and comparable way for examining susceptibility relationships. The traditional index of susceptibility is the LD_{50} value, an estimate of the median lethal dose. LD_{50} values may be expressed in terms of dose per arthropod, which gives an indication of the susceptibility in the field, or in terms of dose per unit body weight, which provides a measure of the intrinsic susceptibility of the organism to the toxicant. Example profiles for the pyrethroid insecticide deltamethrin are given in Figure 35.1. Susceptibility data from laboratory topical bioassays, compiled from a number of sources in the published literature, are presented for 15 arthropod species, including representatives from six different families. The profiles provide useful guidance for ranking species susceptibility, establishing relationships between susceptibility and other factors – such as body weight (e.g. Wiles and Jepson, 1992) or metabolic rate (Chapter 32) – and examining ranges of susceptibilities that occur. The LD_{50} values, expressed in terms of μg a.i./arthropod or in terms of μg a.i./g body

---

body weight (truncated in the graph for scaling purposes). The test species belong to the following taxonomic groups:

- Coleoptera – Carabidae: *B. lampros, B. obtusum, D. atricapillus, T. quadristriatus, A. dorsale, H. rufipes, N. brevicollis, P. melanarius*
- Coleoptera – Staphylinidae: *T. hypnorum*
- Coleoptera – Coccinellidae: *C. septempunctata*
- Aranea – Linyphiidae: *E. atra*
- Hemiptera – Aphididae: *S. avenae*
- Hymenoptera – Apidae: *A. mellifera*
- Lepidoptera – Pieridae: *P. brassicae, P. rapae.*

weight, show a 300-fold range of susceptibilities between the most susceptible species (the linyphiid spider *Erigone atra*) and least susceptible species (the small carabid beetle *Demetrias atricapillus*) tested. It is noticeable that the ranking order changes somewhat when body weight is taken in account, which may highlight physiological differences in coping with the chemical between the species.

From a practical point of view, further development of this approach may require a considerable research effort. Data sets will need to be generated and extended to cover additional species, carefully selected to represent different taxonomic or morphological groups. It would also be interesting to explore where the species used in regulatory testing fit into the ranking, and how near the extremes they fall. By profiling a range of selected compounds (e.g. representatives from different chemical classes), with a similar set of species it is possible to examine relationships between interspecies variability patterns and intercompound variability patterns.

Characterization of species life history patterns and habitat associations

A number of qualitative (e.g. de Snoo *et al.*, 1994) and semi-quantitative approaches (e.g. Jepson, 1989) have been suggested to estimate the likelihood and degree of exposure of terrestrial arthropods to pesticide sprays. The most simplistic approach is to undertake a simple categorization according to whether the invertebrate lives within or outside the treated area. In a more detailed form, factors such as dispersal rate, diet range, voltinism, fecundity, diurnal activity pattern and whether the organism is associated with plant foliage or the soil may be taken into account. Using this information a scoring system may then be employed to determine risks of exposure for any given spraying scenario based upon knowledge of the timing and frequency of spray applications (e.g. Jepson, 1993). This offers a robust approach to extrapolation but requires considerable knowledge of species ecology and biology. Certainly an extensive literature is available for many key arthropod groups, but currently such an approach would at best require considerable time and expert judgement.

Future research efforts may provide useful tools upon which to develop a more formal system for categorizing exposure risk, based upon species phenology or life history patterns. One approach would be to attempt to categorize the range of annual rhythms that occur and then formulate a table of relative risk predictions for different spray application scenarios. Thus organisms may be split into groups which represent, for example: species that are spring breeding and have summer larvae and adults that hibernate; species with winter larvae and reproduction in the summer and/or autumn; species with winter larvae and young adults in spring that undergo aestivation prior to reproduction; species that have a flexible reproduction and development period (i.e. spring and/or autumn); and species that take more than one year to reach sexual maturity. Another promising approach for developing extrapolation tools for the future, recently promoted by Calow *et al.* (1997), involves (theoretical) exploration of how life history variables may influence population growth for different life history strategies.

Characterization of chemical bioavailability patterns

Although the important physicochemical properties of most compounds, such as solubility in water, octanol/water partition coefficient, vapour pressure, stability in relation to pH, dissociation constant pKa, photodegradation rates, volatilization rates, hydrolysis and transformation rates are well characterized, and enable predictive modelling of the fate and persistence of chemicals, our understanding of patterns of bioavailability remains somewhat limited. The reason for this is that bioavailability is mediated not only by complex substrate-dependent interactions (associated, for example, with soil particle composition or plant cuticular architecture) but also by environmental factors, such a temperature and humidity, and species-specific factors, such as morphological characteristics and activity patterns. These all contrive to make precise predictions difficult.

In situ bioassays provide a robust approach for investigating and determining patterns of bioavailability, within and between substrate types. For example, Wiles and Jepson (1994) exposed adult *Coccinella septempuncatata* and *Tachyporus hypnorum* to fresh deltamethrin residues on barley plant foliage and a sandy loam soil and found a 60-fold difference in toxicity between the two substrates, with residues being less available on the soil than the foliage. In another study, Chowdhury (1996), using a laboratory leaf bioassay test with the collembolan *Folsomia candida*, showed a 12-fold difference in toxicity of deltamethrin to the springtails on six types of plant foliage with differing wax contents (Table 35.1). Likewise Longley (1995), in a set of bioassays exposing second instar *Pieris brassicae* to deltamethrin residues on glass and five different species of hedgerow plant, showed a seven-fold range of toxicity, mediated by differences in chemical availability (Figure 35.2).

For extrapolation purposes there is a clear need to explore the broad patterns and ranges of bioavailability that may occur on different substrates. If established, such patterns may have considerable value for allocating weighting factors within extrapolation procedures. If nothing else, this would improve predictions taking account of species habitat associations on the two main receptors of spray deposits: plant foliage and soil.

Table 35.1 Toxicity of deltamethrin residues to *Folsomia candida* on six different leaf types (unpublished data from Chowdhury, 1996)

Plant species	*Mean wax content of leaves assayed* ($\mu g/cm^2$)	*72 h LD_{50} for F. candida* (*g a.i./ha*)
Barley	51.3	6.3
Cabbage	36.1	8.9
Pear	21.2	14.5
Sugarcane	5.4	20.9
Maize	2.8	66.1
Dwarf bean	1.5	77.6

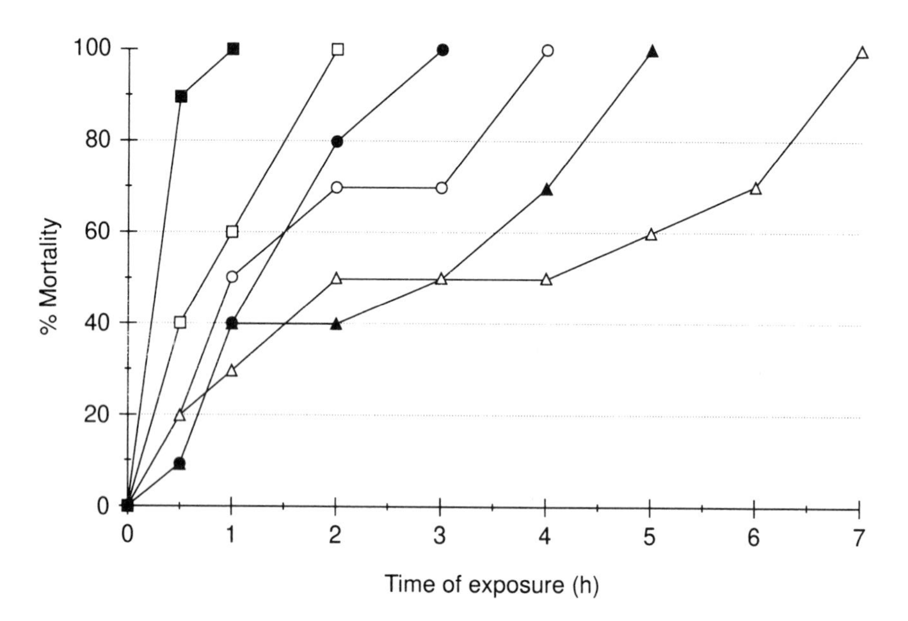

Figure 35.2 Mortality of second instar *Pieris brassicae* larvae exposed to deltamethrin residues on glass and hedgerow plant foliage in laboratory bioassays: ■, glass; □, nettle; ● charlock; ○, grass; ▲, dock; ◇, ivy. (Unpublished data from Longley, 1995.)

An example of integrating approaches using scenario-driven weighting factors

In an ideal world it would be possible to integrate both susceptibility and exposure data for specific groups of organisms, using estimates of relative species susceptibilities and realistic weighting values based on data for chemical availability and toxicity on given substrates. In reality these kind of data are only likely to be available for a limited number of scenarios. The example given in Table 35.2 for a summer application of deltamethrin in cereals indicates how this type of approach may be used. The magnitude of differences between risk estimates for each species provides a general insight into possible adverse side-effects. In this particular case, noteworthy findings may include the implication that nocturnal foliage-active staphylinid beetles may be at a greater risk of suffering adverse effects than diurnally ground-active carabid beetles.

Table 35.2 A semiquantitative approach to estimate exposure risks of different arthropod species in a temperate cereal crop from a deltamethrin spray

Order: Family *Species*	*Relative susceptibility topical LD_{50} (ng a.i./arthropod)*	*Score for activity and crop distribution pattern*				
		N	*D*	*P*	*G*	*Total risk*
Araneae Linyphiidae						
E. atra	289.5	1			1	289.5
Coleoptera Staphylinidae						
T. hynorum	18.53	1		60		1111.8
Coleoptera Coccinellidae						
C. septempunctata (Adults)	2.33		2	60		279.6
C. septempunctata (Larvae)	7.42		2	60		890.4
Coleoptera Carabidae						
B. obtusum	18.23		2		1	36.5
B. lampros	17.95		2		1	35.9
T. quadristriatus	15.04		2		1	30.1
D. atricapillus	1.00		2	60		120.0
A. dorsale	2.77	1			1	2.8
H. rufipes	1.57	1			1	1.6
N. brevicollis	1.05	1			1	1.0
P. melanarius	1.71	1			1	1.7

Score = score for activity and crop distribution pattern.
For direct contact exposure, organisms active in the day (**D**) score 2 and those active in the night (**N**) score 1.
For residual exposure, those active on the plant (**P**) score 60 and those active on the ground (**G**) score 1 (based on differential bioavailability of deltamethrin on the two substrates shown by Wiles and Jepson, 1994).
Species susceptibility data was obtained from Wiles and Jepson (1992). The susceptibility values were calculated relative to the susceptibility of the least susceptible species, *D. atricapillus*. The species with the highest multiplicative score are likely to be most at risk from a spray application.

DISCUSSION

The quest for new 'tools' for extrapolation lies at the heart of many of the research areas of terrestrial ecotoxicology, not only between species, but between scales of investigation and between levels of biological organization. To some degree it is dissatisfying to think that the ideas and approaches presented in this chapter have probably posed more questions than they have answered; however, in doing so we hope that we have stimulated new thoughts on the subject and to some extent given rise to a research agenda for the future.

The examples used in this chapter have focused largely on natural enemy species rather than on other non-target species. We apologize for this but it was somewhat forced upon us by a combination of two factors: firstly, the paucity of toxicity data for other species in the literature; and secondly, the mechanistic nature of the approaches we have presented. For those wishing to explore these ideas for other species, data is available for butterflies (e.g. Longley and Sotherton, 1997), sawflies (e.g. Sotherton, 1990), some other beetle species (e.g. Kjaer and Jepson, 1995) and Collembola (e.g. Wiles and Frampton, 1996). An alternative approach to that taken in this paper, and one that may yield interesting and useful findings for a wider range of species, is to examine data from extensive monitoring studies that have also included faunal inventories (e.g. Aebischer, 1992; Vickerman, 1992). Analysis of depletion and recovery rates, linked with a detailed examination of phenological patterns of these species, may prove fruitful.

The development of new tools for extrapolation remains research driven, as even now we are still at a stage of data gathering. In saying this, we do not suggest that there is a general need to test more species. The real need is to determine how these data can be used more effectively. Certainly we should not overlook the large data set gathered over the last few decades and one of the most potentially rewarding options remains the development of computer databases (Croft, 1990) and expert systems that may be interrogated to elucidate patterns of susceptibility and toxicity further. Careful examination of observed trends may yield another route for finding appropriate weighting factors for use in extrapolation procedures.

The answer to the question, 'How much do we know?', posed in the title of this chapter is probably: 'More than we realize.' The main problem that we are faced with remains how to extract the information at the correct level of resolution for incorporation into the appropriate extrapolation tools. Our current knowledge base may best be summarized (from most extensive to least extensive) as exposure → susceptibility → bioavailability → population consequences of exposure → community consequences of exposure. In practical terms, it may be easiest to search for general rules of thumb which, when combined with expert judgement, may aid the extrapolation process. One thing that should be born in mind is that, at least initially, extrapolation procedures need to be tailored according to specific uses, whether that be for academic research, resource management or regulatory purposes.

ACKNOWLEDGEMENTS

Drs Martyn Longley and Nasir Chowdhury of the School of Biological Sciences, University of Southampton, UK, are gratefully acknowledged for providing unpublished data for use in this chapter.

REFERENCES

Aebischer, N.J. (1992) Twenty years of monitoring invertebrates and weeds in cereal fields in Sussex, in *The Ecology of Temperate Cereal Fields*, (eds L.G. Firbank, N. Carter, J.F. Derbyshire and G.R. Potts), Blackwell Scientific Publications, Oxford, pp. 305–331.

Aldridge, C. and Carter, N. (1992) The principles of risk assessment for non-target arthropods: a UK perspective, in *Interpretation of Pesticide Effects on Beneficial Arthropods*, (eds R.A. Brown, P.C. Jepson and N.W. Sotherton), *Aspects of Applied Biology* **31**, 149–156.

Baird, D.J., Maltby, L., Greig-Smith, P.W. and Douben, P.E.T. (1996) *Ecotoxicology: Ecological Dimensions*, Ecotoxicology Series 4, Chapman & Hall, London, 89 pp.

Barrett, K.L., Grandy, N., Harrison, E.G. *et al.* (1994) *Guidance Document on Regulatory Testing Procedures for Pesticides with Non Target Arthropods. From the ESCORT workshop, Wageningen, Netherlands, March 1994*, SETAC-Europe, Brussels.

Beard, R.C.W. and Mauremootoo, J.R. (1994) The biodiversity of Coleoptera in arable field boundaries, in *Proceedings of the BCPC Conference – Pests and Diseases*, Vol. 3, pp. 943–950.

Calow, P., Sibley, R.M. and Forbes, V. (1997) Risk assessment on the basis of simplified life-history scenarios. *Environmental Toxicology and Chemistry*, **16**, 1983–1989.

Chowdhury, A.B.M.N.U. (1996) Studies on the effects of different leaf characteristics on the mediation of toxicity of exposed organisms. Unpublished M.Phil. transfer report, University of Southampton, UK, 169 pp.

Croft, B.A. (1990) *Arthropod Biological Control Agents and Pesticides*, John Wiley and Sons, New York, 723 pp.

De Snoo, G.R., Canters, K.J., De Jong, F.M.W. and Cupreus, R. (1994) Integral hazard assessment of side effects of pesticides in The Netherlands – a proposal. *Environmental Toxicology and Chemistry*, **13**, 1331–1340.

EPPO (1994) Decision-making schemes for the environmental risk assessment of plant protection products – arthropod natural enemies, *Bulletin OEPP/EPPO* **24**(1), 17–35.

Greig-Smith, P.W. (1992a) The acceptability of pesticide side-effects on non-target arthropods, in *Interpretation of Pesticide Effects on Beneficial Arthropods*, (eds R.A. Brown, P.C. Jepson and N.W. Sotherton), *Aspects of Applied Biology* **31**, 121–132.

Greig-Smith, P.W. (1992b) A European perspective to ecological risk assessment, illustrated by pesticide registration procedures in the United Kingdom. *Environmental Toxicology and Chemistry*, **11**, 1673–1689.

Hoekstra, J.A., Vaal, M.A. and Notenboom, J. (1992) *Sensitivity Patterns of Aquatic Species to Toxicants: a Pilot Study*, RIVM Report No. 719102016, Bilthoven, The Netherlands.

Jepson, P.C. (1989) The temporal and spatial dynamics of pesticide side-effects on non-target invertebrates, in *Pesticides and Non-target Invertebrates*, (ed. P.C. Jepson), Intercept, Wimborne, Dorset, pp. 95–127.

Jepson, P.C. (1993) Ecological insights into risk analysis: the side-effects of pesticides as a case study. *The Science of the Total Environment*, **Supplement 1993**, 1547–1566.

Kjaer, C. and Jepson, P.C. (1995) The toxic effects of direct pesticide exposure for a non-target weed-dwelling chrysomelid beetle (*Gastrophysa polygoni*) in cereals. *Environmental Toxicology and Chemistry*, **14**, 993–999.

Longley, M. (1995) The effects of pesticides on butterflies inhabiting farmland. Unpublished Game Conservancy Report, Fordingbridge, Hants.

Longley, M. and Sotherton, N.W. (1997) Factors determining the effects of pesticides upon butterflies inhabiting farmland. *Agriculture, Ecosystems and Environment*, **61**, 1–12.

Potts, G.R. (1991) The environmental and ecological importance of cereal fields, in *The Ecology of Temperate Cereal Fields*, (eds L.G. Firbank, N. Carter J.F. Darbyshire and G. R. Potts), Blackwell, Oxford, pp. 3–21.

Sotherton, N.W. (1990) The effects of six insecticides used in UK cereal fields on sawfly larvae (Hymenoptera: Tenthredinidae), in *Proceedings of the Brighton Crop Protection Conference – Pests and Diseases 1990*, pp. 999–1004.

Vickerman, G.P. (1992) The effects of different pesticide regimes on the invertebrate fauna of winter wheat, in *Pesticides and the Environment: the Boxworth Study*, (eds P.W. Greig-Smith, G.K. Frampton and A.R. Hardy), HMSO, London, pp. 82–109.

Wiles, J.A. and Frampton, G.K. (1996) A field bioassay approach to assess the toxicity of insecticide residues on soil to Collembola. *Pesticide Science*, **47**, 273–285.

Wiles, J.A. and P.C. Jepson (1992) The susceptibility of a cereal aphid pest and its natural enemies to deltamethrin. *Pesticide Science*, **36**, 263–272.

Wiles, J.A. and P.C. Jepson (1994) Substrate-mediated toxicity of deltamethrin residues to beneficial invertebrates: estimation of toxicity factors to aid risk assessment. *Archives of Environmental Contamination and Toxicology*, **27**, 384–391.

36

Application of pheromones: toxicological aspects, effects on beneficials and state of registration

Albert K. Minks and P.A. Kirsch

INTRODUCTION

Insects produce volatile compounds which are excreted into the environment and can act as chemical messengers between conspecifics. These compounds are called pheromones and they convey, among other things, information about sexual readiness from female individuals to their male partners, in which case they are called sex pheromones. Their powerful attraction and species specificity give pheromones a great potential for application in agricultural practice. In this chapter we restrict our attention to lepidopterous sex pheromones whose possibilities for practical application have been investigated most extensively. Sex pheromones can be used in the following ways (Kirsch, 1988):

- for monitoring purposes in traps;
- for control by means of the attract-and-kill method, in which insects are lured to baits with a small amount of contact insecticide;
- for control by mass trapping;
- for control by means of mating disruption.

Ecotoxicology: Pesticides and beneficial organisms.
Edited by P.T. Haskell and P. McEwen. Published in 1998 by Chapman & Hall, London. ISBN 0 412 81290 8.

MATING DISRUPTION

Most of our attention will be focused on the effects of the mating disruption technique, as this method releases the highest quantity of pheromone per unit area. The concept of the technique is to permeate the air over the crop to be protected with synthetic pheromone. Male moths are then unable to locate their female mates when using their own pheromone system and mating is therefore reduced or even eliminated. For more detailed information on mating disruption, refer to Jutsum and Gordon (1989), Ridgway *et al.* (1990) and Howse *et al.* (1996).

Slow-release formulations are absolutely essential for pheromones used in mating disruption. They prolong the release and efficacy of highly volatile pheromone compounds and provide in-field stabilization of pheromone remaining in the formulation. While formulations can be sprayable (e.g. microcapsules), at present most commercial formulations are designed for hand application by clipping, hanging or twisting them around stems or branches of the crop.

The following formulations have been used commercially in recent years:

- Laminate flakes (Hercon, USA)
- Twin ampulla dispensers (BASF Doppelampullen, Germany)
- Twist-tie polyethylene dispensers (Biocontrol/Shin-Etsu, Japan)
- Polymer dispensers (TNO, the Netherlands)
- Isagro cellulose fibre dispensers (Donegani, Italy)
- Consep membranes (Consep, USA)
- Biosys polymer dispensers (Biosys, USA)
- Scentry micro-fibres (Ecogen, USA)
- Microcapsules (3M, USA).

The latest development in mating disruption formulation is a device with an intermittent delivery system, which can be timed to discharge pheromone only when necessary according to pest mating behaviour (Mafra-Neto and Baker, 1996). In addition to dispenser characteristics, many factors influence dispenser performance, including the chemical structure of the pheromone, the initial dispenser load rate and pheromone purity, the additive/preservative compounds and climatic conditions. Most dispenser types have a high evaporation rate in the first 5–7 days post-field installation, after which evaporation levels off at a constant but low rate. The TNO dispenser has a more gradual evaporation pattern, which has a favourable effect on its longevity. The quantity of active ingredient used for mating disruption is usually expressed as the total amount of pheromone (g/ha) installed in the field. This varies between 10 and 150 g/ha. Of course, much of the pheromone remains in the dispenser and the whole amount is not released in the air at the same moment. Through chemical analysis of dispensers at intervals during the treatment period it is possible to estimate the evaporation rate of an individual dispenser. In general, commercial formulations release in the order of 10–20 μg/dispenser per hour. Unfortunately, it is at present impossible to measure in absolute figures how much pheromone is present in the air within the field at a certain moment. An important recent breakthrough in the development of portable electroantennography (EAG) equipment now enables quantitative measurement of relative pheromone concentrations in the field.

COMMERCIAL USE OF MATING DISRUPTION

In recent years mating disruption technology has steadily gained commercial acceptance against a wide range of pests in many different crops and regions. While the technology has inherent limitations, it is a very effective pest management tool when cautiously monitored and carefully supervised within an overall integrated pest management programme. The most outstanding applications of mating disruption are summarized in Table 36.1. The total area treated with commercial mating disruption products in 1996 is estimated at 300 000 ha. The largest single pheromone application in 1996 took place in Egypt against the pink bollworm, *Pectinophora gossypiella*, for a total sale of US$5.3 million and a total treated area of 80 000 ha (Kirsch, 1996).

A review of current global implementation of mating disruption suggests successful control of more than 30 target insect species. Over 50 different products have been developed commercially and 10 different companies are active in the field. Research, development and commercial implementation are under way in all geographical regions. Table 36.2 summarizes these activities for the most important areas. Although commercial efficacy has been shown for all the insect species listed,

Table 36.1 World-wide application of commercial pheromone products in 1996

Crop	*Hectares*
Cotton	250 000
Pome fruit	16 500
Grapes	14 000
Stone fruit	6 000
Rice	4 000
Vegetables	3 000
Forestry	5 500

Table 36.2 Number of commercial pheromone products applied in various regions in the world against various insect species (season 1996)

Area	*Products (no.)*	*Insect species*
Australia/New Zealand	5[a]	Oriental fruit moth, codling moth, light-brown apple moth, leafroller moth, currant borer
South Africa	6[a]	Oriental fruit moth, codling moth
Egypt	5[a]	Pink bollworm
Western Europe	13[a]	Oriental fruit moth, codling moth, *Adoxophyes*, *Anarsia*, apple clear-wing, grape leafroller moth, rice stemborer
USA/Canada	13[a] + 11[b]	Oriental fruit moth, codling moth, *Anarsia*, diamondback moth, omnivorous leafroller moth, pink bollworm, tomato pin worm, tufted apple bud moth, grape berry moth, artichoke plume moth, gypsy moth, peach tree borer, Mexican rice stem borer
Japan	5[a]	Tea tortrix, beet armyworm, diamond-back moth, cherry tree borer
Chile/Argentina		Oriental fruit moth, codling moth, shoot moth

[a] Hand-applied products
[b] Sprayable products

commercial products are not yet available in all cases (e.g. diamondback moth, light-brown apple moth and various leafrollers (Kirsch, 1996).

TOXICITY OF LEPIDOPTEROUS PHEROMONES

Review of the pheromone data list published by Arn *et al.* (1992) indicates that lepidopterous pheromones are, with very few exceptions, aliphatic compounds with a chain length of 11–18 carbon atoms, with an unsaturated character, and with a functional group of either an ester, alcohol or aldehyde. Tables 36.3 and 36.4 summarize the toxicological data of this group of compounds, as compiled by Weatherston and Minks in 1995. Earlier summaries are also available in Bedoukian (1992) and Kirsch (1988). It is clear from published mammalian toxicity data that lepidopterous pheromones do not present any mammalian toxicity hazard. The data on non-target organisms – fate and expression for known aldehyde and acetate pheromones – again show little hazardous

Table 36.3 Mammalian toxicity data for saturated and unsaturated acetates, alcohols and aldehydes with a chain length of 11–18 C-atoms[a] (data compiled by Weatherson and Minks, 1995)

Toxicology test	*Range of data*		
	Acetate	*Alcohol*	*Aldehyde*
Acute oral toxicity LD_{50} (g/kg) (rat)	> 5–34.6 (n = 10)	> 3 to > 50 (n = 7)	> 5 (n = 3)
Acute dermal toxicity LD_{50} (g/kg) (rat)	> 2–20.25 (n = 9)	> 2.5 (n = 4)	> 2–5 (n = 3)
Acute inhalation toxicity LD_{50} (mg/l) (rat)	3.3–32 (n = 5)	5.26 (n = 1)	> 5–16.88 (n = 3)

Table 36.4 Toxicity of acetate and aldehyde lepidopterous pheromones to non-target avian, fish and aquatic invertebrate organisms (data compiled by Weatherson and Minks, 1995)

Toxicology test	*Range of data*	
	Acetate	*Aldehyde*
Avian acute oral toxicity LD_{50} (g/kg) (mallard duck)	> 2 to > 10 (n = 3)	> 2 (n = 1)
Avian acute oral toxicity LD_{50} (g/kg) (bobwhite quail)	> 2–2.25 (n = 2)	> 2 (n = 1)
Fish toxicity LC_{50} (ppm) (bluegill sunfish)	> 100–540 (n = 4)	No data available
Fish toxicity LC_{50} (ppm) (rainbow trout)	> 100–270 (n = 4)	320 (n = 1)
Aquatic invertebrate toxicity LD_{50} (ppm) (*Daphnia magna*)	1.3–6.8 (n = 3)	0.45–2.23 (n = 2)

potential. Without exaggeration, it can be stated that pheromones are the safest of all currently available insect control products.

Finally, Spittler *et al.* (1992) and other authors cited therein have not been able to establish any residue of pheromone compounds on produce harvested from fields treated with exaggerated levels of mating disruption pheromone. It can be argued that the inherent low toxicological nature, the low application rates and environmentally unstable chemical nature of pheromone compounds all combine to minimize any toxicological hazard.

EFFECTS OF LEPIDOPTEROUS PHEROMONES ON BENEFICIALS

No detrimental effects have been observed against non-target organisms in pheromone-treated fields. All records show positive effects on beneficial insects in pheromone-treated plots. For example, in an extensive multi-year study in German vineyards treated with mating disruption against *Lobesia botrana*, higher numbers of beneficial insects and spiders were observed as well as a larger variety of other arthropod species (Louis *et al.*, 1996). In a four-year study in apple orchards in Pennsylvania, total parasitism was higher for summerbrood *Platynota idaeusalis* larvae in mating disruption orchards than in conventionally treated orchards (Biddinger *et al.*, 1994). A significant increase in egg parasitization of the leafroller moth *Archips fuscocupreanus* was reported from pheromone-treated apple orchards in Japan (Ohira and Oku, 1993). Trials in other crops have shown no adverse effects on:

- parasitization by *Cardiochiles* of *Heliothis virescens* larvae in pre-harvest tobacco or subsequent populations on post-harvest tobacco and weed hosts (Tingle and Mitchell, 1982);
- parasitization of two parasites, *Chelonus insularius* and *Temelucha difficilis*, of the larvae of the noctuid *Spodoptera frugiperda* (Mitchell *et al.*, 1984);
- total percentage of parasitization of *Rhyacionia zozana* in pheromone-treated plots of pine forest (although the abundance of the moth parasite *Glypta zozanae* was reduced, numbers of the pupal parasite *Mastrus aciculatus* were much higher; Niwa and Daterman, 1989).

The data in all these examples were obtained from a comparison between pheromone-treated and conventionally treated plots. Although no evidence exists for a direct effect, the above-mentioned favourable effects are likely to be due to enhanced conservation of beneficial arthropods within a selective pest management system where control of the key pest is achieved using low-toxicity pheromone-based mating disruption strategies.

REGISTRATION OF PHEROMONE PRODUCTS IN THE USA

In most countries, including the USA, pheromone products must be registered as crop protection agents if they are used to control noxious insects by means of mass trapping or mating disruption. In 1978, the US Environmental Protection Agency (EPA) granted the first registration of a pheromone product, called Gossyplure HF, to be used for the suppression of *Pectinophora gossypiella* on cotton. This registration came into effect after a long and laborious procedure, which was characterized by great uncertainty on the part of both applicant and the agency in how to deal with this new category of crop protection agents. A striking difference with other pesticides is that pheromones are not toxicants, but agents that simply influence behaviour. In 1984, the EPA recognized the unique nature of pheromones and classified them in a separate category called 'biochemical pesticides', together with repellents, attractants, plant extracts, etc. Specific guidelines were developed that were significantly different from those for conventional insecticides. This was in agreement with the Food and Agriculture Organization of the United Nations (FAO) guidelines on the registration of biological pest control agents, issued in 1988. Basically, registration data requirements are arranged in a so-called tier system for biochemicals. Data requirements are arranged in four levels progressing from simple and relatively inexpensive to complex and more costly tests. Such tier testing is designed to ensure that only the minimum data are necessary to make a scientifically sound regulatory decision. Data are required in four areas: residue chemistry, toxicology, hazard to non-target organisms and environmental fate. As for all pesticide products, chemistry data are required to characterize the commercial product. In practice, pheromone products have only required submission of the least demanding Tier I data set due to their very low toxicity. This Tier I data set is even further reduced and currently only the data indicated in italics in Table 36.5 are usually required by the EPA to obtain registration of a lepidopterous pheromone product. The US EPA does not require product performance or efficacy data.

Table 36.5 Toxicological data requirements for the registration of pheromone products in the USA in Tier I[a,b]

Mammalian toxicity	*Non-target organism, fate and expression data*
Acute oral toxicity	**Avian acute oral toxicity**
Acute dermal toxicity	Avian dietary toxicity
Primary eye irritation	**Freshwater fish toxicity**
Primary dermal irritation	**Freshwater invertebrate toxicity**
Hypersensitivity study	Non-target plant studies
Hypersensitivity incidents	Non-target insect studies
Genotoxicity studies	
90-day feeding study	
90-day dermal study	
90-day inhalation study	
Teratogenicity study	

[a] From Weatherston and Minks (1995)
[b] The requirements **in bold** are usually required to obtain a registration

RECENT DEVELOPMENTS IN THE USA

The US EPA further streamlined the regulatory process for pheromone products in 1994. Firstly the EPA have granted a generic exemption from the requirement of a food tolerance for arthropod pheromones when used in retrievable polymeric matrix dispensers with an annual application rate of < 150 g a.i./acre (375 g/ha). Secondly, EPA now permits testing of pheromones in solid matrix dispensers on areas up to 250 acres (100 ha) without requesting an experimental use permit (Weatherston and Minks, 1995). In 1995 these exemptions were extended to 'all modes of application', removing the distinction between retrievable (i.e. mostly hand-applied) and non-retrievable (i.e. mostly sprayable) formulations. As a result, companies are no longer required to obtain a food use clearance for each compound or formulation.

The implementation of these new exemptions reflects EPA's conviction that lepidopterous pheromones do not present an 'unreasonable' adverse effect to human health or the environment. EPA hopes these measures will encourage the development and use of environmentally acceptable biological pesticides as alternatives to the more toxic conventional synthetic pesticides. The aim is to ease the testing requirements of these products, to speed their market introduction, and to promote their integration into modern pest management strategies. Finally, EPA established a completely separate division in 1995, the Biopesticide and Pollution Prevention Division (BPPD), to manage the registration of semiochemicals and other biologically based products. This removes such biochemical strategies from the normal pesticide review bureaucracy. EPA now claims that the stepwise reduction of regulatory burdens over the past 10 years has reduced the approval times for pheromone products from over 2 years to as little as 2 months. The new rules have also decreased the cost of bringing new products to market, resulting in the development and introduction of a record number of products used to control the codling moth and other caterpillar pests.

REGISTRATION OF PHEROMONES IN OTHER COUNTRIES

No other country has developed specific regulatory requirements for the registration of pheromones with the exception of Canada, where a US-type policy is in formulation. Most European countries, as well as others such as Australia, New Zealand and South Africa, draw no distinction between chemical and biochemical insecticides. Pheromones are still considered as chemical substances, which in principle have to meet the full set of requirements. However, in practice, pheromone regulation varies widely between countries, ranging from an exemption from regulation (Denmark and New Zealand) and complete exemption from the requirements except for proof of efficacy (Italy), to a case-by-case approach implemented with varying degrees of flexibility. This uncertainty concerning data requirements about which demands can be expected from the registration authorities in the latter group of countries makes applicants hesitant to start any registration procedure. Despite such regulatory impediments, several pheromone products have been granted registration in both more 'flexible' and more difficult countries in Europe and elsewhere (Table 36.2). Regulatory success depends not only on a variable product-by-product interpretation of the rules, but also on the persistence of the applicant, which in turn is determined by market opportunities.

Some countries are very strict in their implementation of regulatory policy. In Germany, public concern about pesticides has grown to such an extent that environmentally friendly agents such as pheromones and microbial agents are now under increasing regulatory pressure. For example, some rumours surfaced in 1995 suggesting that pheromones may have aquatoxicological properties, which is really an amazing consideration given how extremely difficult it is to dissolve these apolar compounds in water. German colleagues from the regulatory office told us that they feel obliged to respond to these statements, in particular if they come from an ecological institute or from members of Parliament. This of course results in extra demands for research and increases the complexity of regulatory requirements. The situation is rather peculiar: the same people who want protection of the environment from pesticides concurrently inhibit the introduction of environmentally friendly agents due to their demands for excessive data, a problem already described in the 1970s (Djerassi *et al.*, 1974). In the Netherlands an impasse exists in pheromone registration, but in this case it has a more bureaucratic character.

Finally, national registration offices in the European Union countries are at present postponing introduction of any regulatory changes due to the imminent harmonization of pesticide legislation in the European Union. We would like to conclude by highlighting a current action of the OECD (Organization for Economic Cooperation and Development) which proposes a harmonization of the pheromone product registration data requirements based on the US/Canadian model within all member countries of that organization. We are very pleased with this initiative, as pheromone researchers and the industry have repeatedly made strong pleas in the past to regulatory authorities in various European countries to adopt the US tier system, rather than focusing resources on developing another unique policy. We sincerely hope that the OECD initiative will be successful and implemented globally in the near future.

REFERENCES

Arn, H., Tóth, M. and Priesner, E. (1992) *List of Sex Pheromones of Lepidoptera and Related Attractants*, 2nd edn, IOBC/WPRS, Montfavet.

Bedoukian, R.H. (1992) Regulation of some classes of phytochemicals: flavour and fragrance ingredients related to insect behaviour-modifying chemicals, in *Insect Pheromones and Other Behaviour-modifying Chemicals: Applications and Regulations* (eds R.L. Ridgway, M.N. Inscoe and H. Arn), BCPC Monograph No.51, British Crop Protection Council, Farnham, UK, pp. 79–92.

Biddinger, D.J., Felland, C.M. and Hull, L.A. (1994) Parasitism of tufted apple bud moth (Lepidoptera: Tortricidae) in conventional insecticide and pheromone-treated Pennsylvania apple orchards. *Environmental Entomology*, **23**, 1568–1579.

Djerassi, C., Shih-Coleman, C. and Diekman, J. (1974) Insect control of the future: operational and policy aspects. *Science*, **186**, 596–607.

Howse, P., Stevens, I. and Jones, O. (1996) *Insect Pheromones and their Use in Pest Management*, Chapman & Hall, New York.

Jutsum, A.R. and Gordon, R.F.S. (1989) *Insect Pheromones in Plant Protection*, John Wiley & Sons, Chichester.

Kirsch, P.A. (1988) Pheromones: their potential role in the control of agricultural insect pests. *American Journal of Alternative Agriculture*, **3**, 87–97.

Kirsch, P.A. (1996) Old wine in new skins: evolution in pheromone commercialisation, in *Proceedings of the 1996 International Symposium 'Insect Pest Control with Pheromones', Suwon, Korea, 18–19 October 1996* (eds K.S. Boo, K.C. Park and J.K. Jung), pp. 165–174.

Louis, F., Schirra, K.J. and Feldhege, M. (1996) Mating disruption in vineyards: determination of population densities and effects on beneficials. Abstract International Conference 'Technology Transfer in Biological Control: From Research to Practice', Montpellier, 9–11 September 1996. *IOBC/WPRS Bulletin*, **19**(8), 63–64.

Mafro-Neto, A. and Baker, T.C. (1996) Timed, metered sprays of pheromones disrupt mating of *Cadra cautella* (Lepidoptera: Pyralidae). *Journal of Agricultural Entomology*, **13**, 149–168.

Mitchell, E.R., Wadill, V.H. and Ashley, T.R. (1984) Population dynamics of the fall army worm (Lepidoptera: Noctuidae) and its larval parasites on whorl stage corn in pheromone-permeated field environments. *Environmental Entomology*, **13**, 1618–1623.

Niwa, C.G. and Daterman, G.E. (1989) Pheromone mating disruption of *Rhyacionia zozana* (Lepidoptera: Tortricidae): influence on the associated parasite complex. *Environmental Entomology*, **18**, 570–574.

Ohira, Y. and Oku, T. (1993) A trial to promote the effect of natural control agents, especially of *Trichogramma* sp., on the apple tortrix, *Archips fuscocupreanus* Waslingham, by disrupting the mating of the pest, in *Proceedings International Symposium 'Use of Biological Control Agents under Integrated Pest Management', Fukuoka, Japan, 4–10 October 1996*, pp. 251–265.

Ridgway, R.L., Silverstein, R.M. and Inscoe, M.N. (1990) *Behavior-Modifying Chemicals for Insect Management*, Marcel Dekker Inc., New York.

Spittler, T.D., Leichtweis, H.C. and Kirsch, P. (1992) Exposure, fate and potential residues in food of applied lepidopteran pheromones, in *Insect Pheromones and Other Behaviour-modifying Chemicals: Applications and Regulations* (eds R.L. Ridgway, M.N. Inscoe and H. Arn), BCPC Monograph No. 51, British Crop Protection Council, Farnham, UK, pp. 93–108.

Tingle, F.C. and Mitchell, E.R. (1982) Effect of synthetic pheromone on parasitization of *Heliothis virescens* (F.) (Lepidoptera: Noctuidae) in tobacco. *Environmental Entomology*, **11**, 913–916.

Weatherston, I. and Minks, A.K. (1995) Regulation of semiochemicals – global aspects. *Integrated Pest Management Reviews*, **1**, 1–13.

37

The importance of the population perspective for the evaluation of side-effects of pesticides on beneficial species

John D. Stark, Julie A.O. Banken and William K. Walthall

ECOTOXICOLOGY: WHERE IS THE ECOLOGY?

Ecotoxicology is an emerging field of study that attempts to combine ecological and toxicological principles so that more realistic estimates of environmental damage by xenobiotics can be made. However, ecotoxicology is dominated by toxicology, and calls to 'put the ecology into ecotoxicology' have been virtually ignored (Baird *et al.*, 1996).

There are inherent problems associated with combining ecological and toxicological methodology to determine the impact of xenobiotics on ecosystems because the two fields of study are so different from one another. Toxicology is a rigidly defined science where most variables are controlled and very narrow endpoints are investigated at or below the individual level (i.e. organ, cellular, molecular). Ecological studies, on the other hand, often involve observations of natural systems where variables are not controlled at all. Levels of organization higher than the individual – such as populations, communities and food webs – are frequently studied by ecologists.

What information is gained from toxicity studies? Laboratory toxicity studies often estimate the susceptibility of one life stage of a species to

Ecotoxicology: Pesticides and beneficial organisms.
Edited by P.T. Haskell and P. McEwen. Published in 1998 by Chapman & Hall, London. ISBN 0 412 81290 8.

a particular chemical. It is possible to compare the relative toxicity of several chemicals to a species and also compare the susceptibility of various species among chemicals. However, how closely is this type of information linked to effects on populations?

To answer this question we have been investigating methods of incorporating more ecologically based approaches into ecotoxicology (Kareiva *et al.*, 1996). One area of our research programme has been to investigate the relationship between acute mortality, reproduction and growth of a population using demography (Stark and Wennergren, 1995), population modelling (Kareiva *et al.*, 1996) and other measures of population growth (Stark *et al.*, 1997; Walthall and Stark, 1997).

In this chapter we base our discussion around a series of questions that are important to understanding the effects of pesticides and other pollutants on organisms.

WHAT ARE THE MOST IMPORTANT TOXICOLOGICAL EFFECTS TO CONSIDER?

At first, this question may seem intuitive. Toxins are, by definition, poisons; thus we typically rely on measures of mortality to answer this question. Do these endpoints consider potential biochemical disruptions that may cause overt or subtle phenotypic or genotypic alterations in an organism? Even if these endpoints are considered in toxicological assessment, what is the importance in relation to population fitness? What toxicological effects will have the biggest impact on population health? For example, some bird species can have brain acetylcholinesterase levels reduced by 50% with no apparent effect on health. Since acetylcholinesterase levels have no long-term effect on survival or reproduction, why measure this endpoint?

The most obvious and easily quantified toxic effect is death and therefore mortality is the most widely reported toxic endpoint, but pesticides can cause a wide array of sublethal effects as well. In fact, a cascade of effects can occur to an organism once it is exposed to a pesticide. The types of sublethal effects that occur will obviously depend on the chemical in question as well as on the organism being studied. Some chemicals will not cause sublethal effects; others will cause a series of effects. Even closely related species can respond differently to the same chemical (Stark and Sherman, 1989).

Some of the common endpoints applied in monitoring for potential sublethal effects of a chemical are the level of a particular enzyme, behavioural changes and short-term reproduction. However, mortality and reproduction studies are usually short term in duration, and rarely are a chemical's effects determined over the life span of the organisms being studied. Furthermore, in almost all toxicological studies, measurements are taken on individuals or components of individuals (tissue, cells, enzymes, etc.).

Traditional toxicological methods may not provide enough useful information for determining the effect of pesticides at the population level or above because toxicologists address processes acting at or below the level of the individual. The assumption of individual-based research is that effects on the individual can be extrapolated to higher-order systems such as populations, communities or ecosystems. Toxicological impacts at the population level or higher often involve additional, fundamentally different processes due to density-dependent regulatory mechanisms and these processes can substantially alter the population-level or higher consequences of individual-level impacts. In attempting to protect beneficial and non-target species from pesticide and xenobiotic exposure, we have become prisoners of human toxicology, where the individual is of great importance (McNair *et al.*, 1995). We must recognize that, in ecotoxicology, the health of a population of animals is more important than the health of any particular individual.

One situation that may occur at the population level that can never be estimated from toxicological measurements of individuals is 'population compensation'. When a percentage of individuals dies after exposure to a pesticide, survivors have more resources available and may reproduce at a greater rate, thus compensating for the loss of individuals. The offspring of these survivors may be larger and more vigorous. Thus, losses even as great as 25% may have no long-term impact on a population, while losses of 50% may result in only a slight change (Nicholson, 1954; Slobodkin and Richman, 1956).

'Magnification of effect' may also occur, where effects on individuals are magnified at the population level for some xenobiotics and species. Subtle effects that reduce individual reproduction, but that are difficult to detect, are at work here. The result is that susceptibility is greater at the population level than is indicated by individual-level effects.

Some life stages may be much more susceptible to a particular chemical than others. If this is the case, a population may be more, or less, susceptible than would be indicated by individual-level assessments. For example, Stark and Wennergren (1995) found that if one life stage of a species is less susceptible than another and the xenobiotic in question is degraded quickly (a definition of many modern pesticides), the less susceptible stage may compensate for the more susceptible stage, with the overall impact on the population being negligible. Thus, attempts to predict the effects of pesticides, and other xenobiotics, in the field might fail if based on laboratory toxicity tests conducted on a single life stage or on measurements of individual responses. Other complex ecological and behavioural factors not evaluated in laboratory studies also reduce the predictability of laboratory toxicity tests (Çilgi and Jepson, 1992; Bakker and Jacas, 1995; Kjaer and Jepson, 1995).

The health of a population depends on life span, time to reproductive maturity and reproductive rate. As such, these are the important parameters to be measured in toxicological studies. As mentioned above, xenobiotics can cause both lethal and sublethal effects, but methods for combining lethal and sublethal effects into a single meaningful parameter have not yet been adequately developed. For example, a delay in the time to reproduction can cause a greater reduction in population growth capacity than reductions in either fecundity or longevity (Cole, 1954; Allan, 1976; Daniels and Allan, 1981). Therefore time to first reproduction should be given serious consideration as a toxicity endpoint, but it has rarely been utilized as such.

HOW DO WE DETERMINE THE TOTAL EFFECT (LETHAL AND SUBLETHAL) OF PESTICIDES AND OTHER POLLUTANTS ON POPULATIONS?

Considering the cumulative impact from both lethal and sublethal effects, this is the second question we want to consider.

Overmeer and van Zon (1982) recognized that the total effect of a pesticide was more important than measurements of separate endpoints. They proposed the use of the total effect index (E) which incorporates survival, reproduction and performance, but their approach has not been widely adopted. Other researchers have suggested that demographic parameters, such as the intrinsic rate of increase, should be used to estimate the total effects of xenobiotics on organisms (Daniels and Allan, 1981; Bechmann, 1994; Stark and Wennergren, 1995; Kareiva *et al.*, 1996).

Demography and its role in evaluation toxicity

Demographic toxicological analysis incorporates age-specific survivorship and reproduction (including age at first reproduction, brood size and frequency, and longevity) of test organisms into one endpoint: the intrinsic rate of growth of the population (Daniels and Allan, 1981; Allan and Daniels, 1982; Bechmann, 1994). The intrinsic rate of growth (r_{max}, or r_m) is defined as the innate capacity for increase, or the rate of increase in a population growing in optimal conditions. The advantage of this approach is that a total measure of effect is determined that incorporates lethal and sublethal effects into one measure. To achieve 'optimal conditions', test subjects receive an unlimited food supply and offspring are removed daily so that crowding does not occur. Females are exposed from birth to a pesticide and their survivorship and fecundity are recorded daily. Life tables are constructed from which r_m can be estimated using the equation:

$$r_m = \log_e R_o / T \qquad (37.1)$$

where R_o is net replacement rate, and is obtained from the formula $l_x m_x$; T is mean generation time, $l_x m_x X/R_o$; X is the age interval in days, l_x is the proportion of females surviving to age x, and m_x is the average number of female offspring produced per female at age x. Because r_m measures fecundity as well as mortality, it can detect subtle, individual-level effects of contaminants that alter the growth of populations at rates below the lethal concentration limits (Bechmann, 1994).

Even though demographic toxicological data provide perhaps the most complete data about toxicant effects on populations, there are several problems associated with this method. The first and perhaps most important one is that it can be incredibly expensive and time consuming to develop life table data. For some species it is virtually impossible to generate this endpoint because of extremely low reproductive rates, long life spans and difficulty in rearing the animals in captivity. This may be overcome in some cases by using abbreviated (truncated) life tables (Walthall and Stark, 1997). Because neonates are used to develop life table data, differential susceptibility among life stages and differential population structures are ignored (Stark and Wennergren, 1995). In addition, individuals used in life table studies are reared separately under unrealistic conditions.

HOW DO THE EFFECTS OF PESTICIDES IN INDIVIDUALS TRANSLATE TO WHAT HAPPENS AT THE POPULATION LEVEL?

Estimating the total effect of a pesticide on individuals is difficult enough, but another complexity arises when we ask this third question.

To answer it we will discuss another population growth rate measure that simplifies the gathering of population-level data. It is called the instantaneous rate of increase (r_i) (Walthall and Stark, 1997). Unlike the intrinsic rate of increase (r_m), which is based on projected growth in an unlimited environment (Birch, 1948), the instantaneous rate of increase reflects the actual growth of a population (Walthall, 1995) and is determined by measuring a population's change in density during a designated time interval.

To determine r_i, a population with known density is realistically exposed to a pesticide for a length of time that will reflect both lethal and sublethal effects, and is then censused. The instantaneous growth equation is:

$$N_t = N_0 e^{rt} \qquad (37.2)$$

where N_t is the number of individuals in the population over the time period (t) studied, and N_o is the initial number of individuals in the population. This equation can be rearranged so that r_i is calculated as:

$$r_i = \ln(N_t/N_o)/t \quad (37.3)$$

Populations can be studied in the field or maintained within microcosms that closely represent field conditions. Laboratory microcosms have been recommended by Kimball and Levin (1985) and Cairns (1983) as useful tools for the prediction of the effects of a chemical on natural populations. Since this endpoint takes into account the growth of the population, the instantaneous rate of increase can be used to evaluate the impact of a pesticide at the population scale.

The development of this method is a step forward in the effort to understand and predict the population- and community-level consequences of xenobiotic exposure. Because actual populations are measured, r_i accounts for the impacts of density-dependent feedback mechanisms, such as decreased population growth under limited environmental conditions. Microcosms can be designed that closely simulate realistic environmental conditions so that test populations receive natural levels of direct, residual and oral chemical exposure. Finally, test populations can be monitored in the presence of predator or competitor species to assess modifications to normal interspecific competition, interactions among the trophic levels and differential susceptibility between predator and prey species (Banken, 1996).

The benefits of the instantaneous rate of increase method are not without certain costs. Less detailed information about specific population growth parameters is gained from studies where the instantaneous rate of increase is the endpoint determined, compared with life table studies. If, for example, you only record the total number of individuals and do not note the number of original survivors, offspring, population structure over time and new adults, you will not know whether population declines are due to mortality, reduced reproductive capacity or a delay to first reproduction. Therefore caution must be taken when using the instantaneous rate of increase. However, the level of realism and the ability to do these studies with predators and competitors makes r_i a very realistic endpoint of toxicological effect.

Acute toxicity versus population growth: a case study

Stark *et al.* (submitted) investigated the relationship between individual endpoints of effect, acute mortality and reproduction, and population growth rates. Acute LC_{50}s (72 h) were estimated for two mite species. The first was a herbivore,, *Tetranychus urticae*, and the second a predator of the herbivore, *Iphiseus degenerans*. The two species exhibited almost identical acute susceptibility to the pesticide dicofol, but they displayed completely different susceptibility when population growth was the evaluated endpoint. Populations of *I. degenerans* became extinct after exposure to 140 ppm dicofol, while *T. urticae* populations became extinct after exposure to 560 ppm (Figure 37.1). The NOEC (no observable effect concentration) for population growth rates after dicofol exposure was 17 ppm for *I. degenerans*, and 35 ppm for *T. urticae*. These NOEC values were equivalent to the acute LC_3 for the immature stage of *I. degenerans* and the acute LC_{23} for the immature stage of *T. urticae*. Thus, populations of *T. urticae* were able to compensate for high losses of individuals while

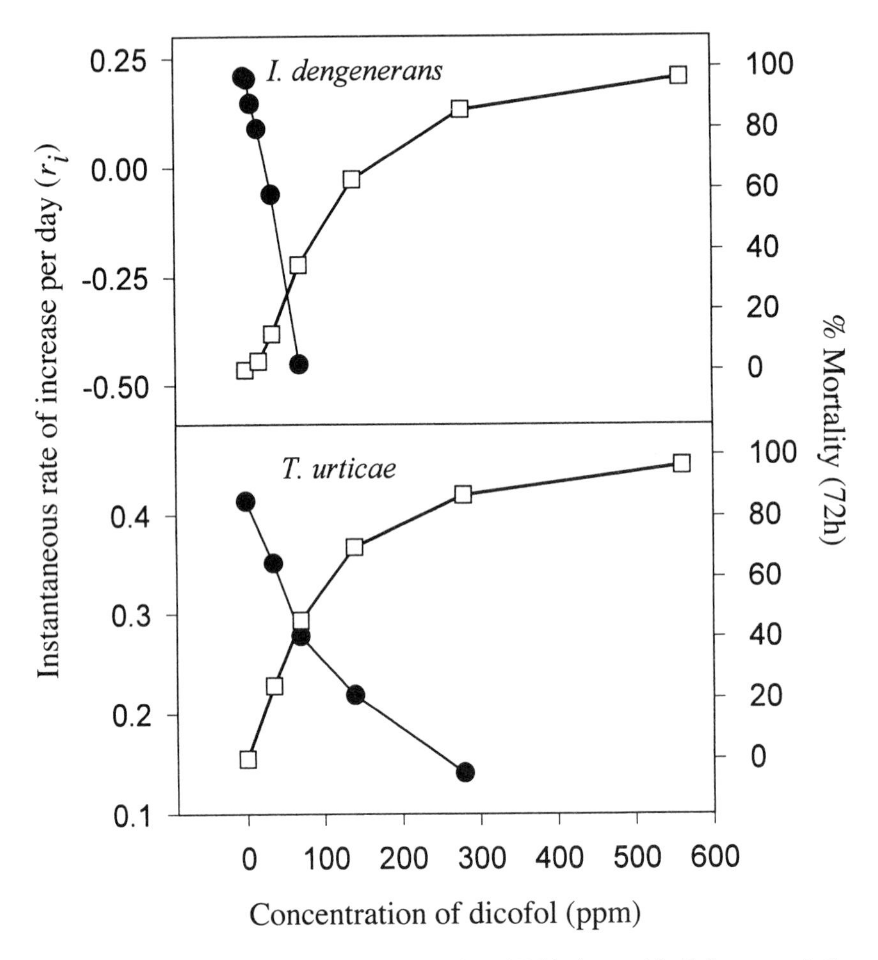

Figure 37.1 Comparison of acute mortality (72 h) data with 7-day population growth data over the same range of concentrations (●, r_i; □, mortality). Note that some r_i values are negative for *Ipheseus degenerans* but not for *Tetranychus urticae*.

I. degenerans populations could not compensate for losses. An analysis of reproduction data indicated that unexposed *T. urticae* produced four to five times more offpsring than *I. degenerans*. This in itself was important because it indicated that *I. degenerans* was intrinsically more susceptible than *T. urticae* because similar effects on reproduction would be more devastating to the species with a lower reproductive rate. Pesticide exposure reduced reproduction in both species, but *I. degenerans* reproduction was reduced over a much narrower range of concentrations, indicating greater susceptibility. Therefore the difference in population growth exhibited between the two species was due to greater reproductive potential of *T. urticae* over the time frame studied, and a greater pesticide effect on reproduction in *I. degenerans* than in *T. urticae*.

Our data suggest that although the acute LC_{50} is a valuable measure of effect, it may be misleading because it does not take into account differential reproductive potential among species and effects on reproduction. Also, reproduction data alone may overestimate effects on populations.

HOW DOES POPULATION STRUCTURE INFLUENCE THE IMPACT OF PESTICIDES ON POPULATIONS?

This fourth question deals with differential susceptibility among life stages and ages. To answer it, we initiated another project to determine whether the starting structure of a population at the time of initial xenobiotic exposure influenced the impact that pesticides, or other xenobiotics, have on population growth rates. The two-spotted spider mite, *T. urticae*, served as our model organism. Populations of *T. urticae* were exposed to dicofol. Three structured populations were tested: the first consisted entirely of eggs; the second consisted of the stable age distribution comprising eggs, immatures and adults; the third consisted entirely of young adult females. Instantaneous rates of increase (r_i) for the three structured populations were determined over time without exposure to pesticides (control) and after exposure to a pesticide concentration that did not kill all individuals. Population growth rates for the three control populations of *T. urticae* converged 16 days after the start of the study (Figure 37.2).

Unlike the three control populations, the r_i of the pesticide-treated populations did not converge by day 16 (Figure 37.2). The r_i of populations that started as eggs remained significantly lower than those of the other two populations ($P < 0.05$), while the r_i of treated adult and of mixed-age populations were not significantly different.

Thus, we conclude that the initial structure of a population does affect the impact of xenobiotics on populations. These results have implications for estimating the effects of xenobiotics on beneficial species.

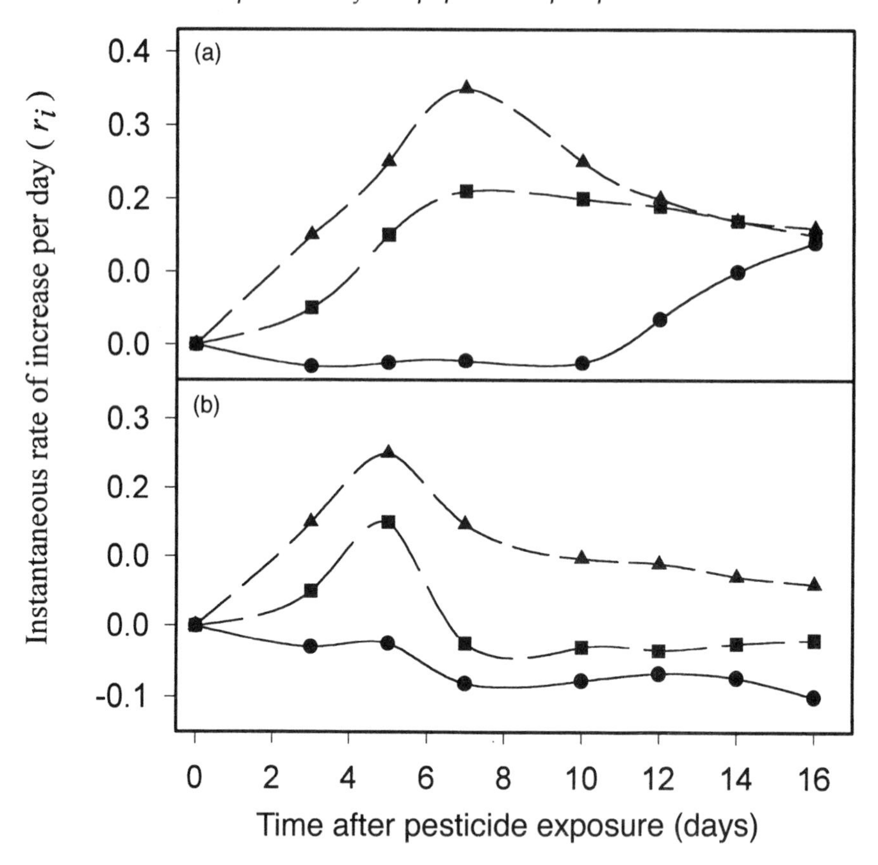

Figure 37.2 Population growth rates of three populations of *Tetranychus urticae* with different initial age/stage structures: ●, all eggs; ■, stable age; ▲, all adults. **(a)** No pesticide exposure; **(b)** exposure to a pesticide.

THE NEED FOR CAUTION WHEN COMPARING LABORATORY AND FIELD-GENERATED ENDPOINTS

When comparing an endpoint that has been generated under both laboratory and field conditions, erroneous conclusions about the impact of the chemical in question can be made. For example, the development of acute toxicity data in the field for a particular pesticide may yield the same results as acute toxicity studies performed under laboratory conditions, but this endpoint does not take into consideration long-term effects on population growth.

IOBC tiered testing system and evaluation of pesticide impact

The International Organization for Biological Control of Noxious Animals and Plants (IOBC) has developed standard protocols for the analysis of the impacts of pesticides on non-target species (Hassan, 1992). It recommends a sequential testing procedure where pesticides are first tested on beneficial organisms in the laboratory. If no harmful effect is observed in the laboratory, the pesticide is considered safe for use in integrated pest management (IPM) programmes. If the pesticide is found to be toxic to beneficial insects in the laboratory, tests are performed in semi-field trials, where non-target organisms are exposed to treated, caged plants in simulated or natural field conditions. Pesticides that cause no toxic effects in semi-field trials are then recommended for use in IPM, and those that still harm beneficial insects are tested again in field trials. The advantage of the IOBC laboratory approach is that a measure of reproduction is often recorded, as well as mortality. Our work has indicated that reproduction is a more sensitive measure of effect than mortality (Stark *et al.*, submitted). Also, the tiered approach reduces the chances of making mistakes. However, mistakes do occur (Bakker and Jacas, 1995) and after our previous discussion it is easy to see why a laboratory test based on individual mortality could fail to predict effects in the field.

The new generation of insecticides and why it will be more difficult to evaluate them in the laboratory

A new generation of insecticides that are slower acting and that produce a greater degree of sublethal effects than the previous generation is being developed and registered. For some of these products, sublethal effects may be as important as lethal effects. The slow-acting nature of these new products and the important sublethal effects they produce make prediction of their potential environmental impact more complicated. Some of them have a much greater impact on one life stage than on others (Stark and Wennergren, 1995). Thus, to estimate the overall toxicity of a chemical in the laboratory to a species exposed to these new insecticides may prove difficult at best.

CONCLUSIONS

More ecological approaches must be adopted to estimate adequately the impact of pesticides and other xenobiotics on populations and communities. This will become increasingly important as new insecticides that have non-neurotoxic modes of action are used. Laboratory toxicity studies might be improved by using population growth rate studies in place of acute mortality and reproduction studies, but it must be realized that use of population growth may be very difficult for some species. Abbreviated population studies that are short in duration are being investigated (Walthall and Stark, 1997): they may provide a greater understanding of pesticide effects than traditional bioassay, but require less work than developing life tables or instantaneous rate of increase data. Modelling pesticide impacts with limited but key data may also result in better predictive power of the impacts that pesticides have on beneficial and non-target populations.

REFERENCES

Allan, J.D. (1976) Life history patterns in zooplankton. *American Naturalist*, **110**, 165–180.

Allan, J.D. and Daniels, R.E. (1982) Life table evaluation of chronic exposure of *Eurytemora affinis* (Copepoda) to kepone. *Marine Biology*, **66**, 179–184.

Baird, D.J., Maltby, L. and Douben, P.E.T. (1996) *Ecotoxicology: Ecological Dimensions*, Chapman & Hall, London.

Bakker, F.M. and Jacas, J.A. (1995) Pesticides and phytoseiid mites: strategies for risk assessment. *Ecotoxicology and Environment Safety*, **32**, 58–67.

Banken, J.A.O. (1996) An ecotoxological assessment of the neem insecticide, Neemix, on the pea aphid, *Acyrthosiphon pisum* Harris, and the seven-spotted lady beetle, *Coccinella septempuntata* L. Masters thesis.

Bechmann, R.K. (1994) Use of life tables and LC50 tests to evaluate chronic and acute toxicity effects of copper on the marine copepod *Tisbe furcata* (Baird.). *Environmental Toxicology and Chemistry*, **13**(9), 1509–1517.

Birch, L.C. (1948) The intrinsic rate of natural increase of an insect population. *Journal of Animal Ecology*, **17**, 15–26.

Cairns, J. (1983) Are single species toxicity tests alone adequate for estimating environmental hazard? *Hydrobiologia*, **100**, 47–57.

Çilgi, T. and Jepson, P.C. (1992) The use of tracers to estimate the exposure of beneficial insects to direct pesticide spraying in cereals. *Annals of Applied Biology*, **121**, 239–247.

Cole, L.C. (1954) The population consequences of life history phenomena. *Quarterly Review of Biology*, **29**, 103–137.

Daniels, R.E. and Allan, J.D. (1981) Life table evaluation of chronic exposure to a pesticide. *Canadian Journal of Fisheries and Aquatic Science*, **38**, 485–494.

Hassan, S.A. (1992) Guidelines for testing the effects of pesticides on beneficial organisms: description of test methods. *IOBC/WPRS Bulletin*, **XV**(3), 1–3.

Kareiva, P., Stark, J.D. and Wennergren, U. (1996) Using demographic theory, community ecology and spatial models to illuminate ecotoxicology, in *Ecological Dimensions*, (eds L. Maltby and P. Grieg-Smith), Chapman & Hall, London, pp. 13–23.

Kimball, K.D. and Levin, S.A. (1985) Limitations of laboratory bioassays: the need for ecosystem level testing. *Bioscience*, **35**, 165–171.

Kjaer, C. and Jepson, P.C. (1995) The toxic effects of direct pesticide exposure for a nontarget weed-dwelling chrysomelid beetle (*Gastrophysa polygoni*) in cereals. *Environmental Toxicology and Chemistry*, **14**(6), 993–999.

McNair, J.N., Goulden, C.E. and Ziegenfuss, M.C. (1995) Is there a place for ecotoxicology? *SETAC News*, **15**(6), 18–21.

Nicholson, A.J. (1954) Compensation reactions of populations to stresses, and their evolutionary significance. *Australian Journal of Zoology*, **2**, 1–8.

Overmeer, W.P.J. and van Zon, A.Q. (1982) A standardized method for testing the side-effects of pesticides on the predacious mite *Amblyseius andersoni* (Acarina: Phytoseiidae). *Entomophaga*, **27**, 357–364.

Slobodkin, L.B. and Richman, S. (1956) The effect of removal of fixed percentages of the newborn on size and variability in populations of *Daphnia pulicaria* (Forbes). *Limnology and Oceanography*, **1**, 209–237.

Stark, J.D. and Sherman, M. (1989) Toxicity, penetration and metabolism of acephate in three fruit fly species (Diptera: Tephritidae). *Journal of Economic Entomology*, **82**, 1564–1571.

Stark, J.D. and Wennegren, U. (1995) Can population effects of pesticides be predicted from demographic toxicological studies? *Journal of Economic Entomology*, **88**(5), 1089–1096.

Stark, J.D., Tanigoshi, L., Bounfour, M. and Antonelli, A. (1997) Reproductive potential: it's influence on the susceptibility of a species to pesticides. *Ecotoxicology and Environmental Safety*, **37**, 273–279.

Walthall, W.K. (1995) An ecotoxicological approach to the assessment of the insecticide Imidacloprid on the pea aphid, *Acyrthosiphon pisum* Harris (Homoptera: Aphididae). Masters thesis, 97 pp.

Walthall, W.K. and Stark, J.D. (1997) Comparison of two population-level ecotoxicological endpoints: the intrinsic (*rm*) and instantaneous (*ri*) rates of increase. *Environmental Toxicology and Chemistry*, **16**, 1068–1073.

Part Five

Global Implications

38

Introduction

Peter T. Haskell

The previous sections show that the main initiative for developing testing techniques on the effects of pesticides on beneficial organisms – mostly beneficial arthropods – has come from Europe, particularly from the International Organization for Biological Control, and that the body of some 15 years of research reviewed in this book has been largely carried out by European government and agrochemical company scientists.

However, the great increase in recent years in public concern about the effect of pesticides in the environment has been global and there are now few countries that do not have a government department dealing with the environment. Historically, concern about pesticides – as exemplified, for example, by Rachel Carson's *Silent Spring* – was one of the first issues to cause global disquiet and a call for action. Action in the pesticide field was, in fact, taken quite quickly in the form of the development of the philosophy and practice of integrated pest management systems, specifically designed to reduce the overuse of and overdependence on pesticides which was characteristic of the 1950s and 1960s. It is an indication of the global importance of integrated pest management, and its consequences for many aspects of pesticide development and usage, that as this book was being prepared news came of the award of the World Food Prize in October 1997 to Ray Smith and Perry Adkinson: these two American scientists played a leading role not only in the technical development of IPM but also in promoting and publicizing the advantages of the system, largely through and with the support of the Plant Protection Service of the UN Food and Agriculture Organization.

Ecotoxicology: Pesticides and beneficial organisms.
Edited by P.T. Haskell and P. McEwen. Published in 1998 by Chapman & Hall, London. ISBN 0 412 81290 8.

Integrated pest management and its successor, integrated crop management, are now received wisdom in many countries and, because of their application, pesticide usage world-wide has been reduced. However, the research that led up to the new approach showed how little was known about the wider environmental effects of pesticide usage – in particular, the ecological interactions in the target crop or vector population.

This was certainly the case in respect of the effect of pesticides on beneficial arthropods – the insects, spiders and mites that prey on or parasitize the target pest complex – and the conference that generated this book was designed to explore and review this area.

Part Two has demonstrated that concern about pesticide usage and its potential environmental consequences is now a global concern, with chapters on official government philosophy (and action plans to reduce pesticide usage) from Israel, Malaysia and Europe. Part Five extends this coverage to consider what action is being taken elsewhere in the world in relation to a range of subjects, all of which have at their core a concern for the non-target effect of pesticides and an interest in quantifying this so that it can be reduced or avoided in the future.

One of the key areas in world food production is Asia, where the typical small farmer still overuses pesticides: most still use 'calendar spraying' and many farmers have no idea that some insects are beneficial.

Williamson (Chapter 39) describes an initiative by IOBC and IIBC, assisted by German and Swiss aid organizations, to alter this state of affairs in Asia by giving an appropriate training course in IPM and ICM systems to pesticide registration officers and extension workers from 11 countries throughout the region. The success of this training, and the interest that it has aroused in Africa and Latin America, emphasizes the need for training and information in this subject and has been a recurring theme in this book. The theme is continued in the final chapter in this section, with Vos's account of farmer field schools (FFS) and training of extension officers as trainers (TOT) in vegetable farming in Southeast Asia. All this training drew heavily on the results of the research reviewed at the Ecotox meeting, as encapsulated in the publication of the ESCORT workshops in 1994 (Barrett *et al.*, 1994).

The remaining chapters in this section deal with specific subjects: locusts and rice.

Locust and grasshopper control has depended on chemical pesticides from the earliest times, with millions of hectares being sprayed annually, but only recently has any serious attempt been made to assess the environmental effects of this global control campaign. Van der Valk (Chapter 40) describes the Locustox project, supported by the governments of Senegal and The Netherlands and organized by the FAO Plant Protection Service. Although this work was carried out in West Africa, it is of considerable interest because, in the arid and fragile desert ecosystems exploited by locusts, even small localized use of pesticides can have severe effects.

In general, and on balance, the project was able to report that current locust control did not result in large irreversible effects on the invertebrate fauna of the desert ecosystem; nevertheless, many unsuspected effects were discovered, which led to the development of improved testing procedures on the effect of pesticides on non-target organisms that are applicable outside the desert ecosystem. The fact that this work was carried out in Africa increases the chance that some of these testing procedures will be incorporated in the pesticide registration regulations that are now being developed or adopted by many African countries.

The world's major food crop is rice and vast amounts of pesticides are used in its production. Because of this a great deal of research, much of it carried out by the International Rice Research Institute in the Philippines, has been undertaken, with increasing emphasis in recent years on IPM/ICM and the reduction of pesticide usage. The brown planthopper of rice, *Nilaparvata lugens*, is a classic case of pesticide-induced outbreaks caused by destruction of natural enemies. In Chapter 41, Heong and Schoenly from IRRI review and discuss the effect of pesticides on natural enemies and, in particular, the capacity of the rice ecosystem to recover from external stresses. This led Way and Heong (1994) to conclude that 'IPM in tropical rice should be based on the contention that insecticides are not needed'.

Chapter 41 also stresses that there is still a great deal of unnecessary pesticide application because of farmer ignorance. The information is there but the problem is how to convince farmers to use it. The chapter reviews some pilot projects where insecticide usage has been halved in 16 months after an intensive media campaign.

This section clearly shows both the need for and the possibilities of research and training in reducing pesticide usage and it underlines the global importance of the developments reviewed in this book in testing and related ecological work, which has a major role to play in bringing about such a reduction.

REFERENCES

Barrett, K.L., Grandy, N., Harrison, E.G. *et al.* (1994) *Guidance Document on Regulatory Testing Procedures for Pesticides with Non Target Arthropods. From the ESCORT workshop, Wageningen, Netherlands, March 1994,* SETAC-Europe, Brussels.

Way, M.J. and Heong, K.L. (1994) The role of biodiversity in the dynamics and management of insect pests of tropical irrigated rice – a review. *Bulletin of Entomological Research,* **84**, 567–587.

39

The Asian initiative in pesticides and beneficials testing

Stephanie F.J. Williamson

BACKGROUND

Under non-Green Revolution farming practices, most insect populations rarely reach levels capable of causing sustained, severe damage to crops. This is largely due to the impact of predators, parasitoids and pathogens which keep pest populations in check. Pest outbreaks occur when the ecological balance is disturbed, most dramatically by the misuse of chemical pesticides. Many pest problems in tropical developing countries in Asia have arisen as a direct consequence of excessive or inappropriate use of pesticides, leading to pest resurgence and pesticide resistance. Well known examples include resurgence outbreaks of brown planthopper on rice and diamondback moth (DBM) resistance to almost all chemicals currently used on crucifers, but the phenomenon has also been documented on fruit, cotton, coffee, cocoa, oilpalm and other vegetables (Lim, 1992).

Successful implementation of alternative integrated pest management (IPM) programmes needs to focus not only on field research and farmer training from a farmer-first perspective (Matteson, 1992) but also on raising IPM awareness among government decision makers responsible for pesticide registration and legislation. One of the prime factors behind Indonesia's impressive achievements in reducing pesticide usage on rice involved first convincing the government of the value of natural enemies and the incompatibility of many widely used pesticides with an IPM approach. Their subsequent decision to ban 57 products, phase

Ecotoxicology: Pesticides and beneficial organisms.
Edited by P.T. Haskell and P. McEwen. Published in 1998 by Chapman & Hall, London. ISBN 0 412 81290 8.

out subsidies and put resources into IPM was the key to the success of the programme (Rengam, 1992).

An important strategic component for minimizing the negative impact of pesticides on natural enemies is to build tests on pesticide effects on beneficials into the registration process for chemical pesticides, with the aim of regulating those compounds that cause most harm to indigenous biological control systems. This approach has been pioneered by the IOBC Working Group on Pesticides and Beneficials, amongst others, over the last decade in Europe and standard regulatory testing procedures are soon to be incorporated into EC-wide legislation for plant protection product registration (Barrett *et al.*, 1994).

In order to share these European experiences with tropical Asian countries, a joint initiative was set up in 1995 by IOBC and the International Institute for Biological Control (IIBC), in collaboration with the German Federal Biological Research Centre (BBA). The main aim is to adapt the testing programmes developed by the IOBC Working Group on Pesticides and Beneficials to the Asian situation. To start this process, a training course/workshop was designed to provide researchers from the region with hands-on skills in how to measure pesticide effects on natural enemies and to develop appropriate protocols for their particular crop systems.

Training in evaluation of pesticide effects on natural enemies and its implications for pesticide registration

The two-week practical training course and workshop was held at the Department of Agriculture Training Centre in Serdang, Malaysia, in March 1995, in close collaboration with research staff from the Malaysian Agricultural Research and Development Institute (MARDI). Fifteen resource people were involved from IOBC, IIBC's UK and Malaysia stations, the University of London's Imperial College, MARDI and Germany's Technical Cooperation Organization (GTZ). The key resource people were Drs Paul Jepson and Pieter Oomen (members of IOBC's Working Group on Pesticide Effects on Beneficial Organisms) and Dr. Rob Verkerk (Imperial College) for his experience on sublethal effects on parasitoids of DBM. The course was sponsored by GTZ, the Swiss Development Corporation (SDC) and IOBC.

Participants were deliberately sought from national registration agencies in the region and from research and extension organizations. Sixteen participants were able to attend from 11 countries (China, Vietnam, Malaysia, Laos, Pakistan, the Philippines, Indonesia, Japan, Korea, Myanmar and Thailand). Approximately half the participants worked as pesticide registration officers and the remainder as IPM researchers or in agricultural extension programmes.

Sessions included practical work, group tasks, discussions and informal lectures, making use of interactive training methods in many cases. Topics covered included:

- the role of biological and chemical control in IPM;
- methodology for testing pesticide effects;
- recognition of natural enemies of insect pests;
- fundamentals of pesticide effects;
- IOBC principles for risk assessment of pesticide effects on natural enemies;
- sublethal effects of pesticides;
- using databases in risk assessment;
- IPM case studies;
- protocol design for field tests;
- sampling techniques for natural enemies;
- communicating biocontrol to farmers and decision-makers;
- risk management, legislation and regulation.

Ten practicals were carried out, both laboratory-based and in the field, giving participants hands-on practice in seven different test methods, reinforcing related skills in sprayer calibration and the collection and rearing of natural enemies. A range of predatory and parasitic natural enemies was used in these sessions, including parasitic wasps, coccinellid beetles, mites and spiders. These were either cultured at MARDI for the course, or field-collected, or purchased from commercial producers and imported.

The test protocols were adapted from IOBC test methods and included exposing test arthropods to locally available pesticides using the following methods:

- hour bioassay for predatory mites attached to treated microscope slides;
- hour bioassay using parasitoids of DBM in treated glass vials and confined in clipcages on leaves of treated whole plants;
- hour bioassay for coccinellids confined on treated cut leaves;
- day laboratory comparison of parasitism rate on DBM by treated and untreated female wasps;
- hour bioassay for spiders confined in plastic pots with treated soil;
- hour bioassay for predatory mites confined on cut leaves;
- day semi-field comparison of DBM parasitoid recruitment on treated and untreated rows of plants.

These practicals were aimed at giving participants experience in a variety of experimental designs and with different taxonomic groups, and for training purposes standard replication was reduced.

The training course succeeded in bringing together a highly enthusiastic group of participants from different backgrounds, all of whom play a role in the implementation of IPM. The topics covered complemented the differing experiences of the group, with pesticide registration officers learning about the role of natural enemies in crop systems, and IPM researchers and extensionists learning about the regulatory processes on pesticide use. In particular it was found that registration officers had very little knowledge of natural enemies and, although they were experienced in chemical analysis or mammalian ecotoxicological testing, the methods for assessing pesticide effects on beneficials were completely new to most of them.

The exercises on risk assessment and management using pesticide labels from the participants' countries were extremely valuable for discussion of IPM compatibility of products and ideas for including natural enemy hazard information on labels. Group work on communication and demonstration of biological control was useful for emphasizing the need for good linkages between regulatory and research approaches to IPM implementation and practical activities at farmer level.

WORKSHOP DISCUSSION

At the workshop, participants presented reports on the situation regarding pesticide use and registration in their respective countries, and the European experience in bringing natural enemy considerations into pesticide regulation was discussed. Participants then divided into job-related interest groups in order to develop a set of recommendations and proposals for standard testing procedures for the Asia region.

The country reports presented by the participants provided insights into the variation in pesticide use and legislation within the region. A few examples will illustrate the challenges faced by those working towards better pesticide regulation and IPM promotion. While some countries already have formal registration schemes in place, only the Philippines includes effects on beneficials in bioefficacy testing. China is now a major pesticide producer (over 200 000 tons of active ingredient per year) with regulation and evaluation of over 300 domestic and imported products. In Laos, pesticide use has increased dramatically, mainly on vegetables, and farmers now buy illegally imported products from Thailand and China. Many of these products are banned in other countries and label instructions are all in foreign languages.

The workshop included fruitful discussions on different approaches to incorporating beneficials testing into registration systems in the Asian context, such as the debate on whether to include natural enemy effects as an integral part of bioefficacy evaluation or within traditional ecotoxicological assessment. The workshop culminated in recommendations which participants agreed to pursue within their organizations and through networking.

DISCUSSION

A summary of the recommendations from the training course/workshop is given below.

- National and regional working groups (including members from regulatory, research and industry organizations) should be set up before April 1996 to coordinate development and use of appropriate methods for testing side-effects on natural enemies, focusing initially on problem crops such as rice and vegetables.
- Standardized test procedures and decision-making schemes following the IOBC/European and Mediterranean Plant Protection Organization/European Standard Characteristics of Beneficial Regulatory Testing (IOBC/EPPO/ESCORT) model should be adapted for use in Asia and incorporated within formal registration procedures, for harmonization region-wide. Priority to be given to crops where alternatives to chemical control have been demonstrated.
- Side-effects data should be collected from the region to draw up a simple Natural Enemy Hazard classification. This system would provide a basis for registration agencies to develop procedures (e.g. via pesticide labelling) to restrict the use of products that pose a hazard to key natural enemies.
- Training inputs should be rapidly developed to raise awareness of natural enemy side-effects and their adverse economic impact, aimed at farmers, extensionists, researchers, industry and regulators.
- A communication network should be set up immediately to encourage continued contact between participants, resource people, sponsors, IOBC and IIBC. It should also facilitate access to side-effects databases and other interested people should be encouraged to join.

Progress to date includes the setting up of national projects for beneficials testing and field research on pesticide effects on natural enemies. Pakistan has a new Ecotoxicology Research Centre which will coordinate natural enemy testing among collaborating institutes. A national Working Group on Side Effects of Pesticides on Natural Enemies and Beneficial Organisms (to include fish) has been set up amongst Malaysian researchers with support from the Department of Agriculture and the Malaysian Pesticide Board. They are collecting data on effects on key natural enemies, looking at fruit, vegetable, oilpalm and rice systems in particular. The Philippines Fertilizer and Pesticides Authority has enthusiastically backed the regional initiative and there is further support from the Agriculture Ministry's Policy and Planning Department. Participants from the Chinese Institute for the Control of Agrochemicals are involved in a new research programme using the IOBC/EPPO guidelines to evaluate pesticide effects. Currently they are studying crop pest and natural enemy distribution in the project zone. Other participants working in extension are using the knowledge gained in their IPM training work with farmers.

For further progress towards the aims of a standardized regional testing procedure, national capabilities need to be strengthened by more training and technical support, information exchange between countries, and access to databases on pesticide effects. There is scope for mutual cooperation between countries, for instance, in the introduction of biological control agents, in which Malaysia and Pakistan have considerable experience. Training on natural enemy rearing methods has been highlighted as an immediate need by many of the registration officers, if they are to carry out comprehensive beneficials evaluation.

The IOBC/IIBC Asian initiative in pesticides and beneficials testing has succeeded in developing a unique practical training curriculum for introducing natural enemy evaluation to researchers, regulators and practitioners in tropical developing countries. It has also raised awareness of the need to look at regulatory approaches that will actively encourage implementation of IPM programmes and how these can complement research and extension efforts. Progress is now under way to collect data specific to the region's crop systems most affected by pesticide-induced problems and to use this information to feed into national registration schemes. Interest has also been raised beyond the region, notably in Latin America and Africa, where the experience gained in Asia could be profitably shared to foster better understanding among decision makers of the negative impact of pesticides on natural pest control and to influence policies in support of IPM.

REFERENCES

Barrett, K.L., Grandy, N., Harrison, E.G. *et al.* (1994) *Guidance Document on Regulatory Testing Procedures for Pesticides with Non Target Arthropods. From the ESCORT workshop, Wageningen, Netherlands, March 1994*, SETAC-Europe, Brussels.

Lim, G.S. (1992) Integrated pest management in the Asia-Pacific context, in *Integrated Pest Management in the Asia-Pacific Region: Proceedings of the Conference on Integrated Pest Management in the Asia-Pacific region, 23–27 September 1991, Kuala Lumpur, Malaysia*, CAB International, Wallingford, Oxon, pp. 1–11.

Matteson, P.C. (1992) 'Farmer First' for establishing IPM. *Bulletin of Entomological Research*, **82**, 293–296.

Rengam, S.V. (1992) IPM: the role of goverments and citizens' groups, in *Integrated Pest Management in the Asia-Pacific Region: Proceedings of the Conference on Integrated Pest Management in the Asia-Pacific region, 23–27 September 1991, Kuala Lumpur, Malaysia*, CAB International, Wallingford, Oxon, pp. 13–19.

40

The impact of locust and grasshopper control on beneficial arthropods in West Africa

Harold van der Valk

INTRODUCTION

Locusts and grasshoppers continue to pose a threat to agriculture in Africa. To date, insecticides are still perceived as the only effective way of combating locusts. Since 1986, on average 3 million hectares have been treated annually with insecticides in Africa and the Middle East. Since locusts are preferably controlled before they reach cultivated areas, almost any type of ecosystem may be affected by the treatments, including desert grassland, wadis and oases in the Sahara, semi-arid pasture lands in the northern Sahel, and crops further south. Sedentary grasshoppers are mainly sprayed in cultivated areas, but not necessarily while on the crops. Most of the insecticides used are organophosphates (e.g. chlorpyrifos, fenitrothion) or pyrethroids (e.g. deltamethrin, λ-cyhalothrin) and have broad-spectrum activity. Recently, more specific chitin synthesis inhibitors (e.g. diflubenzuron, triflumuron) and phenyl pyrazoles (fipronil) are being applied as well.

The scale of the control operations and the wide variety of habitats that may be affected by insecticide treatments gave rise to questions about their environmental impact. This resulted in two groups of research programmes being initiated in the late 1980s. The first focused on developing more environmentally benign alternatives for chemical locust control, such as mycoinsecticides (e.g. Lomer and Prior, 1992) or

Ecotoxicology: Pesticides and beneficial organisms.
Edited by P.T. Haskell and P. McEwen. Published in 1998 by Chapman & Hall, London. ISBN 0 412 81290 8.

semiochemicals (e.g. Torto *et al.*, 1994). The second set of research programmes was to evaluate the environmental impact of the use of synthetic insecticides. Its aim was to generate recommendations on how to minimize the environmental side-effects of chemical locust control (Everts, 1990; Murphy *et al.*, 1994; Peveling *et al.*, 1994; Balança and de Visscher, 1995; Keith *et al.*, 1995).

THE LOCUSTOX PROJECT

In 1989 the Food and Agriculture Organization of the United Nations (FAO) coordinated a pilot study on the environmental impact of desert locust (DL) control in northern Senegal. Based on the results of this study (Everts, 1990) the governments of Senegal and The Netherlands decided to set up a project in Senegal to address these problems further, and FAO was requested to execute it. The Locustox project, as it subsequently became known, started operating in mid 1991.

The principal objective of the project is to assess the side-effects of locust and grasshopper control under Sahelian environmental conditions. This is mainly post-registration research, and aims to influence the choice of insecticides to be used under specific circumstances, as well as the choice of control methods and strategies. Furthermore, and closely linked to the first activity, the project develops ecotoxicological assessment methods appropriate to West African situations. The final objective is to help to set up a research group in West Africa capable of carrying out environmental research and assessments on pesticide use. This implies the training of scientists and technicians in ecotoxicological research as well as the development of relevant infrastructures.

The Locustox project at present consists of five working groups, three of which deal with ecotoxicological research (hydrobiology, vertebrate toxicology, entomology), one with environmental chemistry (pesticide residue analysis), and one specifically with training and information exchange.

RESEARCH ON BENEFICIAL ARTHROPODS

Given the large variety of ecosystems that are treated against locusts and grasshoppers, a choice had to be made on which ones to study. Obviously, not all ecosystems could receive attention, nor all beneficials within a chosen system. Very little quantitative ecological information was available on pests and beneficial arthropods in the (agro-)ecosystems concerned. The work, even though carried out in Senegal, also had to provide an assessment of insecticide risk in other parts of the Sahel and preferably the Sahara desert. Over the last five years the entomological research has focused on three partly related questions.

The first concerns insecticide impact on the natural enemies of grasshoppers, addressing the possibility that chemical control could create resurgence of the same grasshopper species or secondary development of another, through the elimination of natural enemies. This type of resurgence is less likely to occur with migratory species, except possibly in areas where they occur almost permanently (such as the Red Sea Coast in the case of the DL).

A second topic deals with detritivorous soil arthropods in the arid grasslands in northern Senegal, on the fringes of the Sahara desert. It has been suggested that arthropods play a regulating role in the breakdown of organic matter in arid and semi-arid ecosystems because drought limits the importance of free-living micro-organisms. Given the higher susceptibility of arthropods to most insecticides when compared with soil micro-organisms, nutrient cycling in (semi-)arid ecosystems may be more affected by insecticide applications than would be the case in temperate zones. In addition, detritivores form the basis of most food webs in deserts and semi-deserts (van der Valk, 1997).

The third research subject concerns the possibility of the development of secondary pests in millet due to locust and grasshopper control. Of all crops grown in the Sahel, millet is probably the most frequently treated during control operations against locusts and grasshoppers. The creation of secondary pests after elimination of natural enemies would pose large problems for subsistence farmers. Additional insecticide use to control such new pests would be far too expensive to be justifiable for a small farmer. Obviously it would also be undesirable from an environmental point of view.

From the start we have integrated ecotoxicological field studies (on both a large and a small scale), laboratory experiments and semi-field bioassays. This has often not followed the elegant tiered approaches discussed in earlier chapters. More than once, unforeseen findings forced us to abandon what we thought were carefully prepared sequences of hypotheses and expected results. The following example is but one of those cases, and illustrates some of the problems related to ecotoxicological research on beneficial arthropods in relatively unknown agroecosystems.

INSECTICIDES IN MILLET: A CASE STUDY

The millet head miner (*Heliocheilus albipunctella* De Joannis) (Lep.: Noctuidae) is considered one of the most important pests of pearl millet in the Sahel. Given that most of the larval life cycle is hidden under the millet flowers and grains, insecticidal control is often not effective. A large number of natural enemies, many of which are active in millet during the day, are known to attack the head miner (Bhatnagar, 1987).

Insecticide use on millet may therefore reduce natural mortality considerably without much affecting the pest (Waage, 1989).

We set up a number of field studies to assess whether natural enemies were affected by locust control insecticides, and if so, whether this would result in increased head miner densities. Populations of beneficials were monitored using different traps; certain life stages of the head miner were incubated for parasitism rates; and pest densities were assessed (van der Valk and Kamara, 1993; Kamara and van der Valk, 1995, unpublished data).

Figure 40.1 shows that peak densities of millet head miner were on average twice those of the controls for both fenitrothion and chlorpyrifos when spraying occurred at early maturation of the millet. This effect was less pronounced if spraying was earlier, during the panicle extension stage and early flowering. In spite of the fact that deltamethrin is also a wide-spectrum insecticide, no significant effect on millet head miner densities was observed.

We then assessed which groups of natural enemies were affected by the insecticides in the field. We were hoping to identify one or two 'key

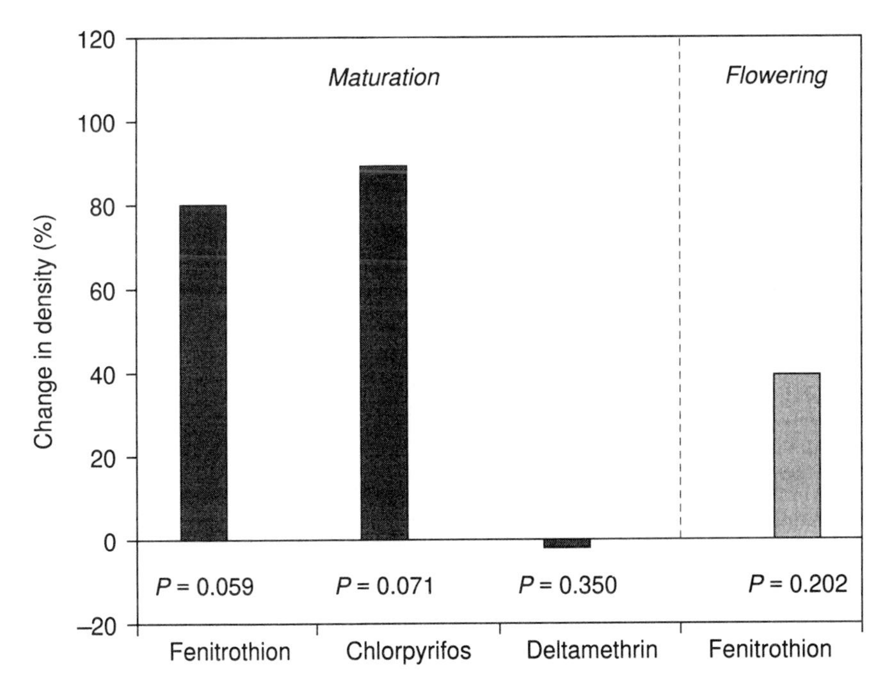

Figure 40.1 Changes in peak densities of *Heliocheilus albipunctella* in millet as compared with untreated control fields in the same study. Each bar is the average of four fields. *P*-values under the bars from paired *t*-tests (control vs. treated). (Sources: van der Valk and Kamara, 1993; Kamara and van der Valk, 1995, unpublished data.)

natural enemies' for which we might develop laboratory screening tests. Here we initially concentrated on those that had been identified by previous authors as being important (Bhatnagar, 1987). This concerned mainly hymenopteran parasitoids (Trichogrammatidae, Encyrtidae, Braconidae). Both the results from our field studies and the life history data of the species concerned suggested that mortality in the first two groups of parasitoids either did not occur or could not explain the observed resurgence in the host. This left us with the braconid parasitoid *Bracon hebetor* as a potential key natural enemy of the head miner. This hypothesis was supported by the fact that ichneumonoid parasitoids were always among the most affected groups of natural enemies in the ecotoxicological field studies.

Subsequently, a number of toxicity tests on an inert substrate were carried out on *B. hebetor* with the different insecticides (van der Valk *et al.*, 1994, unpublished data). A comparison between acute LC_{50}s and initial residues measured on millet leaves after spraying at locust control field rates is shown in Figure 40.2. Even though the organophosphates appear to be more hazardous than deltamethrin, all three insecticides

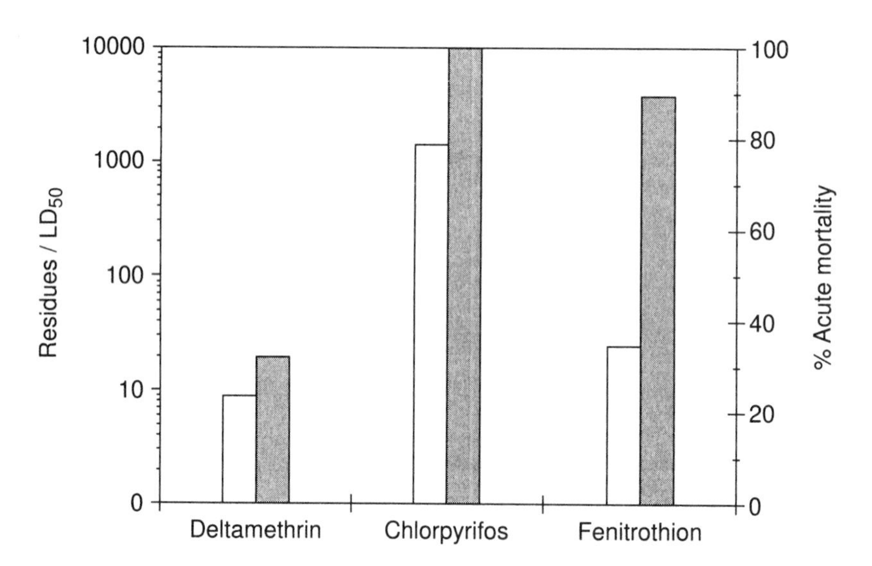

Figure 40.2 Hazard ratios (▭) and mortality (▬) in bioassays of *Bracon hebetor* exposed to different insecticides. The hazard ratio is calculated as the measured initial total residues of the insecticide at recommended field rates against the desert locust (μg a.i./cm^2 millet leaf) divided by the acute 24 h LC_{50} on glass surface (μg a.i./cm^2 glass). The acute mortality in bioassays is determined after 24 hours of exposure of *B. hebetor* on millet leaves sampled from a millet field immediately after treatment at recommended field rates against desert locust. (Sources: van der Valk *et al.*, 1994, unpublished data.)

have the potential to kill the parasitoid. However, the pyrethroid did not cause *Heliocheilus* resurgence, while both organophosphates did.

Bioassays using millet leaves sprayed in the field clarified this apparent inconsistency (Figure 40.2). The high hazard of the organophosphates was confirmed as expected, but deltamethrin only caused 30% mortality. Deltamethrin in highly lipophilic and its bioavailability on vegetation may have been more reduced than that of the organophosphates.

Thus it appeared that the results from laboratory toxicity tests, bioassays and ecotoxicological field experiments, combined with already available knowledge about head miner/natural enemy interactions, could provide a good explanation of the observed *Heliocheilus* resurgence after locust control. Or so it seemed.

The ugly fact slaying the above hypothesis was that observations during our field studies suggested that mortality due to braconid parasitoids was only limited; and that even if insecticides would eliminate these parasitoids, this was unlikely to explain all of the increase in millet head miner densities. Our main problem was that previous research on natural enemies of *Heliocheilus* had been mostly descriptive, and insufficient quantitative data were available to compare properly the contribution of the different natural mortality factors to the generational mortality of the host.

At that point we decided that the toxicological part of our research was to be temporarily put on hold and that the 'eco' in ecotoxicology needed more attention. Two detailed, partial life tables of the millet head miner were established in central Senegal (Thiam and van der Valk, 1996; Sarr, 1996). Using an intensive sampling programme we estimated recruitment into the different life stages from eggs to diapausing chrysalids. The results showed that the largest contribution to the generational mortality occurred in mid-size larvae. This coincided exactly with the period in which insecticide treatments caused millet head miner resurgence, supporting our idea that mortality in natural enemies was involved.

When we tried to identify the different mortality factors, our initial hypothesis about the importance of braconid parasitoids was partly shattered. In spite of the fact that *B. hebetor* turned out to be the most important parasitoid, the parasitoid fraction as a whole contributed only a little to total host mortality. Predators (probably mainly Hemiptera and Hymenoptera) appeared to be the most important factor. So our original idea that *Heliocheilus* resurgence after locust control was caused by insecticide impact on natural enemies is quite probably right, but we had been looking the wrong way – to parasitoids, about which a lot was previously published, rather than to predators, of which we knew very little. Indeed, as Croft and Jepson stress in Chapter 1, the basis of research should be to 'form new conceptions from our misconceptions'.

The next step in the millet programme of the project will be to compare insecticide susceptibility of different predators to *B. hebetor*. This should help to decide which organisms to select for future pre-registration tests in West Africa.

CONCLUSIONS

To a certain extent, research on the impact of pesticides on beneficial arthropods in Africa is caught in a dilemma. Pesticide use is increasing in Africa and will quite certainly continue to do so. As a result, many countries have started setting up pesticide registration and monitoring schemes. This development requires the availability of ecotoxicological tests using locally relevant beneficial arthropods, for both pre- and post-registration evaluation. The example given above shows that a thorough knowledge of the ecosystems concerned is required before we can choose such organisms. Interactions between pests and natural enemies in many African agroecosystems have not yet been studied in much detail, and quantitative data are especially lacking.

Clearly we cannot and should not wait until such ecological data have become available before assessing the risk of pesticides in Africa. Optimal use can be made of existing data from other parts of the world. But this is complicated by the fact that in Africa the access to such data is at present very limited. An easily accessible and regularly updated database containing previous research results would thus be of great value. Furthermore, field research (as opposed to standardized laboratory toxicity tests) will need to be stimulated and supported, especially in the lesser studied agroecosystems that risk being exposed to pesticides. Since ecological and entomological expertise is thinly spread and highly scattered in Africa, ecotoxicologists will need to engage very actively in information exchange and networking to ensure that this knowledge is best applied to their problems. Finally, where ecological knowledge is limited, the resulting increased uncertainty should be taken into account in risk assessment procedures used by pesticide registrators.

ACKNOWLEDGEMENTS

It would be impossible to acknowledge here all people that have been involved in the entomological research of the Locustox project during the last five years. However, I would especially like to thank Jude Andreasen, Lamine Ba, Alioune Beye, Abdoulaye Danfa, James Everts, Baba Gadji, Ousmane Kamara and Abdoulaye Niassy for their very valuable contributions.

Comments on a previous version of the manuscript by Clive Elliott, James Everts and Joost Lahr were highly appreciated.

Work described in this chapter was executed by the Food and Agriculture Organization of the United Nations and the Senegalese Crop Protection Directorate under projects ECLO/SEN/003/NET and GCP/SEN/041/NET, financed by the Governments of The Netherlands and of Senegal.

REFERENCES

Balança, G. and de Visscher, M.-N. (1995) Effets des traitments chimiques antiacridiens sur des coléoptères terrestres au Nord du Burkina Faso. *Ecologie*, **26**(2), 115–126.

Bhatnagar, V.S. (1987) Recherche et développment de la lutte intégrée contre les ennemis des principales cultures vivrières dans les pays du Sahel. Sénégal. Programme de lutte biologique, in *Rapport technique final GCP/RAF/128/CLS*, FAO, Rome.

Everts, J.W. (1990) *Environmental Effects of Chemical Locust and Grasshopper Control – a Pilot Study*, Project ECLO/SEN/003/NET, FAO, Rome.

Kamara, O. and van der Valk, H. (1995) *Side-effects of Fenitrothin and Diflubenzuron on Beneficial Arthropods in Millet in Senegal (the 1992 Study)*, Locustox Report No.95/2, Project GCP/SEN/041/NET, FAO, Rome.

Keith, J.O., Bruggers, R.L., Matteson, P.C. *et al.* (1995) *An Ecotoxicological Assessment of Insecticides Used for Locust Control in Southern Morocco*, Denver Wildlife Research Centre Research Report No. 11–55–005, USDA, Washington DC.

Lomer, C.J. and Prior, C. (1992) *Biological Control of Locusts and Grasshoppers: Proceedings of a Workshop held at the International Institute of Tropical Agriculture, Cotonou, Republic of Benin, 29 April–1 May 1991*, CAB International, Wallingford, Oxon, UK.

Murphy, C.F., Jepson, P.C. and Croft, A. (1994) Database analysis of the toxicity of antilocust pesticides to non-target, beneficial invertebrates. *Crop Protection*, **13**(6), 413–420.

Peveling, R., Weyrich, J. and Müller, P. (1994) Side-effects of botanicals, insect growth regulators and entomopathogenic fungi on epigeal non-target arthropods in locust control, in *New Trends in Locust Control*, Schriftenreihe der GTZ No.245 (eds S. Krall and H. Wilps), GTZ, Eschborn, Germany, pp. 147–176.

Sarr, I. (1996) Détermination de l'impact potentiel des pesticides sur *Heliocheilus albipunctella* (mineuse de l'épe de mil) à partir d'une méthode indirecte, l'étude de la table de survie. MSc thesis.

Thiam, A. and van der Valk, H. (1996) *Impact potentiel des insecticides sur la mortalité naturelle de la chenille mineuse de l'épi de mil (*Heliocheilus albipunctella*): une étude de la table de survie*, Locustox Report No.96/2, Project GCP/SEN/041/NET, FAO, Rome.

Torto, B., Obeng-Ofori, D., Njagi, P.G.N. *et al.* (1994) Aggregation pheromone system of adult gregarious desert locust *Schistocerca gregaria* (Forskal). *Journal of Chemical Ecology*, **20**(7), 1749–1760.

van der Valk, H. (1997) Community structure and dynamics in desert ecosystems: potential implications for insecticide risk assessment. *Archives of Environmental Contamination and Toxicology*, **32**, 11–21.

van der Valk, H. and Kamara, O. (1993) *The Effect of Fenitrothon and Diflubenzuron on Natural Enemies of Millet Pests in Senegal (the 1991 Study)*, Locustox Report No.93/2, Project ECLO/SEN/003/NET, FAO, Rome.

van der Valk, H., van der Stoep, J., Fall, B. and Dieme, E. (1994) *A Laboratory Toxicity Test with* Bracon hebetor *(Say) (Hymenoptera: Braconidae) – First Evaluation of Rearing and Testing Methods*, Locustox Report No.94/1, Project GCP/SEN/041/NET, FAO, Rome.
Waage, J. (1989) The population ecology of pest-insecticide-natural enemy interactions, in *Pesticides and Non-target Invertebrates*, (ed. P. Jepson), Intercept, Wimborne, Dorset, UK, pp. 81–93.

41

Impact of insecticides on herbivore-natural enemy communities in tropical rice ecosystems

K.L. Heong and K.G. Schoenly

INTRODUCTION

Today, rice pest management in tropical Asia is strongly influenced by the agrochemical era of the 1960s and 1970s. Prophylactic insecticide campaigns were components of rice production intensification programmes, like Masagana 99 in the Philippines (Alix, 1978) and BIMAS in Indonesia (Adjid, 1983). Through agricultural subsidy and loan schemes, farmers were encouraged to apply insecticides on regular schedules (Kenmore *et al.*, 1987; Conway and Barbier, 1990; Conway and Pretty, 1991). The agrochemical industry, through its aggressive advertising and marketing campaigns, also played a role in encouraging pesticide use.

The pest problems farmers and researchers witnessed in tropical Asia in the 1970s, particularly insecticide-induced outbreaks of the rice brown planthopper (*Nilaparvata lugens* Stål [BPH]; Heinrichs and Mochida, 1984; Kenmore *et al.*, 1985), led to IPM strategies that emphasized host-plant resistance, biological control and cultural practices, minimizing the use of pesticides. Since the 1970s however, several studies in tropical Asia have concluded that high levels of host-plant resistance for BPH management are unnecessary under certain circumstances (i.e. large areas of irrigated rice production where farmers use of insecticides is low). Experimental

Ecotoxicology: Pesticides and beneficial organisms.
Edited by P.T. Haskell and P. McEwen. Published in 1998 by Chapman & Hall, London. ISBN 0 412 81290 8.

studies conducted in Indonesia, Vietnam and the Philippines, for example, have shown that (a) hopperburn is rare or absent in fields grown with BPH-susceptible varieties not treated with insecticides (Kenmore *et al.*, 1984; Cook and Perfect, 1985; Cuong *et al.*, 1997: but see Sawada *et al.*, 1993); (b) susceptible varieties rarely show yield loss by BPH populations even when outbreaks occur in adjacent plots (Cuong *et al.*, 1997); and (c) moderately-resistant or even BPH-susceptible varieties grown for several years by a large number of farmers are associated with low and stable BPH populations (Gallagher *et al.*,1994; Cohen *et al.*, 1997). Thus a revised IPM strategy for tropical rice, based on these studies, advocates a shift in priority from host-plant resistance to naturally occuring biological control, while minimizing insecticide inputs, for sustainable and durable rice production systems.

Meanwhile, the global market share of pesticides sold in Asia has increased. For example, in 1988, worldwide sales of rice pesticides reached US$2.4 billion, sufficient to nudge out maize and cotton as the single most important crop for pesticides, with 90% of this market located in Asia (Woodburn, 1990). Insecticides accounted for the largest fraction of the total market (40–50%) until 1992, after which herbicide sales exceeded insecticide sales (MacKenzie, 1996). In 1995, herbicides accounted for 39% of the rice pesticide market, followed by insecticides and fungicides at 34% and 27% respectively. Country-by-country comparisons show that Japan leads all other countries in pesticide sales of approximately 50% of the world's total.

Past and ongoing research indicates that most insect pests of tropical rice are controlled by the activity of not just a few natural enemies but a whole array, through a complex and rich food web of generalist and specialist predators and parasites/parasitoids that live above and below the water surface (Heckman, 1979; Heong *et al.*, 1991, 1992; Schoenly *et al.*, 1996a,b; Settle *et al.*, 1996). Farmer interventions impact target and non-target species in different ways because biocontrol mechanisms span multiple trophic levels and act along spatiotemporal gradients. Determining the biocontrol potential of different components of this rich biodiversity (e.g. spiders, beetles, parasitoids, aquatic predators) and their role as stabilizing and buffering agents in rice production systems still remains to be shown through future research.

In this chapter, we highlight and review community-level approaches that have helped ecologists and entomologists to understand better how insecticides affect pest and natural enemy populations in tropical rice ecosystems. We begin with a review of farmer spraying practices and insecticide use patterns and conclude that, in many instances, farmers spray unnecessarily. In the second part of this report, we limit our review of studies to those directed at the community level of biological organization because farmers normally care about the net effect of all pests on their crop, not necessarily about individual species. Methods directed at the community level that have been used by different workers include ecostatistical indices, rank-abundance curves, guild and food web approaches, and multiple regression models. We conclude that much practical knowledge can be gained about how, when and to what degree different classes of insecticides impact different arthropod groups when they are studied at the level of entire communities.

FARMER PERCEPTIONS AND INSECTICIDE USE PATTERNS

Farmer surveys conducted by Heong *et al.* (1994) in the Philippines and Vietnam showed that about 90% of all sprays applied by farmers in 1992 were insecticides. Roughly half of these sprays were organophosphates such as methyl parathion, monocrotophos, methamidophos and chlorpyrifos. Cross-listing these insecticides against World Health Organization (WHO) classification of hazardous pesticides reveals that 37% of these sprays rate as 'extremely' (WHO Ia) or 'highly' (WHO Ib) hazardous to human health. Except in Japan, South Korea and Malaysia, WHO I insecticides are readily available and frequently applied on rice in most Asian countries (Heong and Escalada, 1997a).

In many cases, insecticides for rice are unnecessary. In the Philippines, for example, about 80% of insecticides are misused by farmers because they are applied at the wrong time and to the wrong target; consequently, such sprays are unlikely to result in an economic return on farmer investment (Heong *et al.*, 1995a). Moreover, farmers typically applied their first sprays in the first 40 days after crop establishment (Heong *et al.*, 1994; Heong and Escalada, 1997b) on lepidopterous larvae, commonly called 'worms' by farmers.

In early crop stages, leaf damage that is common and visible to farmers is attributable to rice leaf folders (*Cnaphalocrocis medinalis, Marasmia patnalis, M. exgua*), whorl maggots (*Hydrellia* spp.) or thrips (*Stenchaetothrips biformis*). Among these herbivores, rice leaf folders are the most common target for sprays by farmers in Asia. Under favourable conditions, a leaf folder larva can consume 25 cm^2 (or 25%) of leaf area (Heong, 1990) while leaf folder densities average less than two larvae per hill (Wada and Shimazu, 1978; Guo, 1990). In highly fertilized crops, densities of leaf folders can reach five larvae per hill (deKraker, 1996); however, the fraction of damaged leaves at such densities rarely exceeds 50%. Miyashita (1985) has shown that crops with leaf damage as high as 67% and occurring as late as the tillering stage do not suffer yield loss. Similarly, computer simulation has shown that larval densities need to reach (unrealistic counts of) 15 individuals per hill before yield loss can be detected (Fabellar *et al.*, 1994). Thus, most early-season sprays reflect farmer misperception of insect problems by overestimating yield losses that the larvae are likely to cause.

ECOLOGICAL EFFECTS OF INSECTICIDE SPRAYS

In this section, we review results of studies conducted in the Philippines and Indonesia that evaluated community-wide impacts of insecticides on different functional groups of invertebrates (herbivores, predators, parasitoids, detritivores) in the rice ecosystem.

Species richness in sprayed and unsprayed plots

In an irrigated rice field in Central Luzon, Philippines, species richness (N0) for herbivores and predators was found to be significantly lower in chlorpyrifos-sprayed than unsprayed plots after insecticides were applied at 29 and 43 days after transplanting (DT) (Figure 41.1a,b). After the second spray at 43 DT, species richness in sprayed and unsprayed plots was nearly identical for herbivores and predators at 57 and 63 DT, respectively, suggesting that herbivore richness recovered from chlorpyrifos sprays one week earlier than predators. For herbivores the number of abundant species, measured by N1 = exp[H′], where H′ is the Shannon diversity index (Ludwig and Reynolds, 1988, section 8.2.2), decreased and recovered after each spray (Figure 41.1c). Conversely, the number of abundant predators increased significantly after the second spray (Figure 41.1d) until the end of the trial. Thus, species richness of all herbivores (Figure 41.1a), all predators (Figure 41.1b) and abundant herbivores (Figure 41.1c) recovered relatively quickly (through recolonization) from the cumulative effect of two chlorpyrifos sprays; whereas the sharp rise in number of abundant predators (Figure 4.1d) after the second spray (at 43 DT) suggests overcompensated recovery aided in part by numerical response behaviour.

Species abundance in sprayed and unsprayed plots

In the same twice-sprayed, chlorpyrifos plot in Central Luzon, herbivore abundance was significantly reduced immediately after each spray but climbed to unsprayed abundances one week after the second spray (Figure 41.2a). Over the remaining post-spray dates, differences in herbivore abundance gradually lessened between sprayed and unsprayed plots. Recovery of predators in the sprayed plot reach unsprayed abundances one week after the first spray, but after the second spray predators did not reach unsprayed levels until 63 DT (Figure 41.2b). Herbivore and predator abundances in sprayed and unsprayed plots, when summed over all sampling dates, differed by 1% and 42%, respectively (Table 41.1).

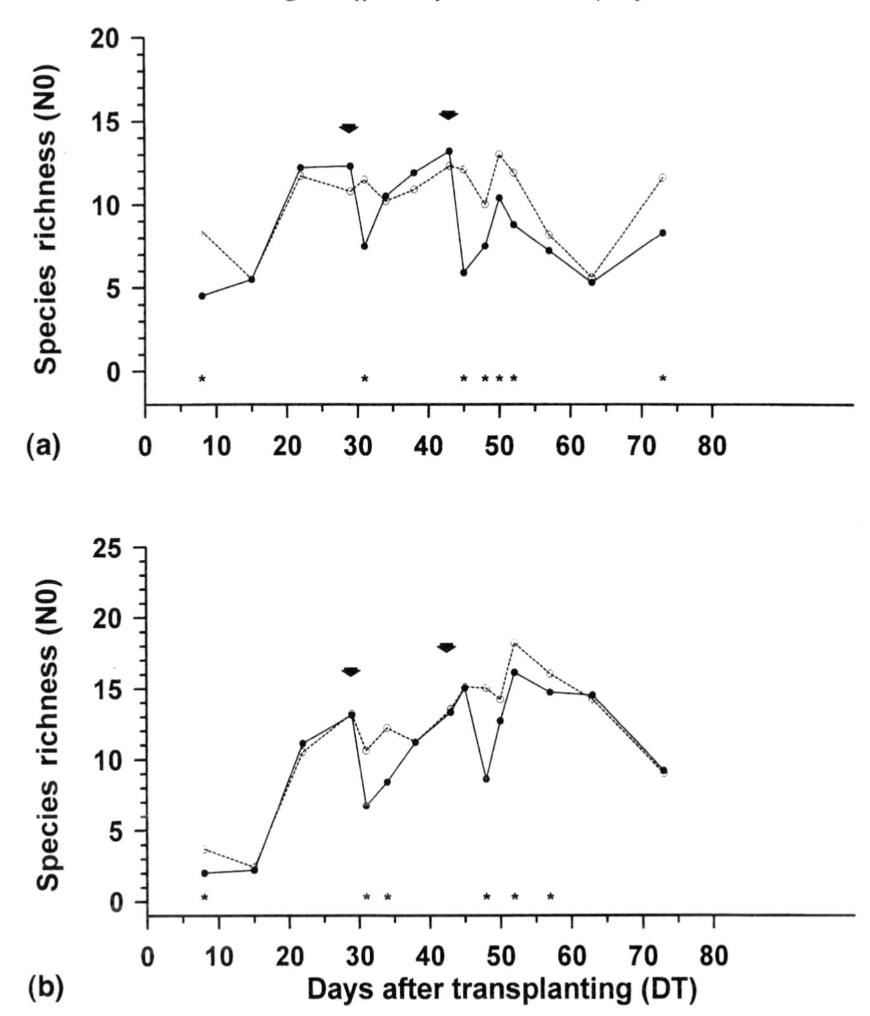

Figure 41.1 (a and b) Mean counts, based on 10 samples, of **(a,b)** species richness and **(c,d)** the most abundant species for **(a,c)** herbivores and **(b,d)** predators in chlorpyrifos-sprayed (●) and unsprayed (○) rice plots in Nueva Ecija, Philippines. Arrows denote spray days; * indicates that mean counts in sprayed and unsprayed plots were significantly different at $P = 0.05$. (See text for details.)

Chlorpyrifos reduced cicadellid homopterans, mainly *Nephotettix virescens* (Figure 41.3a), by 12% and increased delphacid homopterans, mainly *Nilparvata lugens* and *Sogatella furcifera*, by 23% (Figure 41.3b,c). Predators affected most by sprays included lycosid spiders (61% reduction, Figure 41.4a) and veliid bugs (55%, Figure 41.4b). Mirid bugs, particularly *Cyrtorhinus lividipennis*, though reduced after each spray, recovered

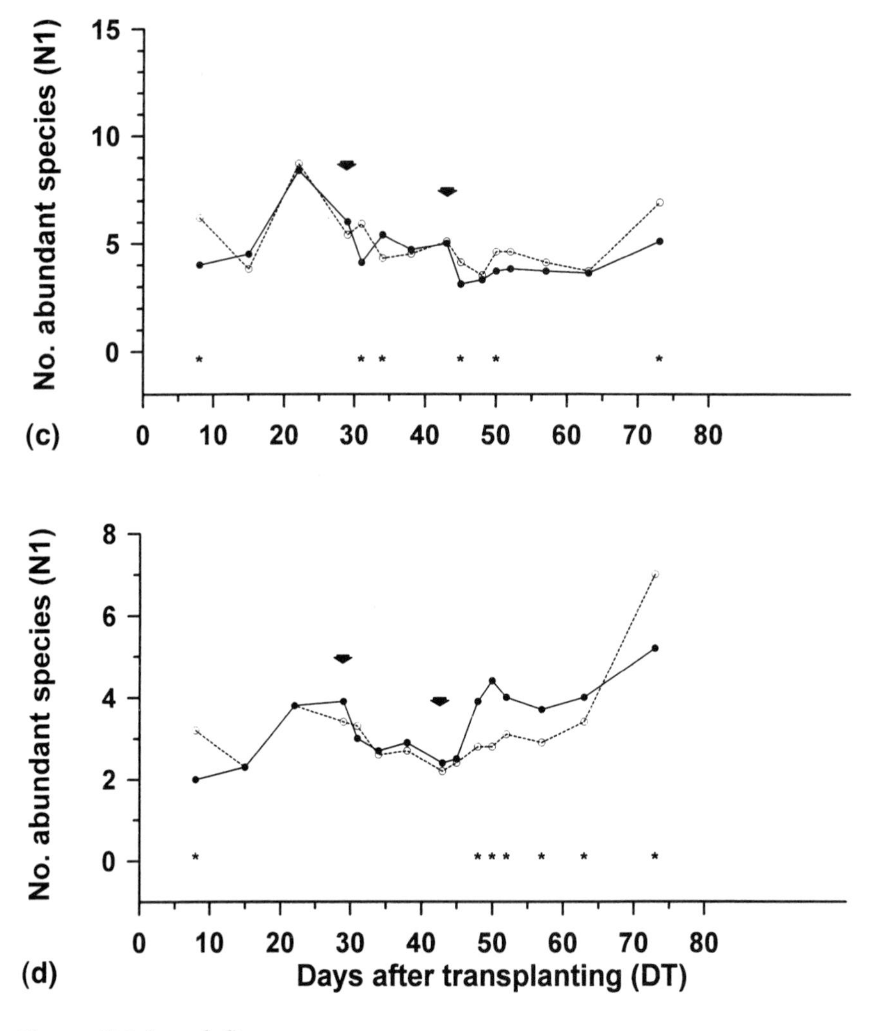

Figure 41.1 (c and d)

quickly to abundances close to or higher than unsprayed abundances; by season's end, total numbers of mirids in sprayed and unsprayed plots differed by only 1% (Figure 41.4c). Numbers of sampled organisms of all taxa, taken over all sampled dates, were higher in the unsprayed plot (75 121 individuals) than the sprayed plot (55 560 individuals).

In the same chlorpyrifos study, rank-abundance curves for herbivore and predator populations manifested different magnitudes of variation before and after the first spray date at 29 DT (Figure 41.5a,b). Interestingly, the rank order of the most common herbivore species remained unaltered between pre- and post-spray dates and between sprayed and

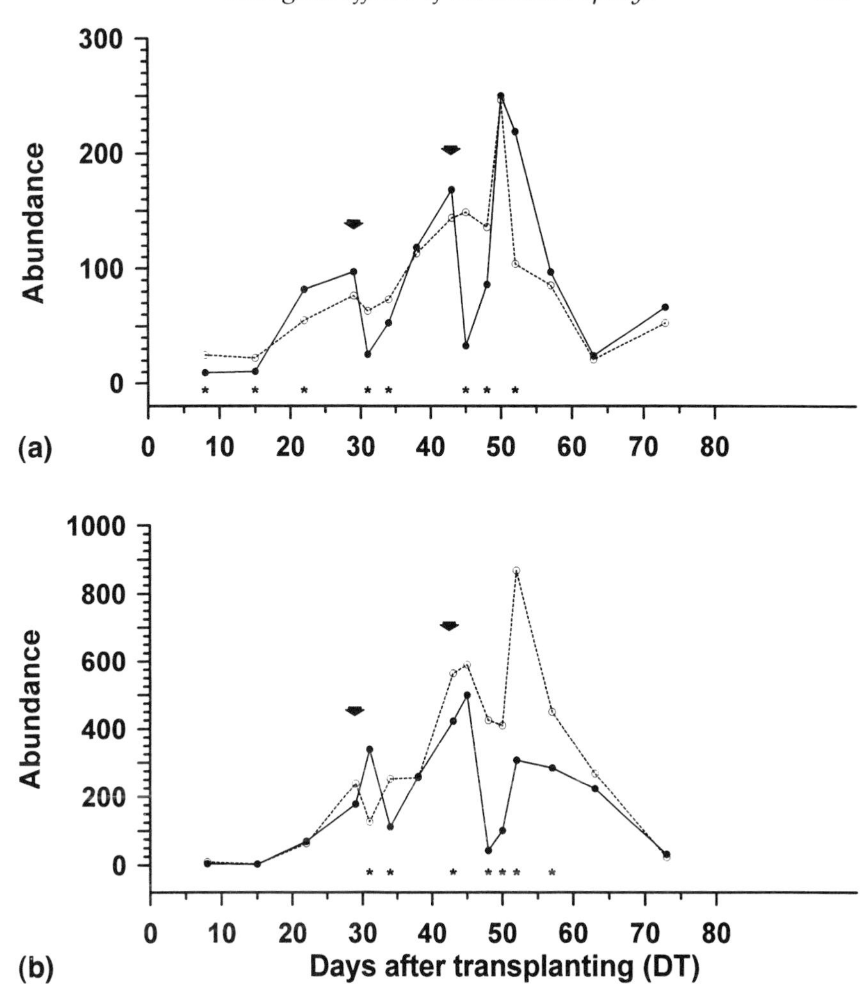

Figure 41.2 (a and b) Mean counts, based on 10 samples, of **(a)** herbivore and **(b)** predators in chlorpyrifos-sprayed (●) and unsprayed (○) plots in farmers' fields in Nueva Ecija, Philippines. Arrows denote spray days; * indicates that mean counts in sprayed and unsprayed plots were significantly different at P = 0.05.

unsprayed plots, with *Nephotettix virescens* and *S. furcifera* retaining numerically dominant positions throughout, followed (less consistently) by *N. lugens* and aleyrodids (Figure 41.5a). On the other hand, the most abundant predators did differ in rank over pre- and post-spray dates and between sprayed and unsprayed plots (Figure 41.5b). For example, between the pre- and post-spray dates, the top three predators at 29DT were *Microvelia douglasi atrolineata*, *Mesovelia vittigera* and *C. lividipennis*,

Table 41.1 Differential effects of two sprays of chlorpyrifos (29 and 43 DT) on arthropod abundances from different functional groups in a farmer's field in Central Luzon, Philippines

Functional group	*Unsprayed plot*	*Sprayed plot*	*% Difference*
All herbivores	14 402	14 221	−1.2
Cicadellidae	8 191	7 229	−11.7
Delphacidae	3 879	4 776	+23.1
All predators	46 248	26 967	−41.7
Miridae	4 255	4 211	+1.0
Lycosidae	1 063	417	−60.8
Veliidae	39 845	18 012	−54.8
All parasitoids	2 359	1 477	−37.4
All detritivores	12 090	12 829	+6.1

followed at 31 DT by *M. d. atrolineata*, *Atypena* (= *Callitrichia*) *formosana* and *Tetragnatha javana* in the sprayed plot.

Population studies of rice brown planthopper (BPH), *N. lugens*, show increases of up to 800-fold after sprays of monocrotophos (Heinrichs and Mochida, 1984; Kenmore *et al.*, 1984). Similar increases in BPH populations have been observed on BPH-resistant varieties after insecticide application (Joshi *et al.*, 1992; Gallagher *et al.*, 1994).

Guild and food web structure in sprayed and unsprayed plots

To gauge the ability of guild and food web concepts to reveal effects of insecticide spraying, another farmer's field in Central Luzon was studied using sprayed and unsprayed plots of equal area (Schoenly *et al.*, 1996a). Three foliar sprays of deltamethrin were applied at 28, 38 and 49 DT. Time-specific food webs for the sprayed and unsprayed plots were constructed from time-series samples and a 546-taxa cumulative Philippines-wide food web.

Triangle graphs are instructive tools for illustrating time-series changes in herbivores (%H), natural enemies of herbivores (%E) and other taxa (detritivore and planktonic) (%O) in sprayed and unsprayed plots (Figure 41.6a). On the three pre-spray dates (8, 15 and 22 DT), sprayed and unsprayed plots had similar levels of %H, %E and %O (Figure 41.6b,c). In both plots, %O declined whereas %H and %E increased over the two-week period. Over the spraying interval (28–49 DT), %H in the sprayed plot exceeded %H in the unsprayed plot. Natural enemy abundances (%E) were much greater in the unsprayed plot than in the sprayed plot (three-date means: 57% and 24%, respectively). Over the remaining eight post-spray dates (50–99 DT), percentage differences in trophic groups gradually lessened between sprayed and unsprayed plots (Figure 41.6b,c). At 78DT, %H,

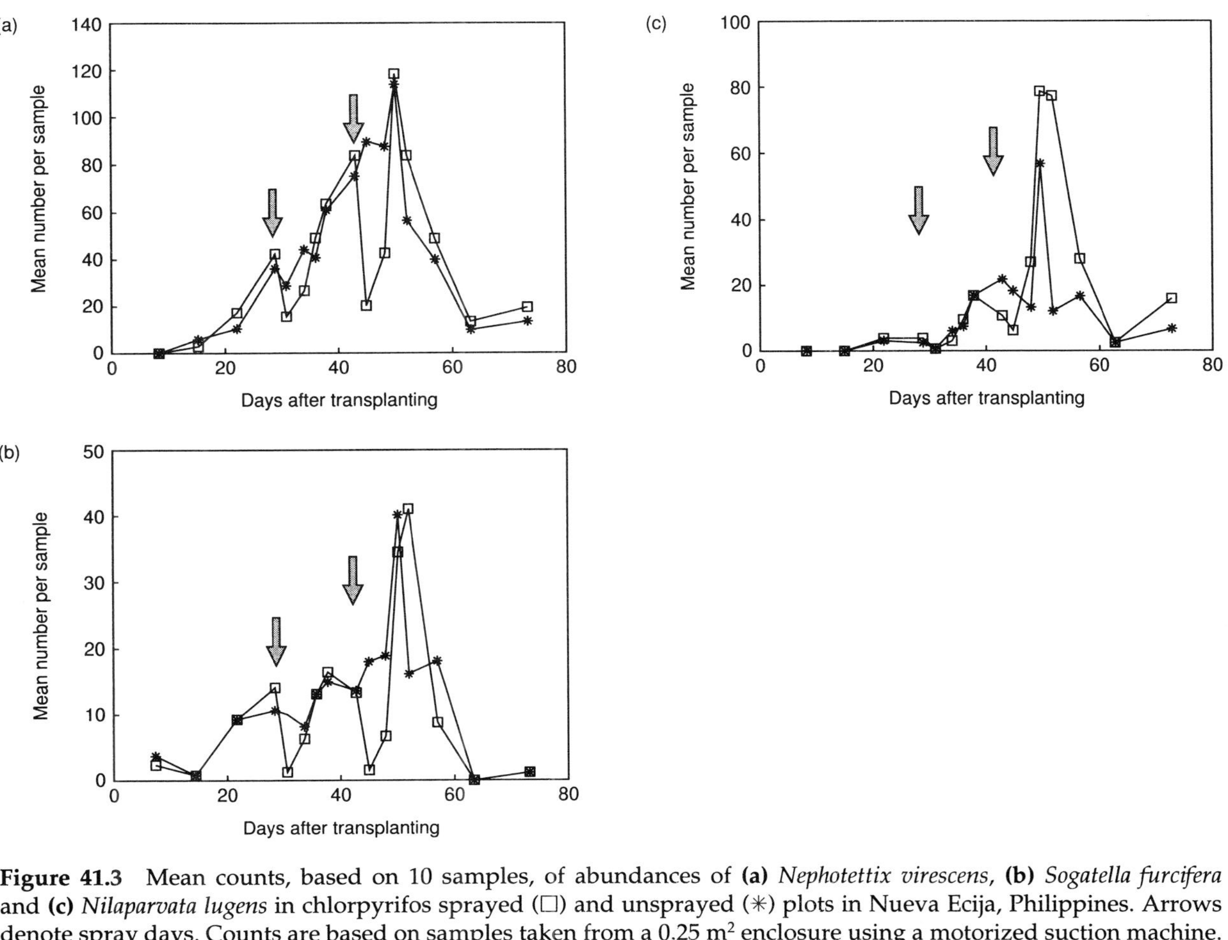

Figure 41.3 Mean counts, based on 10 samples, of abundances of **(a)** *Nephotettix virescens*, **(b)** *Sogatella furcifera* and **(c)** *Nilaparvata lugens* in chlorpyrifos sprayed (□) and unsprayed (✳) plots in Nueva Ecija, Philippines. Arrows denote spray days. Counts are based on samples taken from a 0.25 m^2 enclosure using a motorized suction machine.

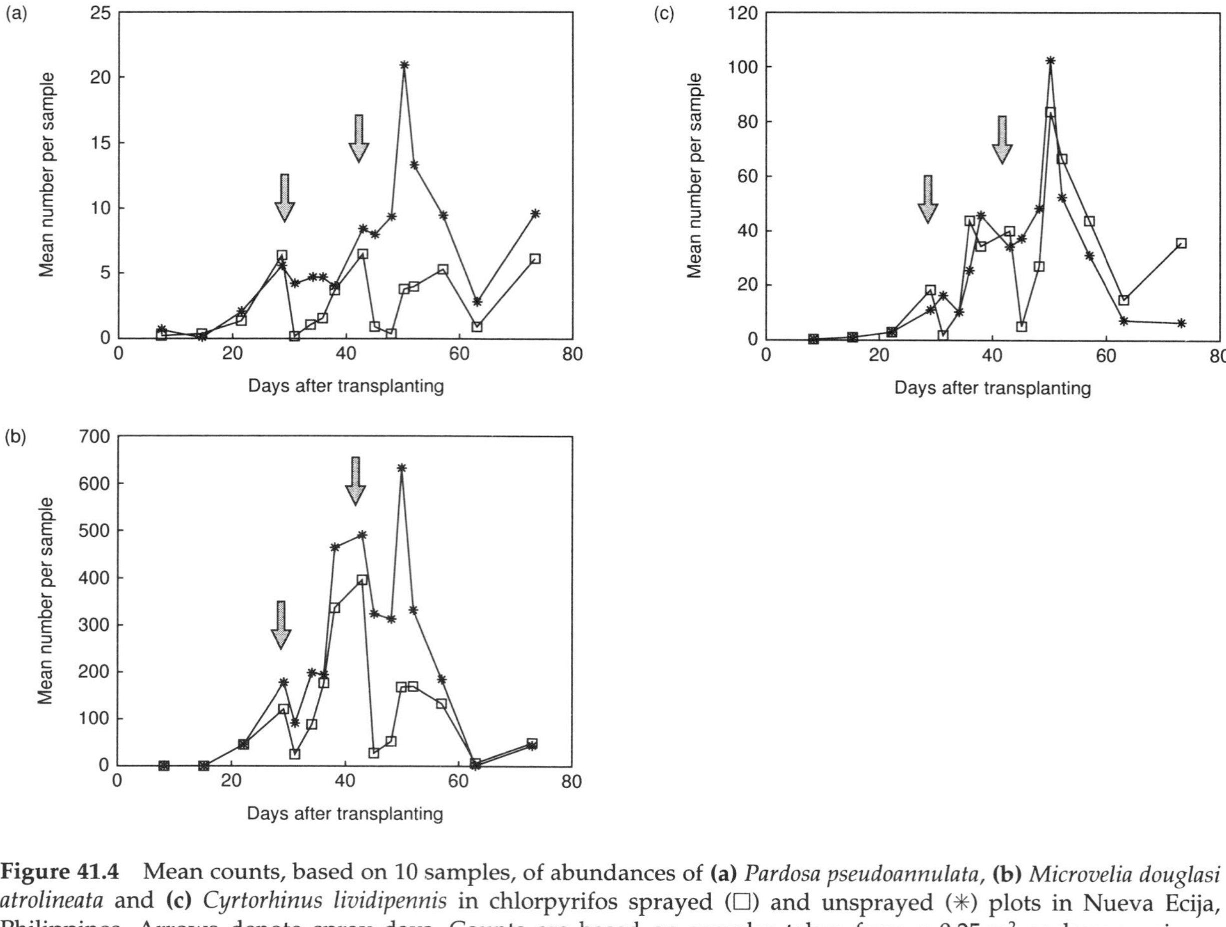

Figure 41.4 Mean counts, based on 10 samples, of abundances of **(a)** *Pardosa pseudoannulata*, **(b)** *Microvelia douglasi atrolineata* and **(c)** *Cyrtorhinus lividipennis* in chlorpyrifos sprayed (□) and unsprayed (✳) plots in Nueva Ecija, Philippines. Arrows denote spray days. Counts are based on samples taken from a $0.25\,m^2$ enclosure using a motorized suction machine.

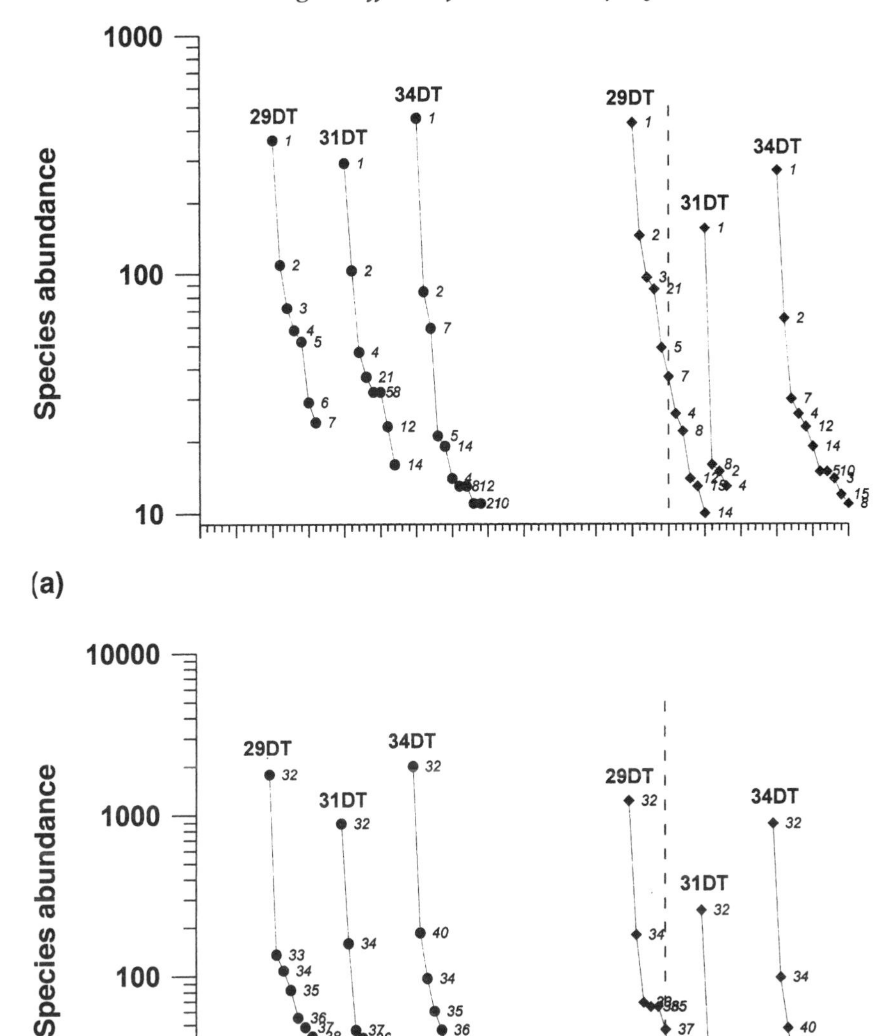

Figure 41.5 Rank abundance curves for **(a)** herbivores and **(b)** predators on chlorpyrifos sprayed (◆) (29 DT) and unsprayed (●) rice plots for common species comprising 10 or more individuals. Vertical dashed lines separate pre- and post-spray dates. Numbers next to data points identify the most common species comprising 10 or more individuals. Herbivores: **1**, *Nephotettix virescens*; **2**, *Sogatella furcifera*; **3**, *Brachydeutera longipes*; **4**, Aleyrodidae; **5**, *Nephotettix nigropictus*; **6**, *Leptocorisa* sp.; **7**, *Nilaparvata lugens*; **8**, Coreidae; **9**, *Tagosodes pusanus*; **10**, *Notiphila similis*. Predators: **32**, *Microvelia douglasi atrolineata*; **33**, *Mesovelia vittigera*; **34**, *Cyrtorhinus lividipennis*; **35**, *Atypena formosana*; **36**, *Pardosa pseudoannulata*; **37**, *Tetragnatha javana*; **38**, *T. maxillosa*; **39**, *Sternolophus* sp.; **40**, *Limnogonus fossarum*; **41**, *Hydrometra lineata*.

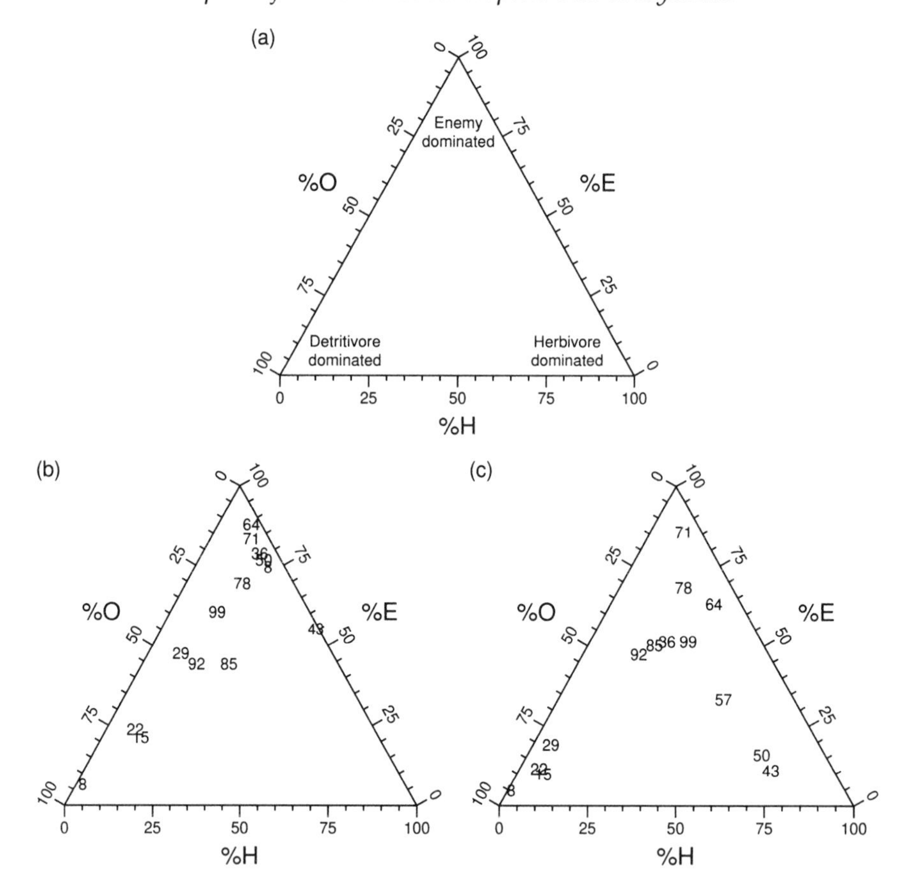

Figure 41.6 Triangle graphs showing temporal variation in percentage abundances of herbivores **(%H)**, natural enemies **(%E)** and other taxa (detritivores and planktonic) **(%O)** from farmers' fields in Zaragosa, Philippines. Percentages of **(a)** herbivore-, enemy- and detritivore-dominated regions of the graph, and of H, E and O abundances for each of 14 sampling dates in **(b)** unsprayed and **(c)** deltamethrin-sprayed plots. Numbers within each graph are days after rice seedling transplantation. Percentages in sprayed and unsprayed plots are means of 10 samples based on counts per 1.6 m^2 per sampling date. Modified after Schoenly *et al.* (1996a).

%E and %O in the sprayed and unsprayed plots were nearly identical, suggesting a (roughly) one-month recovery period of trophic guilds from deltamethrin sprays.

For each time-specific food web, mean chain length (μ) was calculated using the most common taxa that captured 90% of the total abundance (90%A web; Schoenly *et al.*, 1996a). Mean chain length was defined as the average length (counting links, not species) of all maximal food chains from a basal species to a top predator (Cohen, 1978), calculated using the long-way-up algorithm of Cohen and Luczak (1992).

On pre-spray dates, the 90%A webs of the sprayed and unsprayed plots each had food chains up to three links in length: rice/pests/specialist enemies/generalist enemies (Figure 41.7). Following the first deltamethrin spray, μ in the sprayed plot fell from 2.6 at 22 DT to 2.0 at 29 DT, yielding a web that contained only two-linked chains (rice/pests/enemies). Over the same 7-day period, μ in the unsprayed plot increased slightly from 2.4 to 2.6 (Figure 41.7). Subsequent sprays at 38 and 49 DT sustained the treatment effect first seen at 29 DT but did not amplify it.

Over the next five post-spray dates (36–64 DT), six natural enemy species that were present in 90%A webs of the unsprayed plot were absent in 90%A webs of the sprayed plot. In decreasing order of occurrence they were *A. formosana* (six out of six post-spray dates), *C. lividipennis* and *Micraspis* spp. (three out of six dates each) and *Pardosa pseudoannulata*, *Stilbus* sp. and *Tetragnatha* sp. (one out of six dates each). Based on food chain lengths in Figure 41.7, the estimated time to web recovery following three deltamethrin sprays at 28, 38 and 49 DT was 22 days.

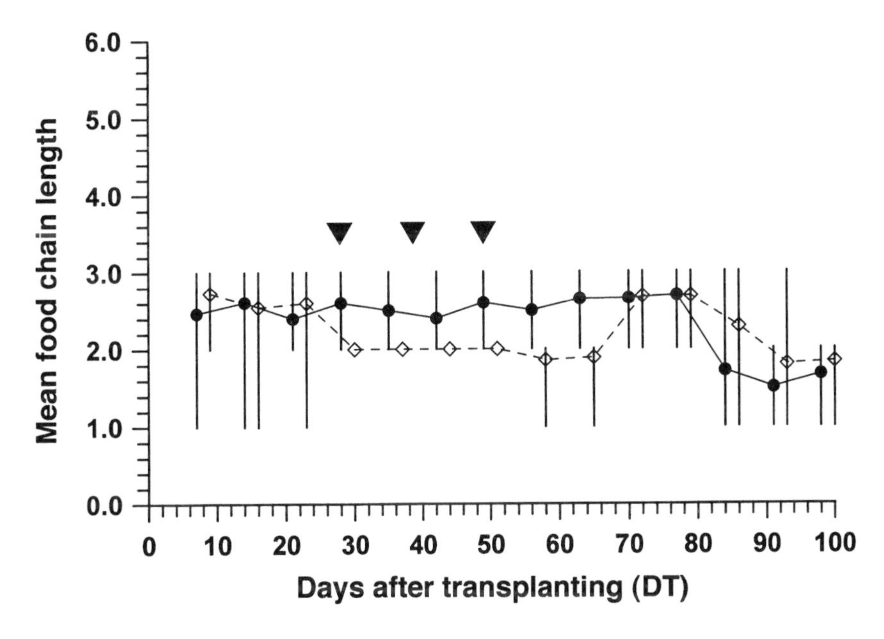

Figure 41.7 Temporal variation in mean food chain lengths (μ) for unsprayed (●) and deltamethrin-sprayed (◇) plots. Vertical lines show range in food chain lengths for sprayed and unsprayed webs on each sampling date. Calculation of μ is based on the most common taxa whose population sizes made up 90% of the total abundance on each sampling date in sprayed and unsprayed plots. ▼, dates of deltamethrin sprays. Modified after Schoenly *et al.* (1996a)

Ecological costs of insecticide sprays to rice farmers

Differences between sprayed and unsprayed plots in herbivore and natural enemy abundances can be translated into ecological costs that farmers can understand (Schoenly *et al.*, 1996a). To assess ecological costs of insecticides, the magnitude and direction of differences in %H (herbivores) and %E (natural enemies) abundances were calculated between sprayed and unsprayed plots and then converted to numbers per hectare per sampling date.

This analysis revealed that deltamethrin sprays brought an additional four million herbivores per hectare per sampling date (Figure 41.8a). Of the 12 putative pest species in both plots, the majority of additional herbivores in the sprayed plot were three delphacids (*S. furcifera*, *N. lugens* and *Tagosodes pusanus*). A second ecological cost attributable to deltamethrin sprays was one million fewer natural enemies per hectare per sampling date in the sprayed plot by mid-season (Figure 41.8b). The timing of enemy losses partly overlapped the hump of extra herbivores at 36–50 DT (Figure 41.8a,b); however, the additional herbivores over this overlapping interval went unchecked in the sprayed plot because there were fewer natural enemy individuals. Of the 34 natural enemy taxa in both plots, *M. d. atrolineata* and *C. lividipennis* comprised the largest fraction of natural enemies affected by deltamethrin sprays. By season's end, the sprayed plot netted an estimated 279 000 more natural enemies per hectare per sampling date than the unsprayed plot. However, early season losses increased the likelihood of pest damage in the sprayed plot.

In Java, Indonesia, Settle *et al.* (1996) used insecticides to demonstrate a link between early-season natural enemies and late-arriving pests. Insecticides reduced natural enemy populations and caused pest populations to resurge, particularly rice brown planthoppers. By season's end, sprayed fields netted higher predator populations than unsprayed fields. However, rebounding populations only partly overlapped the resurgence hump of extra herbivores seen earlier in the season, as in the Schoenly *et al.* (1996a) study.

Multiple regression models

Cohen *et al.* (1994) used data from the Philippines-wide food web and multiple regression models to predict population fluctuations of insect pests in a rice field at the International Rice Research Institute (IRRI) in the Philippines. Independent variables of the seven pest models included the biomass of rice plants in the field, the abundance of each pest, and the abundances of five highly correlated enemies of each pest, all as functions of time.

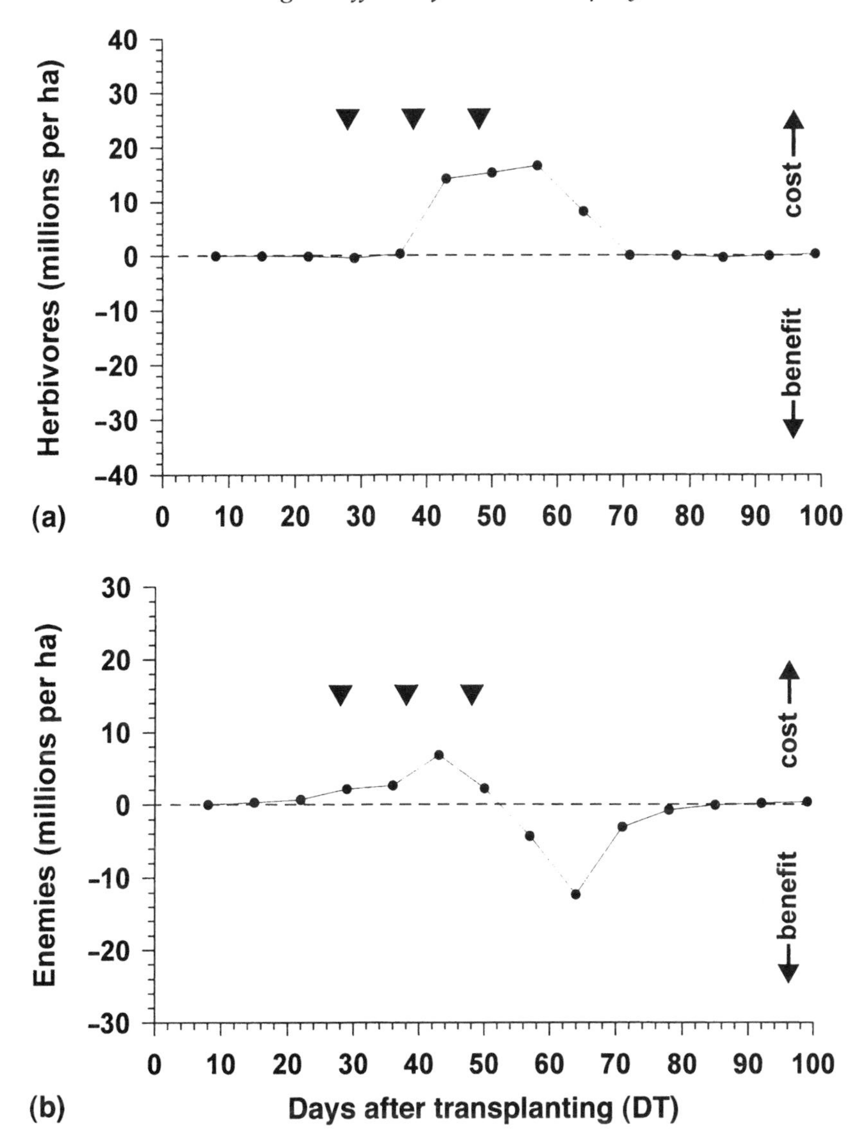

Figure 41.8 Ecological effects of deltamethrin sprays: **(a)** extra herbivores and **(b)** early-season losses of natural enemies. ▼, spray dates. Modified after Schoenly *et al.* (1996a).

In this study, deltamethrin sprays disorganized the population dynamics of insect species feeding in the IRRI field. Multiple regression models were less able in the sprayed plot than in the unsprayed plot to forecast the population fluctuations of pest species on the basis of various numbers and combinations of independent variables. In the unsprayed plot, independent variables were significant predictors of future pest abundance in four of seven initial models, compared with one of seven models in the sprayed plot. Stepwise removal of independent variables in the models enhanced the forecasting power in both the sprayed and unsprayed plots, but significant models in the unsprayed plot nearly always outnumbered those in the sprayed plot. In general, models that included interaction terms and either the presence or absence of sprays among the independent variables did not improve the forecasting power of sprays in either the sprayed or unsprayed plot.

DISCUSSION

The magnitude of the pesticide problem in tropical Asia requires that rice ecologists and IPM practitioners get the most from results of farmer surveys and ecological experiments conducted at the community and ecosystem levels of biological organization. Temporal trends in guild- and community-level measures, and comparisons between sprayed and unsprayed plots, can be translated into ecological costs (increased herbivores, decreased natural enemies and slow recovery times) that farmers can understand. For example, ecostatistical measures (species richness and abundance), guilds and mean food chain length revealed variable recovery times of 1–4 weeks from insecticide sprays which depended on the chemical used (e.g. deltamethrin, chlorpyrifos), its timing of application (e.g. 29 vs. 43 DT) and the functional group (e.g. herbivores, predators). As well as revealing additional economic costs to the farmer, these studies showed that insecticides negatively impact non-target beneficial organisms in predictable ways. Indeed, when rice insect pests are exposed to both natural enemy additions (ie. *P. pseudoannulata*) and insecticide applications (ie. mono crotophos) the treatments cancel each other out (Fagan *et al.*, in press) confirming the incompatible and counterproductive nature of chemical and biological pest control tactics for tropical rice ecosystems. Thus, there is practical value for IPM workers and farmers in concepts and methods that can evaluate the effects of insecticide sprays on entire communities of pests and natural enemies.

Differential effects on guilds

By design, insecticides are biocides with the potential for causing differential mortality in different invertebrate (and vertebrate) species when applied to the rice ecosystem. Besides direct killing action, insecticides alter rates of reproduction, feeding and dispersal, and synchrony of pest and natural enemy interactions. General classes of insecticides such as pyrethroids have been found to increase reproductive rates of the brown planthopper (Heinrichs *et al.*, 1982; Heinrichs and Mochida, 1984), whereas buprofezin reduces searching efficiency of egg predators (e.g. *C. lividipennis*; Suvaparp, 1993). In addition, parasitoids may be more vulnerable to pesticides in the rice ecosystem owing to their smaller size and high mobility. Pyrethroids also have higher toxicity effects on spiders (Kenmore, 1980; Croft, 1990) and hymenopteran parasitoids (Croft, 1990), whereas buprofezin is less toxic to natural enemies (Suvaparp and Heong, 1992).

Because insecticides are often administered over the entire crop cycle, they can evoke whole-community disruptions similar to climatic catastrophes, such as drought or flooding. Such disruptions can re-set predator-prey oscillations and create asynchrony. This effect, referred to as catastrophic asynchronization (Perera *et al.*, 1988), favours herbivore species with high reproductive, survival and recruitment rates and with short life spans (e.g. delphacid homopterans). In rice ecosystems, the delphacids, *S. furcifera* and *N. lugens*, embed their eggs into the rice tissue, which reduces their exposure to insecticide sprays.

When a pesticide is applied to an oscillating predator-prey interaction it kills the most susceptible stages. The result is synchronization of pest population development (Waage, 1989). At one extreme, where only one stage survives, a uniform pest age-structure is created. For a natural enemy species that exhibits a shorter generation time than the pest, such synchronization can cause short-term gaps in availability of the necessary pest stage. Examples of pest resurgences from catastrophic synchronization have been documented for pests of coconut (Perera *et al.*, 1988), tropical plantation crops and coffee (Waage, 1989). On the other hand, when natural enemies have longer generation times than the pest, synchronization creates periods of low natural enemy populations that can markedly reduce the pest's natural control. In the case of *N. lugens* and *S. furcifera* in the rice ecosystem, natural enemies with both longer and shorter generation times exist, suggesting that catastrophic synchronization from insecticide sprays can promote pest outbreaks of a large and persistent nature.

Effects due to recolonization

Ecological disruptions caused by pesticides, unlike those caused by winter, drought and floods, are usually localized. When a farmer sprays his crop, his neighbours may not. Arthropods from neighbouring crops are potential sources of recolonization, rates of which differ for different pest and natural enemy species. For example, post-spray population densities of herbivores *N. lugens*, *S. furcifera* and *N. virescens* in the chlorpyrifos study recovered to densities of the unsprayed plot within one week after each spray. Populations of predators *M. d. atrolineata* and

P. pseudoannulata recovered within two weeks after the first spray but remained significantly lower than unsprayed densities after the second spray. Recolonization rates of the predator *C. lividipennis* were similar to those of several pest species, probably because of similar mobility rates. The generally higher recolonization rates of pest species than predator species, coupled with their high reproductive rates and reduced natural control after insecticide sprays, play important roles in secondary pest developments.

Brown planthopper: a secondary pest caused by early-season insecticide use

Flint and van den Bosch (1981) described two models depicting effects of pesticide impact that illustrate the phenomenon of target pest resurgence and secondary pest outbreaks. It is now well established that some insecticides, when used on rice, induce planthopper increases that can lead to outbreaks (Heinrichs and Mochida, 1984; Kenmore *et al.*, 1984; Joshi *et al.*, 1992; Cohen *et al.*, 1994). In most cases, these insecticides were not targeted for planthopper control. Instead, as indicated earlier, farmers used them to control leaf-feeding insects during early crop stages (Heong *et al.*, 1994). In Vietnam, about 42% of the sprays farmers used were targeted at leaf-feeding insects and 34% at planthoppers (Heong *et al.*, 1994). Organophosphates and pyrethroids accounted for 72% of the sprays used for leaf-feeding insects.

The brown planthopper problem in tropical rice has been attributed to pest resurgence (Heinrichs and Mochida, 1984; Gallagher *et al.*, 1994; Rombach and Gallagher, 1994), implying that farmers were using insecticides to control BPH that resulted in having higher BPH populations. Subsequently the resurgence-causing insecticides were not recommended (and in some cases were banned) for use for BPH control but continued to be promoted for use on other rice pests, such as leaf-feeding insects, stemborers and rice bugs in many countries. Consequently, ecological disruptions that favour BPH development continue to occur, which probably accounts for the few recent BPH outbreaks (Way and Heong, 1994). Farmers' use of these resurgence-causing insecticides, especially in the early crop stages, have been shown to be unnecessary and their removal has had no significant yield consequences (Heong *et al.*, 1995b, IRRI, 1996a,b; Heong and Escalada, 1997a). In addition, natural mortalities and natural enemy abundances build up from the early crop stage to mid season (Kenmore *et al.*, 1984; Cook and Perfect, 1989; Fowler *et al.*, 1992; Heong *et al.*, 1992) and computer simulation shows that early-season spraying is inefficient (Heong, 1989; Cheng *et al.*, 1990). Thus, there are significant policy implications for tropical rice management that depend on whether BPH is regarded as a secondary or a resurgence pest.

Implications for rice pest management

Like other ecosystems, tropical rice agroecosystems have mechanisms for resisting and recovering from external stresses (within finite limits), which led Way and Heong (1994) to conclude that IPM in tropical irrigated rice should be based on the contention that insecticides are not needed, rather than that they are, and that 'pests' should now be critically re-assessed and proved guilty before insecticide use is contemplated. The validity of natural control mechanisms is supported by farmer experiences in Vietnam (Heong *et al.*, 1995b), the Philippines (Way and Heong, 1994; Heong and Escalada, 1997a; IRRI, 1996a) and Indonesia (Matteson *et al.*, 1994). On the other hand, the insecticide-based approach is costly and involves uncertain and complicated choices about which insecticides to use and when and how to use them for different pests. Moreover, some insecticides can create secondary pest problems and ecological and health costs to farmers (Rola and Pingali, 1993; Pingali and Roger, 1995).

The studies discussed above and the work of several others have led to the conclusion that prevention of outbreaks, particularly from planthopper populations, depends on protection of early-season natural enemies through suspension of insecticide sprays in the first 40 days after crop establishment (Way and Heong, 1994). However, there remains widespread, unnecessary insecticide use in tropical rice during early crop stages due to wide gaps in farmers' knowledge and perceptions for rational decisions. One way to bridge this knowledge gap is to change farmer perceptions and risk-averse attitudes towards leaf-feeding pests. When such conflict information ('insecticide spraying for leaf folders in the first 30 days after transplanting is not needed') is presented to farmers as a heuristic (or rule-of-thumb), farmers are motivated to test the rule (Heong and Escalada, 1997a). After participating in such experiments, farmers changed their perceptions about leaf folders and stopped spraying against them. Vietnamese farmers have participated in such experiments and have stopped early-season insecticide spraying (Heong *et al.*, 1995b).

This perception seems strongly held by thousands of farmers and there is a need to explore different ways to communicate this information effectively and as rapidly as possible to change current perceptions. For example, a pilot project was established in Long An Province, Vietnam, to evaluate the use of media materials (leaflet, poster and radio drama) to communicate and encourage farmers to test the heuristic (IRRI, 1996b). Sixteen months after the project was launched, farmers in the project area reduced the number of insecticide sprays per season from 3–4 to 1–2. The proportion of farmers who perceived that leaf folders would cause severe yield loss and sprayed to control them decreased from 67–77% to 17–27%. If farmers resist using insecticides in the early crop stages, then opportunities for insecticide misuse are reduced. The degree to which other herbivores of tropical rice require chemical control measures and under what management options remain open questions for future research.

ACKNOWLEDGEMENTS

The authors are grateful for the support provided by the Swiss Agency for Development and Cooperation (SDC) and the Rockefeller Foundation (through its Environmental Research Fellowship Program in International Agriculture) to the International Rice Research Institute, which made these published and unpublished studies possible. We also thank Professor P.T. Haskell and two anonymous reviewers for their comments on the manuscript.

REFERENCES

Adjid, D.A. (1983) Vertical transfer of agrotechnology in Indonesia: the case of the BIMAS programme, in *Agrotechnology Transfer in Indonesia and the Philippines*, BSP Technical Report 9, HITAHR Res Series 027, Department of Agronomy and Soil Science, University of Hawaii, Manoa, pp. 3–10.

Alix, J.C. (1978) The impact of the Masagana 99: an integrated production drive in the Philippines. Paper presented at the Seminar on Accelarating Agricultural Development and Rural Prosperity, University of Reading, UK, 21 pp.

Cheng, J.A., Norton, G.A. and Holt, J. (1990) A systems analysis to brown planthopper control on rice in Zhejiang Province, China. II: Investigation of control strategies. *Journal of Applied Ecology*, **27**, 100–112.

Cohen, J.E. (1978) *Food Webs and Niche Space*, Princeton University Press, Princeton, NJ.

Cohen, J.E. and Luczak, T. (1992) Trophic levels in community food webs. *Evolutionary Ecology*, **6**, 73–89.

Cohen, J.E., Schoenly, K., Heong, K.L. *et al.* (1994) A food-web approach to evaluating the impact of insecticide spraying on insect pest population dynamics in a Philippine irrigated rice ecosystem. *Journal of Applied Ecology*, **31**, 747–763.

Cohen, M.B., Alam, S.N., Medina, E.B. and Bernal, C.C. (1997) Brown planthopper, *Nilaparvata lugens* resistance in rice cultivar IR64: mechanism and role in successful *N. lugens* management in central Luzon, Philippines. *Entomologia Experimentalis et Applicata*, **85**, 221–229.

Conway, G.R. and Barbier, E.D. (1990) *After the Green Revolution: Sustainable Agriculture for Development*, Earthscan, London.

Conway, G.R. and Pretty, J.N. (1991) *Unwelcomed Harvest: Agriculture and Pollution*, Earthscan, London.

Cook, A.G. and Perfect, T.J. (1985) The influence of immigration on population development of *Nilaparvata lugens* and *Sogatella furcifera* and its interaction with immigration by predators. *Crop Protection*, **4**, 423–433.

Croft, B.A. (1990) *Arthropod Biological Control Agents and Pesticides*, John Wiley & Sons, New York.

Cuong, N.L., Ben, P.T., Phuong, L.T. *et al.* (1997) Effect of host parasite resistance and insecticide on brown planthopper *Nilaparvata lugens* (Stål) and predator population development in the Mekong Delta, Vietnam. *Crop Protection*, **16**, 707–715.

deKraker, J. (1996) The potential of natural enemies to suppress rice leaffolder populations. PhD Dissertation, Wageningen Agricultural University, The Netherlands.

Fabellar, L.T., Fabellar, N. and Heong, K.L. (1994) Simulating rice leaffolder feeding effects on yield using MACROS. *International Rice Research Newsletter*, **19**, 7–8.

Fagan, W.F., Hakim, A.L., Ariawan, H. and Yuliyanthiningsih, S. (in press) Intraguild predation in tropical rice agroecosystems and its consequences for biological control.

Flint, M.L. and van den Bosch, R. (1981) *Introduction to Integrated Pest Management*, Plenum Press, New York.

Fowler, S.V., Claridge, M.F., Morgan, J.C. *et al.* (1991) Egg mortality of the brown planthopper, *Nilparvata lugens* (Homoptera: Delphacidae) and green leafhoppers, *Nephotettix* spp. (Homoptera: Cicadellidae) on rice in Sri Lanka. *Bulletin of Entomological Research*, **81**, 161–167.

Gallagher, K., Kenmore, P.E. and Sogawa, K. (1994) Judicious use of insecticides defer planthopper outbreaks and extend the life of resistant varieties in Southeast Asian rice, in *Planthoppers – Their Ecology and Management*, (eds R. Denno and T.J. Perfect), Chapman & Hall, London, pp. 599–614.

Guo, Y. (1990) Larval parasitization of rice leaffolders (Lepidoptera: Pyralidae) under field and laboratory conditions. PhD Dissertation, University of the Philippines, Los Baños.

Heckman, C.W. (1979) *Rice Field Ecology in Northeastern Thailand: the Effect of Wet and Dry Seasons on a Cultivated Aquatic Ecosystem*, Junk, The Hague.

Heinrichs, E.A. and Mochida, O. (1984) From secondary to major pest status: the case of insecticide-induced rice brown planthopper, *Nilparvata lugens* resurgence. *Protection Ecology*, **7**, 191–218.

Heinrichs, E.A., Reissig, W.H., Valencia, S.L. and Chelliah, S. (1982) Rates and effects of resurgence-inducing insecticides on populations of *Nilparvata lugens* (Hemiptera: Delphacidae) and its predators. *Environmental Entomology*, **11**, 1269–1273.

Heong, K.L. (1989) A simulation approach to evaluating insecticides for brown planthopper control. *Researches on Population Ecology*, **30**, 165–176.

Heong, K.L. (1990) Feeding rates of the rice leaffolder, *Cnaphalocrocis medinalis* (Lepidoptera: Pyrilidae) on different plant stages. *Journal of Agricultural Entomology*, **7**, 81–90.

Heong, K.L. and Escalada, M.M. (1997a) Perception change in rice pest management: a case study of farmers' evaluation of conflict information. *Journal of Applied Communications*, **81**(2), 3–17.

Heong, K.L. and Escalada, M.M. (1997b) Comparative analysis of pest management practices of rice farmers in Asia, in *Pest Management Practices of Rice Farmers in Asia*, (eds K.L. Heong and M.M. Escalada), International Rice Research Institute, Los Baños, Philippines, in press.

Heong, K.L., Aquino, G.B. and Barrion, A.T. (1991) Arthropod community structures of rice ecosystems in the Philippines. *Bulletin of Entomological Research*, **81**, 407–416.

Heong, K.L., Aquino, G.B. and Barrion, A.T. (1992) Population dynamics of plant- and leafhoppers and their natural enemies in rice ecosystems in the Philippines. *Crop Protection*, **11**, 371–379.

Heong, K.L., Escalada, M.M. and Mai, V. (1994) An analysis of insecticide use in rice: case studies in the Philippines and Vietnam. *International Journal of Pest Management*, **40**, 173–178.

Heong, K.L., Escalada, M.M. and Lazaro, A.A. (1995a) Misuse of pesticides among rice farmers in Leyte, Philippines, in *Impact of Pesticides on Farmers' Health and the Rice Environment*, (eds P.L. Pingali and P.A. Roger), Kluwer Press, California, pp. 97–108.

Heong, K.L., Thu Cuc, N.T., Binh, N. *et al.* (1995b) Reducing early season insecticide applications through farmers' experiments in Vietnam, in *Vietnam and IRRI: a Partnership in Rice Research*, IRRI and Ministry of Agriculture and Food Industry, Hanoi, Vietnam, pp. 217–222.

IRRI (1996) *Program Report for 1995*, IRRI, Los Baños, Philippines.

IRRI (1996) *IRRI 1995–1996: Listening to the Farmers*, IRRI, Los Baños, Philippines.

Joshi, R.C., Shepard, B.M., Kenmore, P.E. and Lydia, R. (1992) Insecticide-induced resurgence of brown planthopper (BPH) on IR 62. *International Rice Research Newsletter*, **17**, 9–10.

Kenmore, P.E. (1980) Ecology and outbreaks of a tropical insect pest of the Green Revolution, the rice brown planthopper, *Nilparvata lugens* (Stal.). PhD thesis, University of California, Berkeley.

Kenmore, P.E., Carino, F.O., Perez, C.A. *et al.* (1984) Population regulation of the rice brown planthopper (*Nilparvata lugens* Stal.) within rice fields in the Philippines. *Journal of Plant Protection in the Tropics*, **1**, 19–38.

Kenmore, P.E., Litsinger, J.A., Bandong, J.P. *et al.* (1987) Philippine rice farmers and insecticides: thirty years of growing dependency and new options for change, in *Management of Pests and Pesticides – Farmers' Perceptions and Practices*, (eds J. Tait and B. Napompeth), Westview Press, Boulder, Colorado, pp. 98–108.

Ludwig, J.A. and Reynolds, J.F. (1988) *Statistical Ecology: a Primer on Methods and Computing*, John Wiley & Sons, New York.

Mackenzie, W. (1996) *The Rice Pesticide Market*. Wood Mackenzie Consultants Ltd., London.

Matteson, P.C., Gallagher, K.D. and Kenmore, P.E. (1994) Extension of integrated pest management for planthoppers in Asian irrigated rice: empowering the user, in *Planthoppers: Their Ecology and Management*, (eds R.F. Denno and T.J. Perfect), Chapman & Hall, London, pp. 656–685.

Miyashita, T. (1985) Estimation of the economic injury level in the rice leaf roller, *Cnaphalocrocis medinalis* Guenee (Lepidoptera: Pyralidae). I. Relations between yield loss and injury of rice leaves at heading or in the grain filling period. *Japan Journal of Applied Zoology*, **29**, 73–76.

Perera, P.A.C., Hassell, M.P. and Godfray, H.C.J. (1988) Population dynamics of the coconut caterpillar, *Opisina aeronsella* Walker (Lepidoptera: Xylorctidae) in Sri Lanka. *Bulletin of Entomological Research*, **78**, 479–492.

Pingali, P.L. and Roger, P.A. (1995) *Impact of Pesticides on Farmers' Health and the Rice Environment*, Kluwer Academic Publishers, Norwell, Mass., USA.

Rola, A.C. and Pingali, P.L. (1993) *Pesticides, Rice Productivity and Farmers' Health: An Economic Assessment*, IRRI, Los Baños, Philippines.

Rombach, M.C. and Gallagher, K.D. (1994) The brown planthopper: promises, problems and prospects, in *Biology and Management of Rice Insects*, (ed. E.A. Heinrichs), Wiley Eastern Ltd, New Delhi, India, pp. 613–656.

Sawada, H., Kusmayadi, A., Gaib Subroto, S.W. *et al.* (1993) Comparative analysis of population characteristics of the brown planthopper, *Nilparvata lugens* Stål, between wet and dry rice cropping seasons in West Java, Indonesia. *Researches in Population Ecology*, **35**, 113–137.

Schoenly, K.G., Cohen, J.E., Heong, K.L. *et al.* (1996a) Quantifying the impact of insecticides on food web structure of rice-arthropod populations in a Philippine farmer's irrigated field: a case study, in *Food Webs. Integration of Patterns and Dynamics*, (eds G. Polis and K. Winemiller), Chapman & Hall, London, pp. 343–351.

Schoenly, K.G., Cohen, J.E., Heong, K.L. *et al.* (1996b) Food web dynamics of irrigated rice fields at five elevations in Luzon, Philippines. *Bulletin of Entomological Research*, **86**, 451–466.

Settle, W.H., Ariawan, H., Tri Astuti, E. *et al.* (1996) Managing tropical rice pests through conservation of generalist natural enemies and alternate prey. *Ecology*, **77**, 1975–1988.

Suvaparp, R. (1993) Effects of sublethal doses of insecticides on the mirid predator, *Cyrtorhinus lividennis* (Reuter) (Heteroptera: Miridae). MS thesis, Kasetsant University, Thailand.

Suvaparp, R. and Heong, K.L. (1992) Relative potency of three insecticides on *Cyrtorhinus lividipennis* and brown planthopper (BPH) *Nilaparvata lugens*. *International Rice Research Newsletter*, **17**, 28–29.

Waage, J. (1989) The population ecology of pest-pesticide-natural enemy interactions, in *Pesticides and Non Target Invertebrates*, (ed P. Jepson), Intercept, Wimborne, Dorset, UK, pp. 81–94.

Wada, T. and Shimazu, M. (1978) Seasonal population trends of the rice leaf roller, *Cnaphalocrocis medinalis* Guenee (Lepidoptera: Pyralidae), in the paddy fields at Chikugo in 1977. *Proceedings Association Plant Protection, Kyushu*, **24**, 77–80.

Way, M.J. and Heong, K.L. (1994) The role of biodiversity in the dynamics and management of insect pests of tropical irrgated rice – a review. *Bulletin of Entomological Research*, **84**, 567–587.

Woodburn, A.T. (1990) The current rice agrochemicals market, in *Pest Management in Rice*, (eds B.T. Grayson, M.B. Green and C.L. Copping), Elsevier Applied Science, New York, pp. 15–30.

42

Development of decision-making tools for vegetable farmers in Southeast Asia

Janny Vos

VEGETABLE CULTIVATION IN SOUTHEAST ASIA

Vegetable production and constraints

In Southeast Asia, a wide range of vegetables is cultivated, such as crucifers (cabbage, cauliflower, broccoli), solanaceous crops (tomato, pepper, eggplant), cucurbits (cucumber, melons, gourds), alliums (shallot, garlic) and others. Production is aimed at local as well as export markets. Vegetable farmers are often smallholders and have little access to information services by research and extension organizations. Crop management practices are variable. In general, vegetable cultivation requires high inputs and is labour intensive. One of the main constraints to the production of high yields and high quality products is crop health. Pests (including diseases and weeds) affect crop yield, quality and appearance of the product, and therefore the income of the farmer. Farmers often lack information about symptoms caused by pests and their biology. Storage of the fresh, perishable products is limited and local market prices often drop during harvest peaks, causing vegetable farmers to be risk-averse. For these reasons, vegetable farmers have become vulnerable when exposed to the currently still intensive promotion of chemical pesticides. Monitoring of crop health problems is hardly practised and most vegetable farmers apply pesticides on a calendar basis.

Ecotoxicology: Pesticides and beneficial organisms.
Edited by P.T. Haskell and P. McEwen. Published in 1998 by Chapman & Hall, London. ISBN 0 412 81290 8.

Disadvantages of chemical control as the sole method of crop health management

There are many disadvantages to the intensive use of chemicals to manage crop health. First of all, there is the risk for farmers themselves of becoming exposed to poisonous chemicals when they apply pesticides. An estimated 20% of all farmers in developing countries suffer from pesticide intoxication at least once in their working life. In addition, records of acute poisoning of farmers during spray applications need to be correlated with the cases of illness of those who regularly handle pesticides. Such figures are not easily traced. Another health problem is posed by pesticide residues in the produce. The contamination is usually not recognized because residues in food are only immediately harmful in extreme cases. Nevertheless, daily intakes of pesticide-contaminated vegetables cause health problems to the consumers in the long run. In discussions with farmers, they often admit that they do not consume their own farm produce as they are worried about the residues. Instead, they grow their own - unsprayed - homegarden.

When pesticides are applied in the field, they also become dispersed into the environment. Generally, pesticides reach the soil either through application to the soil or through runoff. Gaseous chemicals can escape into the air. In the soil, pesticides can bind to soil particles or move into groundwater. When a pesticide is highly persistent in the environment, undesirable biological effects may be caused, such as disruption of soil flora and fauna, negative effects on aquatic life, reduction of ecological diversity and air pollution.

In addition to the target pest, pesticides kill beneficials such as natural enemies and antagonistic fungi. When natural enemies are wiped out, resurgence of pests can be expected, especially in cases of pests with a short life cycle. Also, bird and fish populations are indirectly affected as they may feed on poisoned prey. When pesticides are used on a frequent basis, there is a risk of the build-up of pest resistance. Serious outbreaks of pests, e.g. diamondback moth on cabbage and brown planthopper on rice, have been documented in several Southeast Asian countries after the intensive use of chemicals resulted in build-up of pest resistance to pesticides.

Last but not least, farmers tend to increase the frequency and dosage of pesticide applications when crop health problems persist. As farmers get caught in the 'pesticide treadmill', costs of production escalate.

INTRODUCTION TO A SYSTEMS APPROACH

A major reason for failure of pest control programmes in developing countries is the 'tendency of excessively concentrating on insect pest problems alone, rather than on a broader spectrum of crop parasites and at the same time overlooking major agronomic, economic or social constraints' (Zelazny *et al.*, 1985). The overall condition of the crop should become the major issue rather than the control of a single pest. This systems approach is incorporated in the concept of integrated crop management (ICM). ICM has been defined (El-Zik and Frisbie, 1985) as:

> ... a system whereby all interacting crop production and pest control tactics aimed at maintaining and protecting plant health are harmonised in the appropriate sequence to achieve optimum crop yield and quality and maximum net profit, in addition to stability in the agro-ecosystem, benefiting society and mankind.

Integrated pest management (IPM), integrated nutrient management and integrated water management are components of ICM (FAO, 1991). The impact of ICM is expected to be not only the reduction of excessive use of agrochemicals in agriculture, but also the avoidance of pest resistance, pest resurgence and environmental pollution. Research on ICM of hot pepper in tropical lowlands in SE Asia has shown a general applicability of the technology under a wide range of production situations (Vos, 1994).

TRAINING OF EXTENSION OFFICERS AND FARMERS

Farmer field schools and training of trainers courses

Farmer field schools (FFS) are conducted for the purpose of helping farmers to discover and learn about field ecology and integrated crop management. Before the part-time, season-long FFS are conducted, extension officers are trained in full-time courses (training of trainers – TOT). They grow the crops that farmers grow in season-long training courses. In that way, extension officers learn about the problems that farmers face throughout a cropping season and gain confidence to discuss practical farming. The field is the primary classroom, in the TOT courses as well as in the FFS. The four major principles within the training courses are:

1. Grow a healthy crop.
2. Observe field weekly.
3. Conserve natural enemies.
4. Farmers understand ecology and become experts in their own fields.

There are no standard recommendations or packages of technology offered. In the FFS, farmers collect data in the field and undertake action based on their findings in their own fields (discovery-based decision making). Farmers become active learners and independent decision makers through learning by doing. They compare studies in their own fields and become field-level experts in ecology management. Conservation and utilization of local natural enemies and other beneficials play important roles. In ecology-oriented crop management, pesticides (with a preference for biopesticides rather than chemical pesticides) are only applied in case of emergency or when no other management methods are available. As a general rule, crop health management actions are conducted only after field observations have shown the need to act. The understanding of the ecological processes in the field is a prerequisite to successful implementation of integrated crop management.

Development of vegetable field school curricula

The field school programme covers a number of basic concepts, among which are:

- monitoring or field inspection;
- understanding biological control;
- understanding life cycles of insects, pathogens and weeds
- understanding crop growth stages, crop needs and crop compensation.

During a field school, trainees become acquainted with monitoring or field inspection methods. Weekly observations are done in subgroups divided over a conventional practice or calendar-sprayed and an ICM field plot. For each subgroup, plants are sampled and carefully observed. Observations include the crop growth stage and pest and beneficial populations. In addition, the weather, soil and overall crop conditions are recorded. Plant material with disease symptoms and insects is collected in the field, to facilitate drawing later on. After the field inspection, each subgroup draws a plant at the observed crop stage with the complex of pests and beneficials associated with it on a large piece of paper. In the field collected plant material and insects help to make accurate drawings. Pest and natural enemy population sizes are recorded as numbers of, say, insects or leaf spots in the case of a leaf spot disease. Observations on weather, soil and crop conditions are included as well as the recommendation of the subgroup on what action needs to be taken in the observed field plot. After the drawing, each subgroup presents their agro-ecosystem analysis (AESA) for discussion.

As the field school progresses, the weekly AESA shows that farmers improve their skills to observe and recognize pests and beneficials. The decisions on what action needs to be taken in the field become more and more based on actual findings by farmers themselves through 'exercises'. Exercises have been developed to show, for example, biological control to farmers and to identify the need for any additional intervention, whether by use of cultural control, mechanical control or biological pesticides. An exercise to show biological control is the 'Insect Zoo'.

Insect Zoo

1. Jars are lined with tissue paper.
2. Leaves with pest insects (e.g. cabbage leaf with aphids) are collected from the farmers' fields.
3. The number of pest insects one leaf is counted; the leaf is inserted in a jar; the jar is covered with a lid.
4. Predators (e.g. ladybird beetles) are collected from the farmers' fields (or refuges on bunds).
5. One predator is inserted per jar; the jar is covered with a lid.
6. Farmers keep the Insect Zoos in a shady place and study them.

Based on the Insect Zoo (many different Zoos can be made during a season-long field school), farmers learn about the beneficials in their fields. They also assess how many pest insects are consumed by one predator. Based on these discoveries, farmers can predict whether or not the X predators they find on their sampled plants can control the Y pests, and thus whether or not additional pest control actions need to be taken (discovery-based decision making).

Using the same discovery-based learning approach as in the Insect Zoo, exercises have been developed to show the negative effects of pesticides on beneficials, and on farmers themselves (e.g. 'Spraying Dye') as well as the environment.

Spraying Dye

1. One participant is wrapped up with tissue-paper, including hands, face (except eyes), legs and feet.
2. A dye colour solution is prepared and used to fill a knapsack sprayer.
3. The mummified participant is requested to spray a crop with the dye as if spraying with a pesticide.
4. After spraying, the knapsack is removed and participants are requested to observe and record the spots of dye found on the different body parts: head, torso, back, arms, hands, legs, feet.

Exercises have also been or are being developed to show farmers the life cycles of pests (including diseases and weeds), the impact of cultural control, crop compensation, etc. Concepts that are not easily explained in short-duration exercises are sometimes converted into field studies or fun exercises or games.

IMPACT OF THE SEASON-LONG FIELD SCHOOLS

There is very little information on cost comparisons between chemical (including botanical) and alternative treatments of pests of vegetables in

Southeast Asia. The systems approach, as advocated in this chapter, is itself an important alternative to sole reliance on chemicals. In order to calculate the overall cost of chemical applications, one should assess the monetary value of non-target organisms (including natural enemies and antagonists of pests) that are killed in the treated field, the health risk to farmers and consumers, the contamination of groundwater and the overall long-term negative impact on the environment. In fact, this is not a concern to farmers only, but to humankind in general.

Reduced pesticide use, increased yields and improved farmer net incomes have been achieved through training of farmers in integrated management for rice in Indonesia and the Philippines, cotton in India, sugarcane in Pakistan, cabbage in Taiwan and Malaysia, soybean in Brazil, mango in Pakistan and banana in Costa Rica (ADB, 1994). The impact of an FFS programme for vegetables can be illustrated by the recently completed highland vegetable project in the Philippines (ADB, 1996). After FFS, cabbage farmers cut down on insecticide use, mainly to control diamondback moth, from 14 to 3 l/ha (a reduction of 80%). From an average of 6 days, the spray interval was extended to 31 days. The numbers of sprays on a cabbage crop was reduced from 17 to 3 applications. Due to a reduction in inputs, farmers' net incomes increased. Before FFS, most farmers used highly toxic Category I insecticides. After FFS, farmers only use pesticides when severe crop injury is observed and only apply the biopesticide *Bacillus thuringiensis* or other less toxic Category IV insecticides. Before FFS, farmers considered any crawling organism as a pest, but after FFS they became aware of the beneficials ('friends') in their fields. The FFS programme is currently expanding to many Southeast Asian countries, for rice, cotton, legumes and vegetables.

REFERENCES

ADB (1994) *Handbook for Incorporation of Integrated Pest Management in Agriculture Projects*, Asian Development Bank, Manila, Philippines.

ADB (1996) *Integrated Pest Management for Highland Vegetables* Vol. 1, Final TA report, 9 May 1994–30 September 1996. Asian Development Bank, Manila, Philippines, 291 pp.

El-Zik, K.M. and Frisbie, R.E. (1985) Integrated crop management systems for pest control, in *CRC Handbook of Natural Pesticides: Methods – Volume 1: Theory, Practice, and Detection* (ed. N.B. Mandava), CRC Press, Boca Raton, pp. 21–122.

FAO (1991) *Sustainable Crop Production and Protection. Background Document 2*, FAO/Netherlands conference on agriculture and the environment, Hertogenbosch, The Netherlands, 15–19 April 1991.

Vos, J.G.M. (1994) Integrated crop management of hot pepper (*Capsicum* spp.) in tropical lowlands. Evaluation of components of integrated crop management of hot pepper (*Capsicum* spp.) in various production situations. General discussion. PhD thesis, pp. 149–165.

Zelazny, B., Chiarappa, L. and Kenmore, P.E. (1985) Integrated pest control in developing countries. *FAO Plant Protection Bulletin*, **33** (4), 147–158.

Index

Note: page numbers in *italics* refer to tables, those in **bold** refer to figures.